Digital Signal Processing and Applications

M

Digital Signal Processing and Applications

2nd edition

Dag Stranneby

William Walker

AMSTERDAM • BOSTON • HEIDELBERG • LONDON • NEW YORK • OXFORD
PARIS • SAN DIEGO • SAN FRANCISCO • SINGAPORE • SYDNEY • TOKYO

Newnes is an imprint of Elsevier

ELSEVIER

Newnes

Newnes
An imprint of Elsevier
Linacre House, Jordan Hill, Oxford OX2 8DP
30 Corporate Drive, Burlington, MA 01803

First published 2001
Second edition 2004

British Library Cataloguing in Publication Data
Stranneby, Dag
Digital signal processing and applications. 2nd ed.
1. Signal processing – Digital techniques
I. Title   II. Walker, William
621.3′822

Library of Congress Cataloguing in Publication Data
A catalogue record for this book is available from the Library of Congress

ISBN 0 7506 6344 8

For information on all Newnes publications visit our website
at: www/newnespress/com

Typeset by Charon Tec Pvt. Ltd, Chennai, India
Printed and bound in the United Kingdom

Working together to grow
libraries in developing countries

www.elsevier.com | www.bookaid.org | www.sabre.org

ELSEVIER   **BOOK AID** International   Sabre Foundation

Contents

Preface

Dear Reader,

There are many books on digital signal processing (DSP) around these days. The special flavor of this book is to give a broad **overview** of DSP theory and applications in different areas, such as telecommunications, control systems, and measuring and data analysis systems. Such a wide coverage would commonly require at least four or five different books.

The first chapter of this book starts at the novice level with sampling, z-transforms and digital filters, but later chapters treat advanced topics like fuzzy logic, neural networks, information theory and Kalman filters. Since it is our sincere belief that bringing **intuitive understanding** of concepts and systems is by far the most important part of teaching, the text is somewhat simplified at the expense of mathematical rigor. At the end of this book, references for deeper studies of the different topics covered can be found. Some details might be difficult for you to grasp at once, especially if you have novice knowledge of DSP and its techniques; however, there is no cause for alarm. The most important first step in studying any subject is to grasp the **overall picture** and to understand the basics before delving into the details.

As teaching aids, **review questions** and **solved problems** are included in all chapters. Exercises using MATLAB™ are also included. Furthermore, an accompanying **web site** is available where extra teaching material can be found (http://books.elsevier.com/companions/0750663448).

The chapters are organized as follows:

Chapter 1 This chapter starts with an historic overview of DSP and control. The main part of the chapter covers fundamentals of DSP, such as sampling, quantization, convolution, z-transform and transfer functions. It also deals with basic filter structures, finite impulse response (FIR), infinite impulse response (IIR) and with filter synthesis methods, e.g. Butterworth, Chebyshev, impulse invariance, the bilinear transform, Remez exchange algorithm, etc. At the end of this chapter, a brief presentation of digital closed-loop control systems can be found. If you have prior knowledge in DSP and control, you can probably disregard most of this chapter.

Chapter 2 Most signals in real life are certainly not digital. This chapter deals with different types of digital-to-analog (D/A) and analog-to-digital (A/D) conversion techniques and hardware.

Chapter 3 An adaptive filter is a filter which has the property of "tuning" itself. In this chapter, basic models for adaptive filters are presented together with algorithms for adaptation, for instance, the well-known least mean square (LMS) algorithm. Some applications are discussed, such as adaptive interference canceling, equalizers and adaptive beamforming.

Chapter 4 There are an infinite number of non-linear systems. In this chapter, three commonly used non-linear techniques are treated. Starting with the median filter, we continue with artificial neural networks, which is a crude way of trying to mimic the functions in the human brain. The chapter is concluded with a presentation of fuzzy control.

Chapter 5 Spectral analysis and modulation is the title of this chapter. In the first part of this chapter, Fourier transform using fast Fourier transform (FFT) is presented along with correlation-based signal analysis methods, for instance, parametric analysis and wavelets. The second part deals with digital modulation, i.e. ASK, FSK, PSK, QPSK, QAM and so on. The Hilbert filter is also briefly presented.

Chapter 6 This chapter presents recursive least square (RLS) estimation, the pseudo-inverse and the Kalman filter. The Kalman filter technique is illustrated by an "everyday" example. In reading this chapter, it is an advantage if you are used to working with vectors and matrices.

Chapter 7 The main topic of this chapter is information theory and source coding. Some common data compression algorithms are studied, such as Huffman coding, adaptive pulse code modulation (ADPCM), linear predictive coding (LPC), the joint photographics expert group (JPEG), Lempel–Ziv–Welch (LZW) and layer-3 of MPEG-1 (audio compression algorithm)(MP3). We deal with speech compressors and vocoders and make a quick tour through the land of speech and image recognition.

Chapter 8 This chapter is about error-correcting codes and channel coding. We start with a discussion of channel capacity and error probabilities and proceed with block codes, cyclic redundancy check (CRC) codes, convolution codes and Viterbi decoding. Interleaving and turbo codes are also addressed.

Chapter 9 The last chapter of this book is dedicated to practical problems when working with DSP and control. Topics like hardware architecture, numerical problems, execution speed, and data and computer program structures are discussed. We also highlight some problems inherent in the software development process that are significant in DSP environments. Practical code examples for FIR filters, IIR filters and state machines, written in assembly language and in C, are included.

Appendix 1 In this part, you can find solutions to all the problems in the book. Complete calculations, MATLAB™ code and plots are supplied.

Appendix 2 This is not a complete tutorial to MATLAB™ or Simulink™, but a quick guide to getting started using these common software tools.

Finally, a *Glossary* is provided as an aid to the terms used within the book and their abbreviations.

This is the second edition of *Digital Signal Processing* (Stranneby, 2001). The book was originally written to be used in undergraduate courses at Örebro University in Sweden, but has also been used for company on-site training. In addition to the material covered in the following chapters, practical hands-on projects are also an integral part of these courses. The projects consist of the design and implementation of signal processing algorithms on a DSP system.

Welcome to an adventure in the world of DSP!

Dag Stranneby and *William Walker*

1 Introduction

Background All processes and signals commonly found in the real world are "analog" by their nature. Sometimes it may feel a bit awkward trying to represent and process these signals digitally. There are, however, many advantages in digital signal processing (DSP).

In this introductory chapter, we will first give a broad overview of the usage of digital signal processing and digital control. As you will realize, digital signal processing and control systems are all over the place. We will continue by defining digital signals and show some standard digital models and algorithms to process these signals. At the end of the chapter, some methods related to the design of digital signal processing and control applications will be presented.

Objectives In this chapter we will discuss:

- Analog and digital signals, sampling and quantization
- Linearity, difference equations, state–space models, impulse response and convolution
- Transfer functions using the z-transform, the frequency response of a digital system
- Some filter architectures: finite impulse response (FIR), infinite impulse response (IIR) and lattice filters
- Filter approximations: Butterworth, Chebyshev, Cauer and Bessel
- Filter synthesis methods: impulse invariance, the bilinear transform, the Fourier method and Remez exchange algorithm
- Digital control, proportional–integral–derivative (PID) controllers, pole placement controllers and dead-beat controllers.

1.1 The history of digital signal processing

Since the Second World War, if not earlier, technicians have speculated on the applicability of digital techniques to perform signal processing tasks. For example at the end of the 1940s, Shannon, Bode and other researchers at the Bell Telephone Laboratories discussed the possibility of using digital circuit elements to implement filter functions. At this time, there was unfortunately no appropriate hardware available. Hence, cost, size and reliability strongly favored conventional, analog implementations.

During the middle of the 1950s, Professor Linville at Massachusetts Institute of Technology (MIT) discussed digital filtering at graduate seminars. By then, control theory, based partly on works by Hurewicz had become established as a discipline, and sampling and its spectral effects were well understood. A number of mathematical tools, such as the z-transform, which had existed since Laplace's time, were now used in the electronics engineering community. Technology at that point, however, was only able to deal with low-frequency control problems or low-frequency seismic signal processing problems. While seismic scientists made notable use of digital filter concepts to solve problems, it was not until the middle of the 1960s that a more formal theory of digital signal processing (DSP) began to emerge. During this period, the advent of the silicon integrated circuit technology made complete digital systems possible, but still quite expensive.

The first major contribution in the area of digital filter synthesis was made by Kaiser at Bell Laboratories. His work showed how to design useful filters using the bilinear transform (BLT). Further, in about 1965 the famous paper by Cooley and Turkey was published. In this paper, fast Fourier transform (FFT), an efficient and fast way of performing the discrete Fourier transform (DFT), was demonstrated.

At this time, hardware better suited for implementing digital filters was developed and affordable circuits started to be commercially available. Long finite impulse response (FIR) filters could now be implemented efficiently, thereby becoming a serious competitor to the infinite impulse response (IIR) filters, having better passband properties for a given number of delays. At the same time, new opportunities emerged. It was now possible to achieve time varying, adaptive and non-linear filters that could not be built using conventional analog techniques. One such filter is the Kalman filter named after R.E. Kalman. The Kalman filter is a model-based filter that filters the signal according to its statistical rather than its spectral properties.

In the area of adaptive filters, B. Widrow is an important name, especially when talking about the least mean square (LMS) algorithm. Widrow also made significant contributions in the area of neural networks as early as in the 1960s and 1970s.

Today, there are many commercial products around which utilize the advantages of digital signal processing, namely:

- an essentially perfect reproducibility
- a guaranteed accuracy (no individual tuning and pruning necessary)
- well suited for large volume production.

To conclude this section, we will give some everyday examples where digital signal processing is encountered in one way or another. Applications can be divided into two classes. The first class consists of applications that **could** be implemented using ordinary analog techniques but where the use of digital signal processing increases the performance considerably. The second class of applications are these that **require** the use of digital signal processing and cannot be built using entirely "analog" methods.

1.1.1 Measurements and analysis

Digital signal processing traditionally has been very useful in the areas of measurement and analysis in two different ways. One is to precondition the measured signal by rejecting the disturbing noise and interference or to help interpret the properties of collected data by, for instance, correlation and spectral transforms. In the area of medical electronic equipment, more or less sophisticated digital filters can be found in electrocardiograph (ECG) and electroencephalogram (EEG) equipment to record the weak signals in the presence of heavy background noise and interference.

As pointed out earlier, digital signal processing has historically been used in systems dealing with seismic signals due to the limited bandwidth of these signals. Digital signal processing has also proven to be very well suited for air and space measuring applications, e.g. analysis of noise received from outer space by radio telescopes or analysis of satellite data. Using digital signal processing techniques for analysis of radar and sonar echoes are also of great importance in both civilian as well as military contexts.

Another application is navigational systems. In global positioning system (GPS) receivers (RXs) today, advanced digital signal processing techniques are employed to enhance resolution and reliability.

1.1.2 Telecommunications

Digital signal processing is used in many telecommunication systems today; for instance, in telephone systems for dual-tone multi-frequency (DTMF) signaling, echo canceling of telephone lines and equalizers used in high-speed telephone modems. Further, error-correcting codes are used to protect digital signals from bit errors during transmission (or storing) and different data compression algorithms are utilized to reduce the number of data bits needed to represent a given amount of information.

Digital signal processing is also used in many contexts in cellular telephone systems, for instance speech coding in groupe speciale mobile or global system for mobile communication (GSM) telephones, modulators and demodulators, voice scrambling and other cryptographic devices. It is very common to find five to ten microcontrollers in a low-cost cellular telephone. An application dealing with high frequency is the directive antenna having an electronically controlled beam. By using directive antennas at the base stations in a cellular system, the base station can "point" at the mobile at all times, thereby reducing the transmitter (TX) power needed. This in turn increases the capacity of a fixed bandwidth system in terms of the number of simultaneous users per square kilometer, i.e. increases the service level and the revenue for the system operator.

The increased use of the Internet and personal computers (PCs) implies the use of digital processing in many layers. Not only for signal processing in asymmetric digital subscriber loop (ADSL) and digital subscriber loop (DSL) modems, but also for error correction, data compression (images and audio) and protocol handling.

1.1.3 Audio and television

In most audio and video equipment today, such as digital video disc (DVD) and compact disc (CD) players, digital audio tape (DAT), digital compact cassette (DCC) recorders and MPEG layer 3 (MP3) players, digital signal processing is mandatory. This is also true for most modern studio equipment as well as more or less advanced synthesizers used in today's music production. Digital signal processing has also made many new noise suppression and companding systems (e.g. Dolby™) attractive.

Digital methods are not only used for producing and storing audio and video information, but also for distribution. This could be between studios and transmitters, or even directly to the enduser, such as in the digital audio broadcasting

(DAB) system. Digital transmission is also used for broadcasting of television (TV) signals. High definition television (HDTV) systems utilize many digital image processing techniques. **Digital image processing** can be regarded as a special branch of digital processing having many things in common with digital signal processing, but dealing mainly with two-dimensional image signals. Digital image processing can be used for many tasks, e.g. restoring distorted or blurred images, morphing, data compression by image coding, identification and analysis of pictures and photos.

1.1.4 Household appliances and toys

In most modern dishwashers, dryers, washing machines, microwave ovens, air conditioners, toasters and so on, you are likely to find embedded microcontrollers performing miscellaneous digital processing tasks. Almost every type of algorithm can be found, ranging from simple timers to advanced fuzzy logic systems. Microprocessors executing digital signal processing algorithms can also be found in toys, such as talking dolls, speech recognition controlled gadgets and more or less advanced toy robots.

1.1.5 Automotive

In the automotive business, digital signal processing is often used for control purposes. Some examples are ignition and injection control systems, "intelligent" suspension systems, "anti-skid" brakes, "anti-spin" four-wheel-drive systems, climate control systems, "intelligent" cruise controllers and airbag controllers.

There are also systems for speech recognition and speech synthesis being tested in automobiles. Just tell the car: "Switch on the headlights" and it will, and maybe it will give the answer: "The right rear parking light is not working". New products are also systems for background noise cancellation in cars using adaptive digital filters, and radar assisted, more or less "smart" cruise controllers.

To summarize: digital signal processing and control is here to stay ...

1.2 Digital signal processing basics

There are numerous good references available dealing with basic digital signal processing, such as Oppenheimer and Schafer (1975), Rabiner and Gold (1975), Mitra and Kaiser (1993), Marven and Ewers (1993), Denbigh (1998), Lynn and Fuerst (1998) and Smith (2003). The following sections are a brief summary of underlying important theories and ideas.

1.2.1 Continuous and discrete signals

In this book, we will mainly study systems dealing with signals that vary over time (temporal signals). In "reality" a signal can take on an infinite amount of values and time can be divided into infinitely small increments. A signal of this type is **continuous in amplitude** and **continuous in time**. In everyday language, such a signal is called an "analog" signal.

Now, if we only present the signal at given instants of time, but still allow the amplitude to take on any value, we have a signal type that is **continuous in amplitude**, but **discrete in time**.

The third signal type would be a signal that is defined at all times, but only allowed to take on values from a given set. Such a signal would be described as **discrete in amplitude** and **continuous in time**.

The fourth type of signal is only defined at given instants of time, and is only allowed to take on values from a given set. This signal is said to be **discrete in amplitude** and **discrete in time**. This type of signal is called a "digital" signal, and is the type of signal we will mainly deal with in this book.

1.2.2 Sampling and reconstruction

The process of going from a signal being **continuous in time** to a signal being **discrete in time** is called **sampling**. Sampling can be regarded as multiplying the time-continuous signal $g(t)$ with a train of unit pulses $p(t)$ (see Figure 1.1)

$$g^{\#}(t) = g(t)p(t) = \sum_{n=-\infty}^{+\infty} g(nT)\,\delta(t - nT) \tag{1.1}$$

where $g^{\#}(t)$ is the sampled signal. Since the unit pulses are either one or zero, the multiplication can be regarded as a pure switching operation.

The time period T between the unit pulses in the pulse train is called the **sampling period**. In most cases, this period is constant, resulting in "equidistant sampling". There is however no theoretical demand for the sampling period to be constant. In some systems, many different sampling periods are used ("multi-rate sampling") (Åström and Wittenmark, 1984). Yet, in other applications the sampling period may be a stochastic variable. This results in "random sampling" (Papoulis and Pillai, 2001) which complicates the analysis considerably. In most systems today, it is common to use one or more constant sampling periods. The sampling period T is related to the **sampling rate** or **sampling frequency** f_s such that

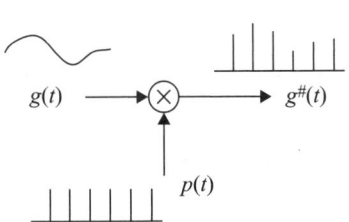

$g(t)$ ⊗ $g^{\#}(t)$

$p(t)$

Figure 1.1 *Sampling viewed as a multiplication process*

$$f_s = \frac{\omega_s}{2\pi} = \frac{1}{T} \tag{1.2}$$

The process of sampling implies reduction of knowledge. For the time-continuous signal, we know the value of the signal at every instant of time, but for the sampled version (the time-discrete signal) we only know the value at **specific points** in time. If we want to reconstruct the original time-continuous signal from the time-discrete sampled version, we therefore have to make more or less qualified interpolations of the values in between the sampling points. If our interpolated values differ from the true signal, we have introduced distortion in our reconstructed signal.

Now to get an idea of our chances of making a faithful reconstruction of the original signal, let us study the effect of the sampling process in the frequency domain. First, referring to equation (1.1), the pulse train $p(t)$ can be expanded in a Fourier series

$$p(t) = \sum_{n=-\infty}^{+\infty} c_n e^{jn(2\pi/T)t} \tag{1.3}$$

where the Fourier coefficients c_n are

$$c_n = \frac{1}{T} \int_{-T/2}^{+T/2} p(t)\, e^{-jn(2\pi/T)t} \, \mathrm{d}t = \frac{1}{T} \tag{1.4}$$

Hence, the sampling equation (1.1) can now be rewritten as

$$g^{\#}(t) = p(t)\,g(t) = \left(\frac{1}{T} \sum_{n=-\infty}^{+\infty} e^{jn(2\pi/T)t} \right) g(t) \tag{1.5}$$

The Laplace transform of the sampled signal $g^{\#}(t)$ is (using the multiplication-shift property)

$$G^{\#}(s) = \frac{1}{T} \sum_{n=-\infty}^{+\infty} G\left(s + jn\frac{2\pi}{T} \right) \tag{1.6}$$

To get an idea of what is happening in the frequency domain, we investigate equation (1.6) following the $s = j\omega$ axis

$$G^{\#}(j\omega) = \frac{1}{T} \sum_{n=-\infty}^{+\infty} G(j(\omega + n\omega_s)) \tag{1.7}$$

From this, we see that the effect of sampling creates an infinite number of copies of the spectrum of the original signal $g(t)$. Every copy is shifted by multiples of the sampling frequency ω_s. Figure 1.2 shows a part of the total (infinite) spectrum.

The spectrum bandwidth of the original signal $g(t)$ is determined by the highest-frequency component f_{max} of the signal. Now, two situations can occur. If $f_{max} < f_s/2$, then the copies of the spectrum will **not** overlap (Figure 1.2(a)). Given only one spectrum copy, we have a limited, but correct knowledge of the original signal $g(t)$ and can reconstruct it using an inverse Fourier transform. The sampling process constitutes an unambiguous mapping.

If, on the other hand, $f_{max} \geq f_s/2$, the spectra will overlap (Figure 1.2(b)) and the too-high-frequency components will be aliased (or "folded") into the lower part of the next spectrum. We can no longer reconstruct the original signal, since aliasing distortion has occurred. **Hence, it is imperative that the bandwidth of the original time-continuous signal being sampled is smaller than half the sampling frequency, also called the "Nyquist frequency".**

To avoid aliasing distortion in practical cases, the sampling device is always preceded by some kind of low-pass filter ("anti-aliasing" filter) to reduce the bandwidth of the incoming signal. This signal is often quite complicated and contains a large number of frequency components. Since it is impossible to build perfect filters, there is a risk of too-high-frequency components leaking into the sampler, causing aliasing distortion. We also have to be aware that high-frequency interference may somehow enter the signal path after the low-pass filter, and we may experience aliasing distortion even though the filter is adequate.

In some literature, the concept of "relative frequency" (or "fnosq") is used to make calculations simpler. The relative frequency is defined as

$$q = \frac{f}{f_s} = \frac{\omega}{\omega_s} \tag{1.8a}$$

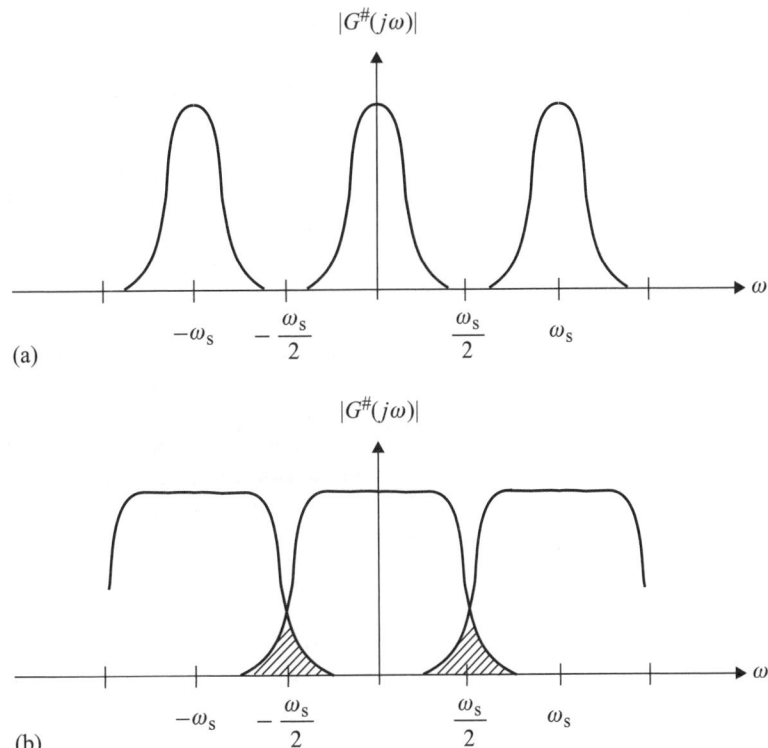

Figure 1.2 *Part of spectrum of sampled signal. In (a) the bandwidth of the signal is less than the Nyquist frequency $\omega_s/2$ and no aliasing takes place. In (b) the bandwidth is greater than $\omega_s/2$ and aliasing takes place, hence the original signal cannot be reconstructed*

Another way of defining frequency in sampled systems is to use the time discrete ("digital") angular frequency

$$\Omega = 2\pi\frac{f}{f_s} = 2\pi\frac{\omega}{\omega_s} = 2\pi q \tag{1.8b}$$

Note that ω (lower case omega) is used for "analog" angular frequency in radians per second, while Ω (upper case omega) is used for "digital" angular frequency in radians per sample period.

Hence, to avoid aliasing distortion: $|q_{max}| < 0.5$ or $|\Omega_{max}| < \pi$ or $|f_s| > 2|f_{max}|$. **This is important!**

If the Nyquist criteria is met and hence no aliasing distortion is present, we can reconstruct the original bandwidth limited time-continuous signal $g(t)$ in an unambiguous way. This is achieved using a low-pass **reconstruction filter** to extract only **one** copy of the spectrum from the sampled signal. It can be shown as a consequence of "Shannon's sampling theorem" or the "cardinal reconstruction formula" (Mitra and Kaiser, 1993) that the ideal low-pass filter to use for reconstruction has the impulse response of a sinc function

$$h(t) = \frac{\sin\left(\pi(t/T)\right)}{\pi(t/T)} = \operatorname{sinc}\left(\frac{t}{T}\right) \tag{1.9}$$

This is a **non-causal** filter (the present output signal depends on future input signals) having an infinite impulse response, and an ideally sharp cut-off at the frequency $\pi/T = \omega_s/2$ radians per second, i.e. the Nyquist frequency. While of theoretical interest, this filter does not provide a practical way to reconstruct the sampled signal. A more realistic way to obtain a time-continuous signal from a set of samples is to hold the sample values of the signal constant between the sampling instants. This corresponds to a filter with an impulse response

$$h(t) = \begin{cases} 1/T & \text{if } 0 \leq t < T \\ 0 & \text{otherwise} \end{cases} \tag{1.10}$$

Such a reconstruction scheme is called a **zero-order hold** (ZOH) or **boxcar hold** that creates a "staircase" approximation of the signal. The zero-order hold can be thought of as an approximation of the ideal reconstruction filter.

A more sophisticated approximation of the ideal reconstruction filter is the **linear point connector** or **first-order hold** (FOH), which connects sequential sample values with straight-line segments. The impulse response of this filter is

$$h(t) = \begin{cases} (T+t)/T^2 & \text{if } -T \leq t < 0 \\ (T-t)/T^2 & \text{if } 0 \leq t < T \end{cases} \tag{1.11}$$

This filter provides a closer approximation to the ideal filter than the zero-order hold (Åström and Wittenmark, 1984).

1.2.3 Quantization

The sampling process described above is the process of converting a continuous-time signal into a discrete-time signal, while quantization converts a signal **continuous in amplitude** into a signal **discrete in amplitude**.

Quantization can be thought of as classifying the level of the continuous-valued signal into certain bands. In most cases, these bands are equally spaced over a given range and undesired non-linear band spacing may cause harmonic distortion. Some applications called companding systems use a non-linear band spacing, which is often logarithmic.

Every band is assigned a code or numerical value. Once we have decided to which band the present signal level belongs, the corresponding code can be used to represent the signal level.

Most systems today use the binary code, i.e. the number of quantization intervals N are

$$N = 2^n \tag{1.12}$$

where n is the word length of the binary code. For example with $n = 8$ bits we get a resolution of $N = 256$ bands, $n = 12$ yields $N = 4096$ and $n = 16$ gives $N = 65536$ bands. Obviously, the more bands we have, i.e. the longer the word length, the better **resolution** we obtain. This in turn renders a more accurate representation of the signal.

Another way of looking at resolution of a quantization process is to define the **dynamic range** as the ratio between the strongest and the weakest signal level that can be represented. The dynamic range is often expressed in decibels. Since every new bit of word length being added increases the number of bands by a factor of 2 the corresponding increase in dynamic range is 6 dB. Hence, an 8-bit system has a dynamic range of 48 dB, a 12-bit system has 72 dB, etc. (This of course only applies for linear band spacing.)

Now, assuming we have N bands in our quantizer, this implies that the normalized width of every band is $Q = 1/N$. Further, this means that a specific binary code will be presented for all continuous signal levels in the range of $\pm Q/2$ around the ideal level for the code. We hence have a random error in the discrete representation of the signal level of $\varepsilon = \pm Q/2$. This random error, being a stochastic variable, is independent of the signal and has a uniformly distributed probability density function ("rectangular" density function)

$$p(\varepsilon) = \begin{cases} \frac{1}{Q} & \text{for } -\frac{Q}{2} < \varepsilon < \frac{Q}{2} \\ 0 & \text{else} \end{cases} \tag{1.13}$$

This stochastic error signal will be added to the "true" discrete representation of our signal and appear as **quantization noise**. The root mean square (RMS) amplitude of the quantization noise can be expressed as (Pohlmann, 1989)

$$\varepsilon_{\text{rms}} = \sqrt{\int_{-\frac{Q}{2}}^{+\frac{Q}{2}} \varepsilon^2 p(\varepsilon)\, d\varepsilon} = \frac{Q}{\sqrt{12}} = \frac{2^{-n}}{\sqrt{12}} \tag{1.14}$$

The ratio between the magnitude of the quantization noise and the maximum signal magnitude allowed is denoted the **signal-to-error ratio**. A longer word length gives a smaller quantization noise and hence a better signal-to-error ratio.

1.2.4 Processing models for discrete-time series

Assume that by proper sampling (at a constant sampling period T) and quantization we have now obtained a "digital" signal $x(n)$. Our intention is to apply some kind of linear signal processing operation, for instance a filtering operation, to the signal $x(n)$ thereby obtaining a digital output signal $y(n)$ (see Figure 1.3).

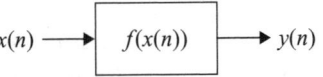

Figure 1.3 A discrete-time linear digital signal processing operation, input x(n)*, output* y(n) *and transfer function* f(x(n))

1.2.4.1 Linear systems

The processing model is said to be **linear** if the **transfer function**, i.e. $y = f(x)$, the function defining the relationship between the output and input signals, satisfies the **principle of superposition**, such that

$$f(x_1 + x_2) = f(x_1) + f(x_2) \tag{1.15}$$

For clarity, let us illustrate this using a few simple examples. If k is a constant, the function (a perfect amplifier or attenuator)

$$f(x) = kx \tag{1.16}$$

is obviously linear since

$$f(x_1 + x_2) = k(x_1 + x_2) = kx_1 + kx_2 = f(x_1) + f(x_2) \tag{1.17}$$

Now, if we add another constant m (bias), what about this function

$$f(x) = kx + m \tag{1.18}$$

It looks linear, doesn't it? Well, let us see

$$f(x_1 + x_2) = k(x_1 + x_2) + m = kx_1 + kx_2 + m \tag{1.19a}$$

but

$$f(x_1) + f(x_2) = kx_1 + m + kx_2 + m = kx_1 + kx_2 + 2m \tag{1.19b}$$

Obviously, the latter function is **not** linear. Now we can formulate one requirement on a linear function: it has to pass through the origin, i.e. $f(0) = 0$.

If we consider a multiplier "circuit" having two inputs y_1 and y_2, this can be expressed as a function having two variables

$$f(y_1, y_2) = y_1 y_2 \tag{1.20}$$

Let us now connect two composite signals $y_1 = x_1 + x_2$ and $y_2 = x_3 + x_4$ to each of the inputs. We now try to see if the function satisfies the principle of superposition (equation (1.15)), i.e. if it is linear or not

$$\begin{aligned} f(x_1 + x_2, x_3 + x_4) &= (x_1 + x_2)(x_3 + x_4) \\ &= x_1 x_3 + x_1 x_4 + x_2 x_3 + x_2 x_4 \end{aligned} \tag{1.21a}$$

while

$$f(x_1, x_3) + f(x_2, x_4) = x_1 x_3 + x_2 x_4 \tag{1.21b}$$

Hence, a multiplication of two variables (composite signals) is a **non-linear** operation. For the special case, we connect the same signal to both inputs, i.e. $y_1 = y_2$, we get

$$f(y_1, y_2) = y_2^2$$

which is obviously a non-linear function. If, on the other hand, we multiply a signal $y_1 = x_1 + x_2$ by a **constant** $y_2 = k$, we get

$$f(x_1 + x_2, k) = (x_1 + x_2)k \quad \text{and} \quad f(x_1, k) + f(x_2, k) = x_1 k + x_2 k$$

as in equations (1.16) and (1.17), then the operation is **linear**. A two input function that is linear in each input if the other is held constant is called a **bilinear** function. Usually we call it a **product**.

The next observation we can make is that the first derivative $f'(x)$ of the function $f(x)$ is required to be a constant for the function to be linear. This also implies that higher-order derivatives do not exist.

An arbitrary function $f(x)$ can be approximated using a Taylor or MacLaurin series

$$f(x) \approx f(0) + x f'(0) + \frac{x^2}{2!} f''(0) + \frac{x^3}{3!} f'''(0) + \cdots \tag{1.22}$$

By inspection of the two first terms, we can find out if the function is linear or not. To be linear we demand (see above)

$$f(0) = 0 \quad \text{and} \quad f'(x) = k$$

It might be interesting to note that many common signal processing operations are indeed non-linear, e.g. rectifying, quantization, power estimation, modulation, demodulation, mixing signals (frequency translation), correlating, etc. In some cases, it is a matter of how fast the signals in question change over time compared to each other. Filtering a signal using a filter with **fixed** (or "slowly varying") coefficients can be regarded as linear, while using an adaptive filter, having variable coefficients, may be regarded as a non-linear operation.

If the parameters of the transfer function are constant over time, the signal processing operation is said to be **time invariant**. Sometimes such an operation is referred to as an **linear time invariant (LTI) processor** (Lynn and Fuerst, 1998; Chen, 1999). Quite often, such processors are also assumed to be **causal**, i.e. the present output signal only depends on present and past input signals, but not on future ones.

The operation of a time-discrete LTI processor can be expressed using a number of different mathematical models, e.g. as a difference equation model, a state–space model, a convolution model or a transfer function model.

1.2.4.2 The difference equation model

The difference equation describing the behavior of a linear time-discrete system can be regarded as the cousin of the linear differential equation describing a continuous-time ("analog") system. An example difference equation of the order of two is

$$y(n) - 0.8 y(n-1) + 0.2 y(n-2) = 0.1 x(n-1) \tag{1.23}$$

To obtain a first output value, $y(0)$, the initial conditions $y(-1)$ and $y(-2)$ have to be known. The difference equation can of course be solved stepwise by inserting the proper values of the input signal $x(n)$. In some cases, this type of direct solution may be appropriate. It is however far more useful to obtain a closed form expression for the solution. Techniques for obtaining such closed forms are well described in the literature on difference equations (Spiegel, 1971; Mitra and Kaiser, 1993). The general form of the difference equation model is

$$\sum_{i=0}^{N} a_i \, y(n-i) = \sum_{j=0}^{M} b_j x(n-j) \tag{1.24}$$

where given $x(n)$, $y(n)$ is to be found. The order of the difference equation equals N, which is also the number of initial conditions needed to obtain an unambiguous solution.

When designing and analyzing digital signal processing algorithms, the difference equation model is commonly not the first choice. Quite often, for instance, the transfer function model described below is a better tool. However, when it comes down to writing the actual computer program code to implement the algorithm, it has to be done using a difference equation model.

1.2.4.3 The state–space model

The state–space model can be seen as an alternative form of the difference equation, making use of matrix and vector representation. For example, if we introduce the variables

$$\begin{cases} y_1(n) = y(n) \\ y_2(n) = y(n+1) \end{cases} \quad \text{and} \quad \begin{cases} x_1(n) = x(n) \\ x_2(n) = x(n+1) \end{cases} \tag{1.25}$$

our example difference equation (1.23) can be rewritten as

$$y_2(n-1) = 0.8\, y_2(n-2) - 0.2\, y_1(n-2) + 0.1\, x_2(n-2) \tag{1.26}$$

or in matrix equation form

$$\begin{bmatrix} y_1(n-1) \\ y_2(n-1) \end{bmatrix} = \begin{bmatrix} 0 & 1 \\ -0.2 & 0.8 \end{bmatrix} \begin{bmatrix} y_1(n-2) \\ y_2(n-2) \end{bmatrix} + \begin{bmatrix} 0 & 0 \\ 0 & 0.1 \end{bmatrix} \begin{bmatrix} x_1(n-2) \\ x_2(n-2) \end{bmatrix} \tag{1.27}$$

If we introduce the vectors

$$\mathbf{Y}(n) = \begin{bmatrix} y_1(n) \\ y_2(n) \end{bmatrix} \quad \text{and} \quad \mathbf{X}(n) = \begin{bmatrix} x_1(n) \\ x_2(n) \end{bmatrix}$$

and the matrices

$$\mathbf{A} = \begin{bmatrix} 0 & 1 \\ -0.2 & 0.8 \end{bmatrix} \quad \text{and} \quad \mathbf{B} = \begin{bmatrix} 0 & 0 \\ 0 & 0.1 \end{bmatrix}$$

the difference equation can now be rewritten in the compact, matrix equation form

$$\mathbf{Y}(n+1) = \mathbf{A}\mathbf{Y}(n) + \mathbf{B}\mathbf{X}(n) \tag{1.28}$$

This is a very useful way of expressing the system. Having knowledge of the system state $\mathbf{Y}(n)$ and the input signal vector $\mathbf{X}(n)$ at time instant n, the new state of the system $\mathbf{Y}(n+1)$ at instant $n+1$ can be calculated. The system is completely specified by the transition matrix $\mathbf{A}$ and the input matrix $\mathbf{B}$. The state–space model is common when dealing with control system applications (Åström and Wittenmark, 1984; Chen, 1999).

1.2.4.4 The convolution model

The convolution model expresses the relationship between the input and output signals as a **convolution sum** (Denbigh, 1998)

$$y(n) = \sum_{k=-\infty}^{+\infty} h(k)x(n-k) = h(n) * x(n) \tag{1.29}$$

The star symbol is an alternative way to express convolution. If we choose the input signal $x(n)$ as the unit pulse $\delta(n)$

$$\delta(n) = \begin{cases} 1 & n = 0 \\ 0 & n \neq 0 \end{cases} \tag{1.30}$$

the output from the system will be

$$y(n) = \sum_{k=-\infty}^{+\infty} h(k)\,\delta(n-k) = h(n) \tag{1.31}$$

where $h(n)$ is denoted as the **impulse response** of the system. Figure 1.4 shows the impulse response of our example system described by the difference equation (1.23). If $h(n)$ of the system is known (or obtained by applying a unit pulse), the output $y(n)$ can be calculated for a given input signal $x(n)$ using convolution as in equation (1.29). The basic idea of convolution is that an arbitrary digital input signal $x(n)$ can be expressed as a sum of unit pulses with different amplitudes and delays

$$x(n) = x_0\,\delta(n) + x_1\,\delta(n-1) + x_2\,\delta(n-2) + \cdots \tag{1.32}$$

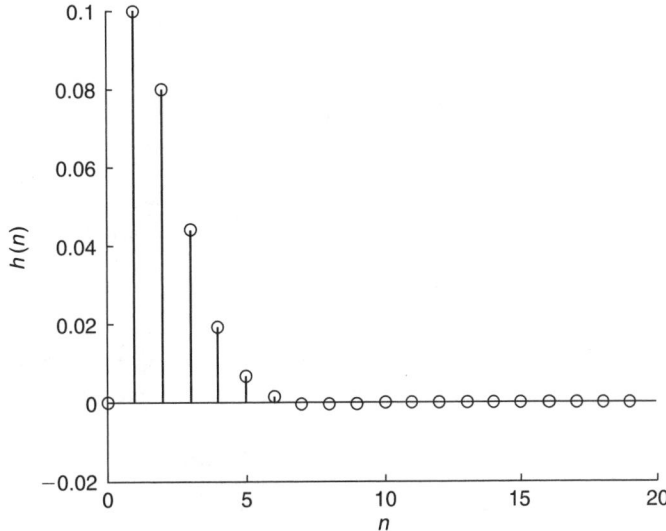

Figure 1.4 *The impulse response of the example system*

Now, we know that when applying a unit pulse to the input of the system, the output will be exactly the impulse response $h(n)$. Hence, if we apply a sequence of unit impulses with different amplitude, i.e. $x(n)$, the output should be a sequence of impulse responses having different amplitude. Summing up all these impulse responses, we obtain the total output, i.e. $y(n)$. This is exactly what the convolution sum, equation (1.29), does. Thanks to the system being linear, this approach works. According to the principle of superposition, instead of calculating the total output when applying a complicated input signal, the input signal can be broken down into sub-parts having known output responses. These output responses are then added to obtain the total output signal, which would give the same result as calculating the total output directly when applying all the sub-parts of the complicated input signal simultaneously. In real world, causal systems, there are of course limits to the index k in equation (1.29).

1.2.4.5 The transfer function model

Another way of representing a digital signal processing system is to use the **transfer function model** or **z-transform model**. If we assume that we are dealing with a causal system

$$h(n) = 0 \quad \text{and} \quad x(n) = 0 \qquad \text{for} \quad n < 0$$

or in other words, the input signal is zero for "negative time" sampling instants, the summation interval of equation (1.29) can be reduced to

$$y(n) = \sum_{k=0}^{+\infty} h(k)x(n-k) \tag{1.33}$$

Taking the z-transform (Oppenheimer and Schafer, 1975; Rabiner and Gold, 1975; Denbigh, 1998), i.e. multiplying by z^{-n} and summing over $0 \le n < \infty$ of both sides of the equation (1.33) we obtain

$$\sum_{n=0}^{+\infty} y(n)z^{-n} = \sum_{n=0}^{+\infty}\sum_{k=0}^{+\infty} h(k)x(n-k)z^{-n}$$

$$= \sum_{k=0}^{+\infty} h(k)z^{-k} \sum_{n=0}^{+\infty} x(n-k)z^{-(n-k)} \tag{1.34a}$$

$$Y(z) = H(z)X(z) \tag{1.34b}$$

where $X(z)$ is the z-transform of the input signal, $Y(z)$ is the transform of the output signal and the transform of the impulse response is

$$H(z) = \sum_{k=0}^{+\infty} h(k)z^{-k} \tag{1.35}$$

Hence, the causal system can be fully characterized by the **transfer function** $H(z)$ which in the general case is an infinite series in the polynomial z^{-k}.

For many series, it is possible to find a closed form summation expression. Note, for a system to be stable and hence useful in practice, the series must converge to a finite sum as $k \to \infty$.

When working in the z-plane, a multiplication by z^{-k} is equivalent to a delay of k steps in the time domain. This property makes, for instance, z-transformation of a difference equation easy. Consider our example of the difference equation (1.23). If we assume that the system is causal and that the z-transforms $X(z)$ and $Y(z)$ of $x(n)$ and $y(n)$, respectively exists, it is straightforward to transform equation (1.23) as

$$Y(z) - 0.8Y(z)z^{-1} + 0.2Y(z)z^{-2} = 0.1X(z)z^{-1} \tag{1.36}$$

Rearranging equation (1.36) we obtain

$$H(z) = \frac{Y(z)}{X(z)} = \frac{0.1z^{-1}}{1 - 0.8z^{-1} + 0.2z^{-2}} = \frac{0.1z}{z^2 - 0.8z + 0.2} \tag{1.37}$$

The **zeros** are the roots of the numerator, while the **poles** are the roots of the denominator. Hence, in equation (1.37) above, the zero is

$$0.1z = 0$$

having the root $z_1 = 0$, and the poles are

$$z^2 - 0.8z + 0.2 = 0$$

having the complex roots $z_1 = 0.4 + j0.2$ and $z_2 = 0.4 - j0.2$.

The study of the locations of the poles and zeros in the complex z-plane are sometimes referred to as the **root locus**. Figure 1.5 is a plot showing the location of the poles and zeros in the complex z-plane, where a pole is commonly marked with a cross and a zero with a ring.

Stating only the locations of the poles and zeros is an alternative way of specifying a system, containing as much information as the transfer function itself. The location of poles and zeros may affect the behavior of system considerably. This is especially true regarding the issue of **stability**.

For the system described by equation (1.37) to be stable, it is required that the poles of the system lie within the unit circle in the complex z-plane, hence

$$|z_p| < 1 \quad \text{for all } p$$

For our example system above, $|z_1| = |z_2| = \sqrt{(0.4^2 + 0.2^2)} \approx 0.45$, i.e. the system is stable.

There are many mathematical definitions of the term stability. A heuristic approach is to say that an **unstable** system is a system where the output signal tends to run out of bounds or oscillates with an increasing amplitude. Anybody who has experienced feedback in an audio system when bringing a microphone too close to a loudspeaker knows what instability is all about. It is easy to realize that an unstable system, "running wild", is commonly not usable.

1.2.4.6 *The frequency function model*

Under some circumstances the properties of the system in the **frequency domain** are of primary interest. These properties can be found by studying

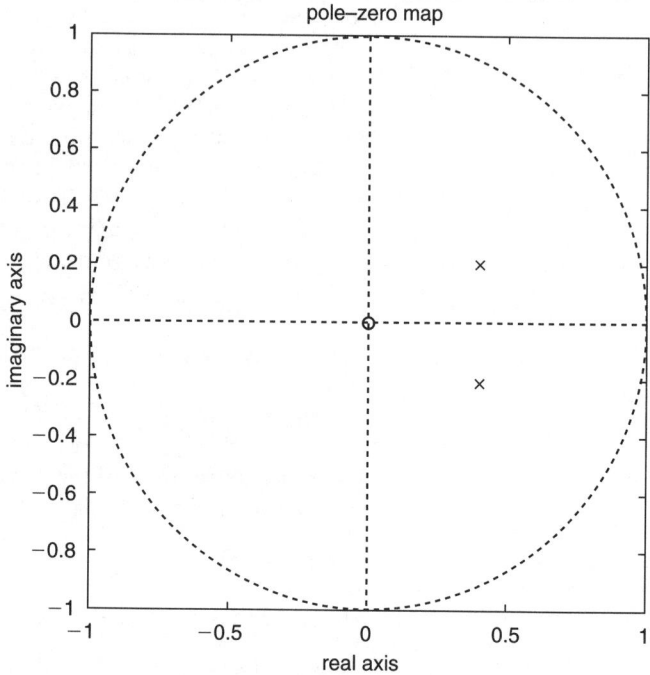

Figure 1.5 *A pole–zero plot showing the location of poles (×) and zeros (0) of the example system in the complex z-plane*

the system for a special class of input signals, which are functionally equivalent to a sampled sinusoid of frequency f

$$x(n) = e^{j2\pi(f/f_s)n} = e^{j\Omega n} = \cos(\Omega n) + j\sin(\Omega n)$$

$$\text{for } -\infty < n < +\infty \tag{1.38}$$

Applying equation (1.38) to equation (1.29) by convolution, we obtain

$$y(n) = \sum_{k=-\infty}^{+\infty} h(k)e^{j\Omega(n-k)} = e^{j\Omega n} \sum_{k=-\infty}^{+\infty} h(k)e^{-j\Omega k}$$

$$= x(n)H(\Omega) \tag{1.39}$$

Thus, for this special class of inputs, we see from equation (1.39) that the output is identical to the input to within a complex, frequency-dependent gain factor $H(\Omega)$, which is defined from the impulse response of the system as

$$\frac{y(n)}{x(n)} = H(\Omega) = \sum_{k=-\infty}^{+\infty} h(k)e^{-j\Omega k} \tag{1.40}$$

This gain factor is often referred to as the **frequency response** or the **frequency function** of the system. Taking the magnitude and argument of the frequency response, **Bode**-type plots, showing gain and phase shift can be

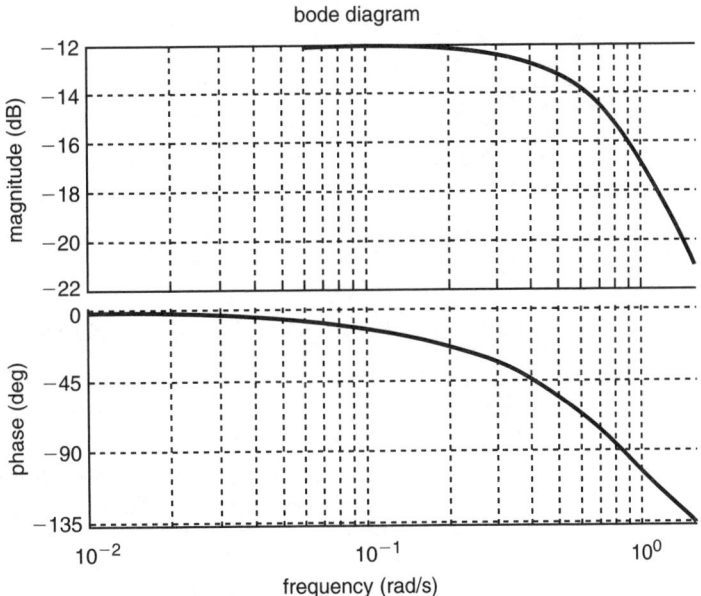

Figure 1.6 *Bode plot of the frequency function of the example system. The upper plot is the gain function (log–log scaling) and the lower is the phase shift function (lin–log scaling)*

obtained (see Figure 1.6). The Bode plot was introduced by H.W. Bode in 1945. The upper diagram shows the **gain function**, i.e. the magnitude of the frequency response, $A(\Omega) = |H(\Omega)|$. Commonly, the gain is plotted using decibels, resulting in a log-gain versus log-frequency scaling. The lower plot is the **phase shift function**, $\varphi(\Omega) = \angle H(\Omega)$ shown in linear-phase versus log-frequency scaling.

Note that the frequency response $H(\Omega) = H(e^{j2\pi(\omega/\omega_s)})$ is essentially the discrete Fourier transform (DFT) (see Chapter 5) of the impulse response $h(n)$. A well-known fact can be seen; that **a convolution in the time domain corresponds to a multiplication in the frequency domain** and vice versa.

In this section, we have demonstrated a number of ways to represent mathematically a signal processing operation (Figure 1.3). It is convenient to work with these models "off-line" in a powerful PC, workstation or minicomputer, using floating-point arithmetic. It is, however, often a challenge to migrate the system into a single-chip, fixed-point, digital signal processor required to perform in real time. "Smart" algorithms and "tricks" are often needed to keep up processing speed and to avoid numerical truncation and overflow problems.

1.3 Common filters

1.3.1 Filter architectures

Since digital filters are nothing but computer program code, there are of course an infinite number of possible filter architectures and variations. In this section, however, some common filter architectures are shown.

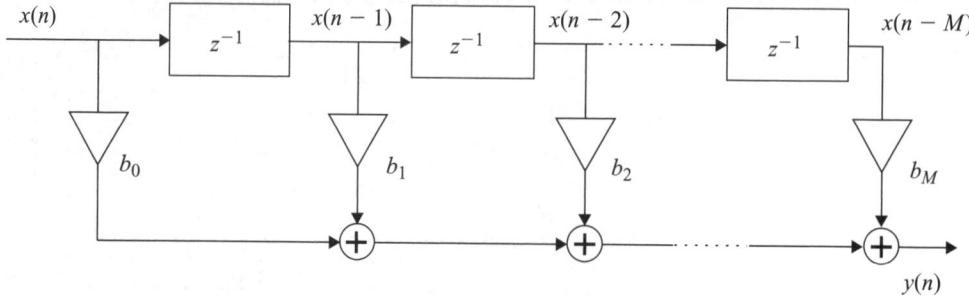

Figure 1.7 *Non-recursive (FIR) filter having length* M *with weights* b_j

1.3.1.1 The non-recursive filter

This filter (Figure 1.7) is sometimes denoted **tapped delay-line filter** or **transversal filter** or **FIR filter** since it has a **finite impulse response**. Using the convolution model, the response can be expressed as

$$y(n) = \sum_{j=0}^{M} h(j)x(n-j) \tag{1.41}$$

where M is the length of the filter. If we apply a unit pulse to the input of the system, i.e. $x(n) = \delta(n)$, it is straightforward to realize that the impulse response $h(j)$ of the system is directly obtained from the weight (gain) of the taps of the filter

$$h(j) = b_j \tag{1.42}$$

If we prefer a z-plane representation of equation (1.41), one of the nice features of the FIR filter is apparent

$$H(z) = b_0 + b_1 z^{-1} + b_2 z^{-2} + \cdots + b_M z^{-M} \tag{1.43}$$

Since the FIR filter does not have any poles, it is always guaranteed to be stable. Another advantage is that if the weights are chosen to be symmetrical (Lynn and Fuerst, 1998), the filter has a **linear-phase response**, i.e. all frequency components experience the same time delay through the filter. There is no risk of distortion of compound signals due to phase shift problems. Further, knowing the amplitude of the input signal, $x(n)$, it is easy to calculate the maximum amplitude of the signals in different parts of the system. Hence, numerical overflow and truncation problems can easily be eliminated at design time (see Chapter 9).

The drawback with the FIR filter is that if sharp cut-off filters are needed so is a high-order FIR structure, which results in long delay lines. FIR filters having hundreds of taps are however common today, thanks to low-cost integrated circuit technology and high-speed digital signal processors.

1.3.1.2 The recursive filter

Recursive filters, sometimes called **IIR** filters use a feedback structure (Figure 1.8), and have an **infinite impulse response**. Borrowing some ideas

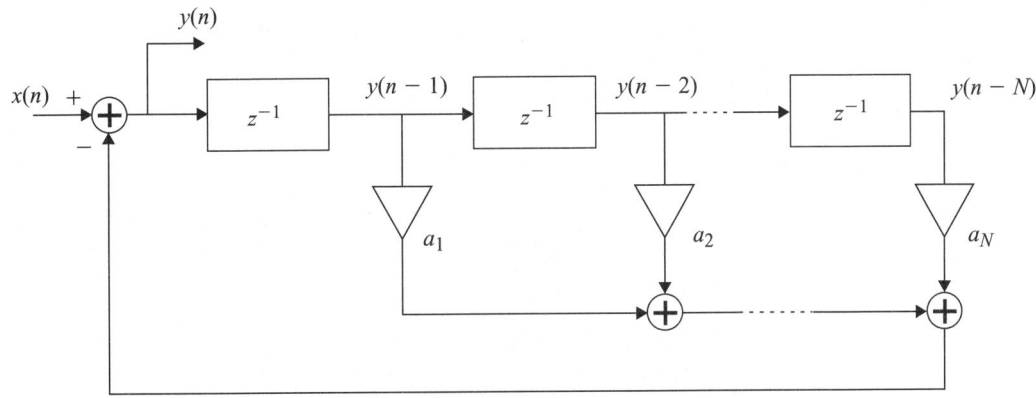

Figure 1.8 *Recursive (IIR) filter having length* N *with weights* a_i

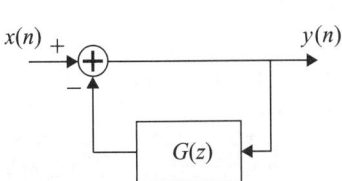

Figure 1.9 *The IIR filter seen as an FIR filter in a feedback loop*

from control theory, an IIR filter can be regarded as an FIR filter inserted in a feedback loop (Figure 1.9). Assume that the FIR filter has the transfer function $G(z)$, the transfer function of the total feedback structure, i.e. the IIR filter is then

$$H(z) = \frac{1}{1 + G(z)} = \frac{1}{1 + a_1 z^{-1} + a_2 z^{-2} + \cdots + a_N z^{-N}} \tag{1.44}$$

The IIR filter only has poles, hence it is of great importance to choose the weights a_i in such a way that the poles stay inside the unit circle to make sure the filter is stable. Since the impulse response is infinite, incoming samples will be "remembered" by the filter. For this reason, it is not easy to calculate the amplitude of the signals inside the filter in advance, even if the amplitude of the input signal is known. Numerical overflow problems may occur in practice.

The advantage of the IIR filter is that it is quite easy to build filters with sharp cut-off properties, using only a few delay elements.

Pure IIR filters are commonly used for alternating current (AC) coupling and smoothing (averaging) but it is more common to see combinations of FIR and IIR filter structures. Such a structure is the second-order combination shown in Figure 1.10, having the transfer function

$$H(z) = \frac{b_0 + b_1 z^{-1} + b_2 z^{-2}}{1 + a_1 z^{-1} + a_2 z^{-2}} \tag{1.45}$$

Of course, higher-order structures than two can be built if required by the passband specifications. It is however more common to combine a number of structures of order two in cascade or parallel to achieve advanced filters. In this way, it is easier to obtain numerical stability and to spot potential numerical problems.

Note! If a filter has zeros **canceling** the poles (i.e. having the same location in the complex plane), the total transfer function may have only zeros. In such a case, the filter will have a finite impulse response (FIR), despite the fact that the structure may be of the recursive type and an infinite impulse response (IIR) is expected. Hence, a non-recursive filter (FIR filter) **always** has a finite impulse

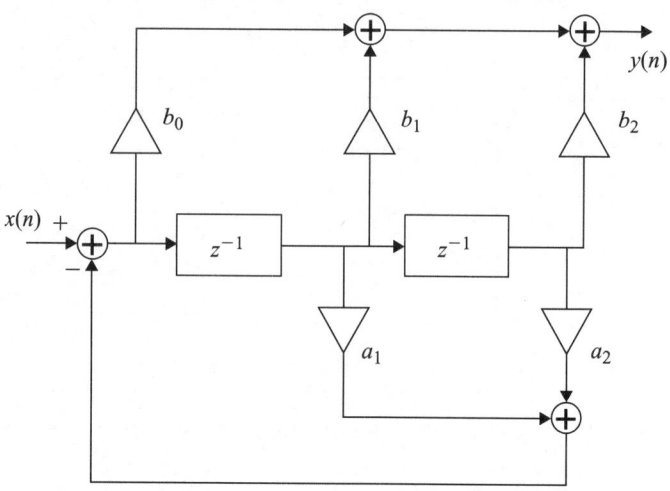

Figure 1.10 *Combined second-order FIR and IIR filter structure. Note: only two delay elements are needed*

response, while a recursive filter commonly has an infinite impulse response, but may in some cases have a finite impulse response.

If for instance a non-recursive filter is transformed into a recursive structure, the recursive filter will have the same impulse response as the original non-recursive filter, i.e. an FIR. Consider for example a straightforward non-recursive **averaging filter** having equal weights

$$y(n) = \frac{1}{N} \sum_{i=0}^{N-1} x(n-i) \tag{1.46}$$

The filter has the transfer function

$$H_{\mathrm{N}}(z) = \frac{Y(z)}{X(z)} = \frac{1}{N} \sum_{i=0}^{N-1} z^{-i} \tag{1.47}$$

Obviously, this filter has a finite impulse response. To transform this FIR filter into a recursive form, we try to find out how much of the previous output signal $y(n-1)$ that can be reused to produce the present output signal $y(n)$

$$y(n) - y(n-1) = \frac{1}{N} \sum_{i=0}^{N-1} x(n-i) - \frac{1}{N} \sum_{i=0}^{N-1} x(n-i-1)$$

$$= \frac{1}{N} x(n) - \frac{1}{N} x(n-N) \tag{1.48}$$

The corresponding recursive filter structure can hence be written

$$y(n) = y(n-1) + \frac{1}{N} x(n) - \frac{1}{N} x(n-N) = y(n-1) + \frac{1}{N} (x(n) - x(n-N)) \tag{1.49}$$

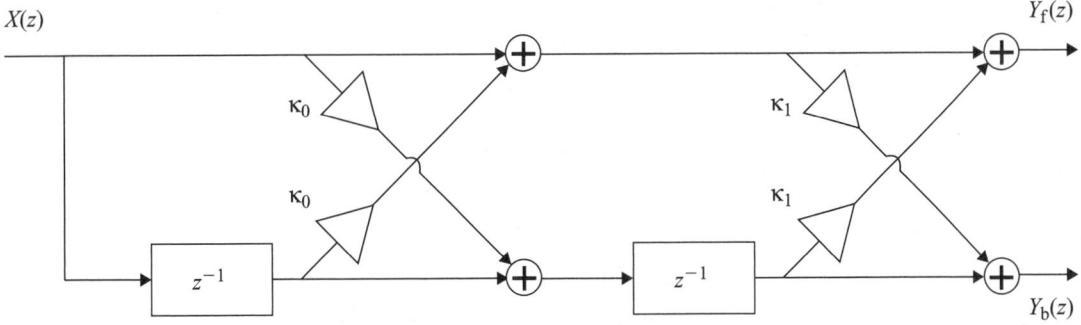

Figure 1.11 *Example of an all-zero lattice filter consisting of two lattice elements*

having the transfer function

$$H_R(z) = \frac{Y(z)}{X(z)} = \frac{1}{N} \cdot \frac{1 - z^{-N}}{1 - z^{-1}} \tag{1.50}$$

This is indeed a recursive filter having the same impulse response as the first filter and hence an FIR. Another interesting observation concerns the implementation of the filter. In the first case, one multiplication and $N - 1$ additions are needed, while in the second case only one multiplication, one subtraction and one addition are required to perform the same task. Since digital filters are commonly implemented as computer programs, less computational burden implies faster execution and shorter processing time (see Chapter 9). In this example, the recursive algorithm seems advantageous, but this is not generally true.

1.3.1.3 The lattice filter

Besides the standard filter structures discussed above, there are many specialized ones for specific purposes. One example of such a structure is the **lattice filter** (Orfandis, 1985; Widrow and Stearns, 1985) which is commonly used in adaptive processing and particularly in linear prediction. An example lattice filter consisting of two lattice elements is shown in Figure 1.11.

There are many variations of lattice structures. This example is an all-zero version, i.e. a structure having a transfer function containing only zeros (no poles), so it may be regarded as a lattice version of a non-recursive filter (FIR). The filter has one input and two outputs. The upper output signal is often denoted the **forward prediction error**, while the lower output signal is called the **backward prediction error**. The reason for this is explained below.

The transfer function, with respect to the forward prediction error, has the general form

$$H_f(z) = \frac{Y_f(z)}{X(z)} = \sum_{i=0}^{L} b_i z^{-i} \tag{1.51}$$

where $b_0 = 1$ and L is the number of lattice elements in the filter. In our example $L = 2$, hence

$$Y_f(z) = X(z) + X(z)b_1 z^{-1} + X(z)b_2 z^{-2} \tag{1.52}$$

The corresponding difference equation is

$$y_f(n) = x(n) + b_1 x(n-1) + b_2 x(n-2) = x(n) - \hat{x}(n) \tag{1.53}$$

Now, if we assume that we can find filter coefficients b_i in such a way that $y_f(n)$ is small or preferably equal to zero for all n, we can interpret $\hat{x}(n)$ as the **prediction** of the input signal $x(n)$ based on $x(n-1)$ and $x(n-2)$ in this example. Hence, we are able to predict the input signal one step **forward** in time. Further, under these assumptions it is easy to see that $y_f(n) = x(n) - \hat{x}(n)$ is the forward prediction error.

In a similar way, it can be shown that the transfer function with respect to the backward prediction error (the lower output signal) has the general form

$$H_b(z) = \frac{Y_b(z)}{X(z)} = \sum_{i=0}^{L} b_{L-i} z^{-i} \tag{1.54}$$

where $b_0 = 1$ and L is the number of lattice elements in the filter. In our example $L = 2$, hence

$$Y_b(z) = X(z)b_2 + X(z)b_1 z^{-1} + X(z)z^{-2} \tag{1.55}$$

The corresponding difference equation is

$$y_b(n) = x(n-2) + x(n)b_2 + x(n-1)b_1 = x(n-2) - \hat{x}(n) \tag{1.56}$$

In this case $\hat{x}(n)$ is the **prediction** of the input signal $x(n-2)$ based on $x(n-1)$ and $x(n)$. Hence, we are able to predict the input signal one step **backward** in time and $y_b(n) = x(n-2) - \hat{x}(n)$ is the backward prediction error.

The procedure for determining the lattice filter coefficients (Orfandis, 1985; Widrow and Stearns, 1985) κ_i is not entirely easy and will not be treated in this book. There are of course other types of filters which are able to predict signal sample values more than only one step ahead. There are also more advanced types, e.g. non-linear predictors. An important application using predictors is data compression (see Chapter 7). The basic idea is to use the dependency between present samples and future samples, thereby reducing the amount of information.

1.3.2 Filter synthesis

So far we have discussed some possible filter architectures, but how to determine the numerical filter coefficients for a specific application? To start with, there are some basic limitations. Consider the following example: using digital signal processing, we would like to build the perfect software radio receiver (RX) having very good selectivity. Such a receiver would have a very good, preferably "brick-wall-type", bandpass filter to allow only signals from the desired transmitter (TX) with frequency f_c, while strongly rejecting signals

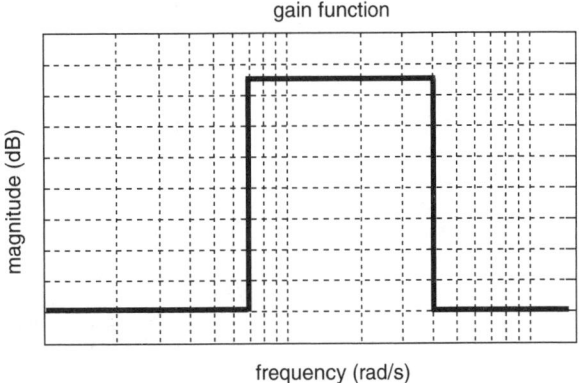

gain function

Figure 1.12 *Gain function of an unrealistic "brick-wall" bandpass filter*

transmitted on nearby frequencies, outside the bandwidth B (we assume that $B \ll f_c$). Figure 1.12 shows the gain function of the bandpass filter of our dreams.

For frequencies $f < f_c - (B/2)$ and $f_c + (B/2) < f$, i.e. the **stopband**, the filter gain is close to zero, i.e. no signals are let through, while in the **passband**, $f_c - (B/2) < f < f_c + (B/2)$ the gain is equal to one. Intuitively, we realize that such a filter can hardly be built using analog, electronic components like capacitors and inductors. But what about digital filters implemented as program code on a computer?

By making an inverse Fourier transform of the gain function A in Figure 1.12, we could obtain the impulse response $h(n)$ of the filter. This would in turn give us all filter coefficients b_n needed to design an FIR filter (compare to equations (1.40) and (1.42)). Unfortunately, we find the impulse response to be a sinc function (compare to equation (1.9)) of the type

$$h(n) = B \operatorname{sinc}\left(\frac{B}{f_s}n\right) \quad \text{for } -\infty < n < \infty \tag{1.57}$$

A part of the impulse response is shown in Figure 1.13. One problem is that the impulse response is of IIR type, i.e. infinite. Since we cannot build an FIR filter with an infinite number of taps, we have to use only a limited number of taps, which will indeed degrade the performance of the filter. Why cannot we build an IIR filter then? The next problem that strikes us is the fact that we have a non-causal impulse response, i.e. we need to know **future** values of the signal which are not available yet. This can of course be amended by introducing an infinitely long delay function in the system. Since there is obviously a very limited market for such filters, we must conclude that the perfect bandpass filter can never be built. We will have to stick to approximations of the gain curve sketched in Figure 1.12. This is mainly what filter synthesis is all about: finding acceptable compromises (not easy) that can be implemented in reality and at a reasonable cost. (Like human life in general?) In this text we have so far only discussed bandpass filter functions, however, the same reasoning applies to low-pass, high-pass and bandstop filters.

So, what parameters should be taken into account when making a filter passband approximation? There are three common aspects: the **slope** of the

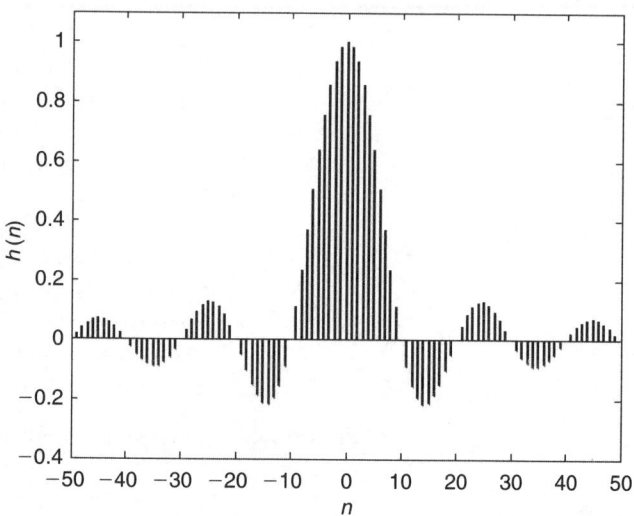

Figure 1.13 *Impulse response (the sinc function) of the unrealistic bandpass filter*

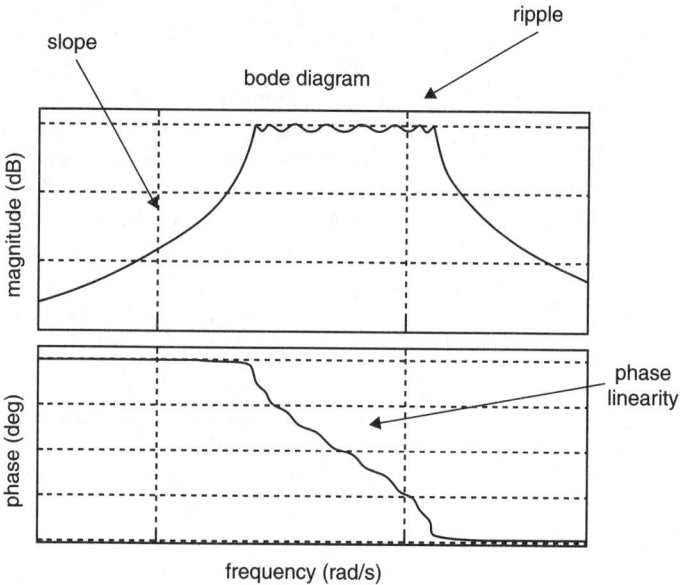

Figure 1.14 *Example of a Bode plot showing a realistic bandpass filter*

passband curve, **ripple** in the passband and **phase linearity** (see the example Bode plot in Figure 1.14).

In the dream filter above, the slopes of the passband curve were infinitely steep, which turned out to be impossible, but in many cases, we would like to get as steep slopes as possible. Ripple is unwanted gain variations in the passband. In the ideal case, the gain is exactly one everywhere in the passband.

Phase linearity, finally, means that the phase shift of the filter in the passband should ideally be proportional to the frequency, i.e.

$$\phi(f) = -kf \tag{1.58}$$

since the group delay τ_g for the signal to pass through the filter is

$$\tau_g = -\frac{1}{2\pi}\frac{d\phi(f)}{df} \tag{1.59}$$

Inserting equation (1.58) into equation (1.59) we realize that having a phase function of the filter according to equation (1.58), the group delay for all frequencies will be the same.

$$\tau_g = -\frac{1}{2\pi}\frac{d}{df}(-kf) = \frac{k}{2\pi} \tag{1.60}$$

This is of course a desirable property of a filter, since sending a composite signal consisting of many frequency components through the filter, all the components will reach the output simultaneously, thus avoiding phase distortion.

It is not surprising that the three good properties above cannot be achieved at the same time. There is always a trade-off. Below, some commonly used approximations for low-pass filters (bandpass filter with $f_c = 0$) are shown. In the expressions below, ω is the normalized, angular frequency defined in such a way that the gain of the filter is $1/\sqrt{2} \approx 0.707$ or -3 dB at $\omega = \omega_0 = 1$. The parameter $\omega_0 = 2\pi f_0$ called the cut-off frequency is commonly used when specifying filters.

- The **Butterworth** approximation is also called the maximally flat approximation, implying that there is no ripple in the passband or in the stopband. On the other hand, neither the slope of the passband curve nor the phase linearity is very impressive. The Butterworth approximation of the gain function is

$$A(\omega) = \frac{1}{\sqrt{1 + \omega^{2k}}}$$

where k is the order of the filter.
- The **Chebyshev** approximation has a steeper slope than the Butterworth type, but has ripple in the passband and poor phase linearity. The Chebyshev approximation of the gain function can be written as

$$A(\omega) = \frac{1}{\sqrt{1 + \varepsilon^2 C_k^2(\omega)}}$$

where ε is a ripple constant and $C_k(\omega)$ is the Chebyshev polynomial defined as

$$C_k(\omega) = \begin{cases} \cos(k \arccos(\omega)) & \text{for } |\omega| \leq 1 \\ \cosh(k \operatorname{arccosh}(\omega)) & \text{for } |\omega| > 1 \end{cases}$$

The magnitude of the passband ripple, i.e. the variations in gain in the passband can be determined by (expressed in decibels)

$$\Delta A = 10 \log(1 + \varepsilon^2)\,\text{dB}$$

- The **Cauer** approximation is an extension of the Chebyshev approximation in the sense that ripple is allowed in the stopband also, resulting in an even steeper slope. The Chebyshev and Cauer approximations belong to the class of **elliptic filters**.
- The **Bessel–Thomson** approximation has good phase linearity, but the slope of the passband is not very impressive.

There are two main ways of synthesizing filters: the indirect method and the direct method.

1.3.2.1 Indirect filter synthesis

The idea of the indirect method is first to design an analog filter using well-known classical methods and then transform the analog filter function into a digital one. The advantage is that there are many well-known design methods for analog filters around and old filter designs can be reused. There are however a couple of drawbacks. Firstly, analog filters are traditionally implemented using electronic components like inductors, capacitors, resistors and operational amplifiers. Therefore, there are limitations to what type of analog transfer functions can be achieved. Typically, an analog transfer function formulated using the Laplace transform would have the structure of a fraction of polynomials

$$G(s) = \frac{Y(s)}{X(s)} = \frac{b_0 + b_1 s + b_2 s^2 + \cdots}{a_0 + a_1 s + a_2 s^2 + \cdots} \tag{1.61}$$

This means that there are only a limited number of filter types that can be designed using the indirect method.

Secondly, the process of transforming the analog transfer function in the s-plane into a digital one in the z-plane is not without problems. Once again we have a compromise situation. There is no perfect transformation method, i.e. there is no method that transforms **all** the properties of an analog filter to a digital counterpart. Below, some common transform methods are presented.

The **impulse invariance method** is based on the idea that the impulse response of the analog filter with transfer function $G(s)$ should be identical to the impulse response of a corresponding digital filter having the transfer function $H(z)$

$$h(t) = L^{-1}[1 \cdot G(s)] = Z^{-1}[1 \cdot H(z)] = h(n) \Rightarrow H(z) = Z\left[L^{-1}[G(s)]\right] \tag{1.62}$$

where the operator L^{-1} is the inverse Laplace transform, Z^{-1} is the inverse z-transform and Z is the z-transform. The following are expressed in equation (1.62): the impulse response of the analog system is the unit impulse 1 times the transfer function $G(s)$. This impulse response, expressed in the time domain, should be equal to the impulse response of the digital system, given by the unit impulse 1 times the transfer function $H(z)$. Solving for $H(z)$ means first applying the inverse Laplace transform to $1 \cdot G(s)$ and then applying the z-transform. The coupling between continuous time and discrete time is

$$t = \frac{n}{f_s} \tag{1.63}$$

As an example, assume that we have an analog transfer function as in equation (1.61). Factorizing the function and splitting it into partial fractions, we obtain

$$G(s) = \frac{r_1}{s - p_1} + \frac{r_2}{s - p_2} + \cdots \tag{1.64}$$

where p_i is the pole of polynomial i and r_i is the corresponding residual, etc. Applying the inverse Laplace transform, every term in equation (1.64) will transform to

$$h(t) = r_1 e^{p_1 t} + r_2 e^{p_2 t} + \cdots \tag{1.65}$$

Replacing the continuous time with discrete time using equation (1.63), and approximating the continuous exponential function with a discrete step function (step width equal to the sampling period $1/f_s$) gives

$$h(n) = \frac{r_1}{f_s} e^{p_1(n/f_s)} + \frac{r_2}{f_s} e^{p_2(n/f_s)} + \cdots \tag{1.66}$$

Taking the z-transform of equation (1.66)

$$H(z) = \frac{r_1/f_s}{1 - e^{p_1/f_s} z^{-1}} + \frac{r_2/f_s}{1 - e^{p_2/f_s} z^{-1}} + \cdots \tag{1.67}$$

After some tedious housekeeping we finally end up with a digital transfer function of the type

$$H(z) = \frac{Y(z)}{X(z)} = \frac{b_0 + b_1 z^{-1} + b_2 z^{-2} + \cdots}{1 + a_1 z^{-1} + a_2 z^{-2} + \cdots} \tag{1.68}$$

which can easily be implemented as a standard IIR filter as in equation (1.45). **Note** that this method only guarantees that the digital filter has the same impulse response as the analog "original" filter. No promises are made about, for instance, the frequency function. This method is used in signal processing applications.

Starting from similar ideas, the **step invariance method** can be formulated as

$$L^{-1}\left[\frac{G(s)}{s}\right] = Z^{-1}\left[\frac{H(z)}{1 - z^{-1}}\right] \Rightarrow H(z) = (1 - z^{-1})Z\left[L^{-1}\left[\frac{G(s)}{s}\right]\right] \tag{1.69}$$

In this case, the analog and digital filters have the same step response. Further, the **ramp invariance method** will be

$$L^{-1}\left[\frac{G(s)}{s^2}\right] = Z^{-1}\left[\frac{z^{-1}H(z)}{(1 - z^{-1})^2}\right] \Rightarrow H(z) = \frac{(1 - z^{-1})^2}{z^{-1}}Z\left[L^{-1}\left[\frac{G(s)}{s}\right]\right] \tag{1.70}$$

where the analog and digital filters will have the same ramp response. The step invariance and ramp invariance methods are commonly used in control system applications.

A commonly used transform, both for digital signal processing and for digital control systems, is the bilinear transform (**BLT**), also called Tustin's approximation. The idea behind the bilinear transform is to find a way of formulating a digital transfer function $H(z)$ that resembles the analog transfer function $G(s)$ as closely as possible considering the frequency response. No guarantees regarding, for instance, the impulse response are given. Omitting the details that can be found in, for instance (Denbigh, 1998), we conclude that the method of the bilinear transform is to replace the Laplace variable s in the analog transfer function by an expression in the z variable as

$$s \rightarrow 2f_{\mathrm{s}}\frac{1 - z^{-1}}{1 + z^{-1}} = 2f_{\mathrm{s}}\frac{z - 1}{z + 1} \tag{1.71}$$

Unfortunately, the shape of the gain function of the analog filter $|G(s)|$ and the digital counterpart $|H(z)|$, obtained by bilinear transformation, differs. This is especially true for higher frequencies, approaching the Nyquist frequency $f_{\mathrm{s}}/2$. To counteract this problem, **pre-warping** is used. Pre-warping means that when designing the initial analog filter, which will later be transformed to a digital one, frequencies used to specify the filter are modified according to

$$f_{\mathrm{A}} = \frac{f_{\mathrm{s}}}{\pi} \tan\left(\pi\left(\frac{f_{\mathrm{D}}}{f_{\mathrm{s}}}\right)\right) \tag{1.72}$$

where f_{D} is the frequency in question for the final digital filter, while the corresponding analog filter should use the pre-warped frequency f_{A}. For small frequencies, the difference between f_{D} and f_{A} is negligible, i.e. $f_{\mathrm{A}} \approx f_{\mathrm{D}}$, while for frequencies approaching the Nyquist frequency, the deviation may be considerable. Let us illustrate the bilinear transformation and pre-warping with an example.

Assume we are asked to design a digital Butterworth filter of order two with cut-off frequency 250 Hz. The sampling frequency of the system is 1 kHz. First, we have to do frequency pre-warping. Inserting f_{D} and f_{s} into equation (1.72) we obtain

$$f_{\mathrm{A}} = \frac{1 \cdot 10^3}{\pi} \tan\left(\pi\frac{250}{1 \cdot 10^3}\right) = 318$$

This means that we need to design the analog filter to have a cut-off frequency of 318 Hz in order to obtain a transformed digital filter with the desired cut-off frequency 250 Hz. The next step is to design the analog Butterworth filter, which can be done using conventional methods, for instance, using MATLAB™ the resulting transfer function will be (see also equation (1.61))

$$G(s) = \frac{b_0}{a_0 + a_1 s + a_2 s^2} \tag{1.73}$$

where the coefficients are $b_0 = 4 \cdot 10^6$, $a_0 = 4 \cdot 10^6$, $a_1 = 2.83 \cdot 10^3$, $a_2 = 1$. Applying the bilinear transform equation (1.71) to the analog transfer function (1.73) we obtain

$$H(z) = \frac{b_0}{a_0 + a_1 2f_s[(z-1)/(z+1)] + a_2(2f_s[(z-1)/(z+1)])^2}$$

$$= \frac{b_0(z+1)^2}{a_0(z+1)^2 + 2f_s a_1(z-1)(z+1) + 4f_s^2 a_2(z-1)^2}$$

$$= \frac{b_0(z^2 + 2z + 1)}{a_0(z^2 + 2z + 1) + 2f_s a_1(z^2 - 1) + 4f_s^2 a_2(z^2 - 2z + 1)}$$

$$= \frac{b_0(z^2 + 2z + 1)}{z^2(a_0 + 2f_s a_1 + 4f_s^2 a_2) + z(2a_0 - 8f_s^2 a_2) + (a_0 - 2f_s a_1 + 4f_s^2 a_2)}$$

$$= \frac{b_0(1 + 2z^{-1} + z^{-2})}{(a_0 + 2f_s a_1 + 4f_s^2 a_2) + z^{-1}(2a_0 - 8f_s^2 a_2) + z^{-2}(a_0 - 2f_s a_1 + 4f_s^2 a_2)}$$

$$= \frac{b_0}{(a_0 + 2f_s a_1 + 4f_s^2 a_2)}$$

$$\cdot \frac{(1 + 2z^{-1} + z^{-2})}{1 + z^{-1}\dfrac{(2a_0 - 8f_s^2 a_2)}{(a_0 + 2f_s a_1 + 4f_s^2 a_2)} + z^{-2}\dfrac{(a_0 - 2f_s a_1 + 4f_s^2 a_2)}{(a_0 + 2f_s a_1 + 4f_s^2 a_2)}}$$

Inserting the filter coefficients, we obtain the final, digital transfer function

$$H(z) = \frac{Y(z)}{X(z)} = 0.293 \frac{1 + 2z^{-1} + z^{-2}}{1 - 0.001z^{-1} + 0.172z^{-2}} \tag{1.74}$$

The filter having the transfer function (1.74) can easily be implemented as a standard second-order IIR filter shown in Figure 1.10 (see equation (1.45)).

Figure 1.15(a) shows the desired gain response (solid line) of a Butterworth filter with cut-off frequency 250 Hz. The dotted line shows the gain function of the digital filter obtained using pre-warping as above. This filter is the bilinear transformation of an analog filter designed for cut-off frequency 318 Hz. In Figure 1.15(b), the dotted line is the gain function of a digital filter designed **without** pre-warping, i.e. the bilinear transform of an analog filter designed with cut-off frequency 250 Hz.

Another, simpler transform is the **Euler transform**, sometimes used in control system applications. The Euler transform consists of variable substitution somewhat simpler than for the bilinear transform above

$$s \to f_s \frac{z-1}{z} = f_s(1 - z^{-1}) \tag{1.75}$$

The idea behind this transform is that s in the Laplace method represents a derivative, while the right side of equation corresponds to a numerical approximation of a differentiation, see also Section 1.4.1. This transform works better the higher the sampling frequency compared to the signal frequencies.

1.3.2.2 Direct filter synthesis

In direct filter synthesis, the digital filter is designed directly from the filter specifications, without using any analog prototype to transform. The advantage

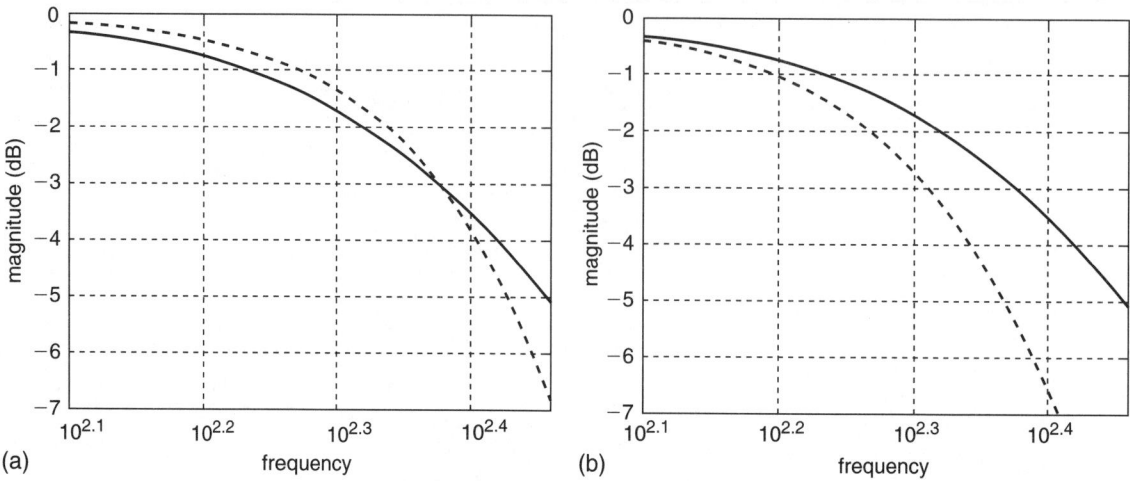

Figure 1.15 *(a) Desired gain response (solid line) of a Butterworth filter with cut-off frequency 250 Hz. The dotted line is the response of the digital filter obtained using pre-warping and bilinear transformation (analog filter designed for cut-off frequency 318 Hz). (b) The dotted line is the response of a digital filter designed **without** pre-warping (analog filter designed with cut-off frequency 250 Hz)*

is that new filter types can be achieved, since we are no longer hampered by the limitations implied by the use of analog, electronic components.

The first direct method presented is the **Fourier method** also called the **Window method**. In this method (Denbigh, 1998) one first determines a desired gain function $A(f)$. Since the transfer function of a filter is the Fourier transform of the impulse response (see equation (1.40)), the **inverse** Fourier transform of the desired gain function would give the impulse response

$$h(n) = \frac{1}{f_s} \int_{-\frac{f_s}{2}}^{\frac{f_s}{2}} A(f) e^{j2\pi n(f/f_s)} \, df \qquad (1.76)$$

Since we know that the impulse response of an FIR filter is exactly the values of the filter taps (equation (1.42)), implementing the filter is straightforward. The problem associated with this method is shown in the beginning of this section. Often the desired gain function results in infinite, non-causal impulse response. Hence, compromises must be made and filter lengths cut using some **windowing** function (see Chapter 5), implying that the resulting gain response will deviate somewhat from the desired one.

A dual of the Window method, briefly presented above, is the **Frequency sampling** method. In this method, the desired frequency response $H(\Omega_k)$ is defined at crucial frequencies. Frequency points in between the specified points are interpolated, and the impulse response is obtained using the inverse DFT. The interpolation frequency points are important, since too abrupt transitions in the desired frequency response function may result in considerable ripple in the frequency response of the final implementation.

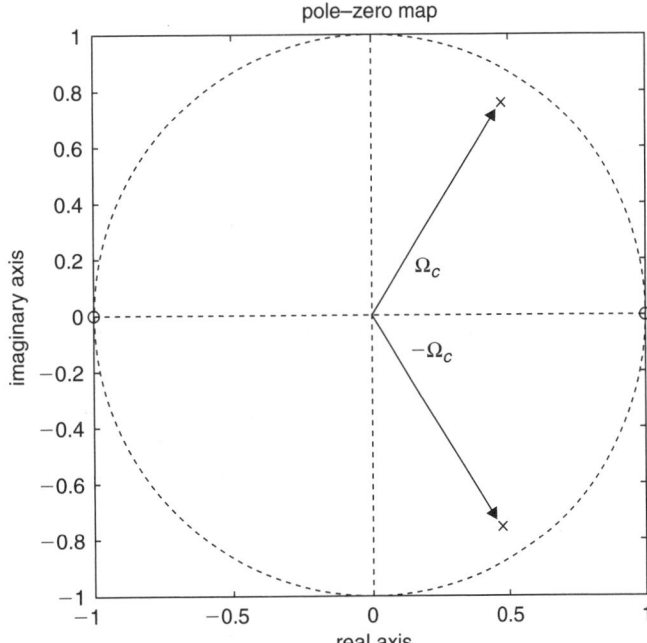

Figure 1.16 *A pole–zero plot for a bandpass filter; poles close to the unit circle result in a narrow passband filter*

Another direct filter design method is the **simulation method**, in which the filter is simulated and poles and zeros are placed manually or automatically in the complex z-plane until the desired gain and/or phase function is obtained. The method may seem a bit *ad hoc*, but it is nevertheless frequently used. When placing the poles and zeros in the z-plane, there are some elementary rules:

- The frequency axis in a gain function plot corresponds to the unit circle in the z-plane (see equation (1.38)). Traversing the positive frequency axis in a gain function plot corresponds to moving counterclockwise on the upper arc of the unit circle (see Figure 1.16).
- All poles have to be placed **inside** the unit circle for stability reasons; the location of zeros does not matter from a stability point of view.
- The distance from a pole to a given position on the unit circle, i.e. a given frequency, is inversely proportional to the gain. The closer a pole is located to the unit circle, the higher the gain for frequencies in vicinity of the pole. A pole should preferably not be placed **on** the unit circle, since that will create an oscillator, or in the worst case, lead to instabilities due to numerical truncation effects.
- The distance from a zero to a given position on the unit circle, i.e. a given frequency, is proportional to the gain. The closer a zero is located to the unit circle, the lower the gain for frequencies in vicinity of the zero. A zero can be placed on the unit circle, resulting in zero gain at the given frequency.

- Single poles or zeros can only appear on the real axis.
- Pairs of poles or zeros appear as complex conjugate pairs, i.e. $p = x \pm jy$.
- Poles or zeros at the origin do not affect the gain function, but contribute to the phase shift function of the filter.
- A pole in an improper location (outside the unit circle) can be cancelled by placing a zero on exactly the same place, thus stabilizing the system. One should be aware, however, that this method is risky. If, for instance, for numerical truncation reasons the zero moves away just a little, we may be stuck with an unstable system.

The example in Figure 1.16 shows a bandpass filter with center frequency $\Omega_c = 2\pi(f_c/f_s)$ which corresponds to the angle $\pm\Omega_c$ in the figure. The closer the poles are placed to the unit circle, the more narrow the passband B, i.e. the higher the Q-factor of the filter.

The **McClellan–Parks method** based on the **Remez exchange algorithm** is a common method for designing optimal linear-phase FIR filters, complying to the passband specification as closely as possible, while requiring as few taps as possible. The basic idea of this method is to approximate the desired gain function by a series of cosines of the form

$$A(\omega) = \sum_k b(k) \cos(\omega k)$$

An error weighting function $W_1(\omega)$ is defined, which governs how to penalize errors between the desired response $A_1(\omega)$ and the approximation $A(\omega)$ at different frequencies. For frequencies where $W_1(\omega)$ is large, a high degree of conformity to the desired gain function is demanded. Using the Remez exchange algorithm, the cosine coefficients $b(k)$ are found by minimizing the maximum error function

$$|W_1(\omega)(A_1(\omega) - A(\omega))| = \left| W_1(\omega) \left(A_1(\omega) - \sum_k b(k) \cos(\omega k) \right) \right|$$

From the cosine coefficients, the impulse response, i.e. the FIR filter taps, can be obtained using the inverse DFT and taking symmetry conditions into account to achieve phase linearity. The presentation above is highly simplified, since the details of the algorithm are quite complicated. A detailed presentation can be found, e.g. Cavicchi (2000).

1.4 Digital control systems

A basic, "analog" closed-loop control system is shown in the block diagram in Figure 1.17 (Tewari, 2002; Wilkie *et al.*, 2002). It consists of a **process** (or plant) having transfer function $G(s)$ (including actuators), a **transducer** with transfer function $H(s)$ (including signal conditioning) and a **controller** with transfer function $K(s)$, expressed in the Laplace domain. The signals appearing in the block diagram are: the **reference signal** or **set point** $R(s)$, the **process error signal** $E(s)$, the **controller output signal** $U(s)$, the **process disturbance signal** $D(s)$, the **process output signal** $Y(s)$, the **measurement disturbance signal** $N(s)$ and the **measured output signal** $Y_m(s)$.

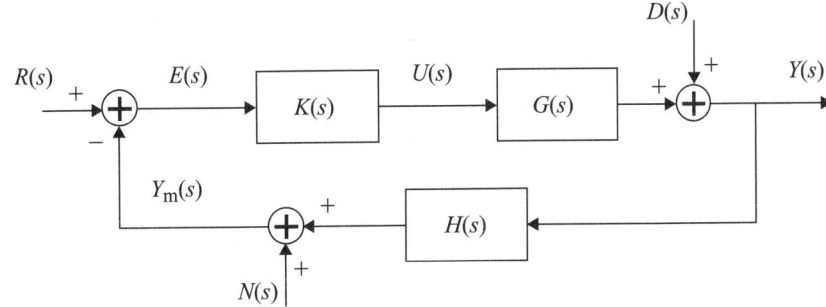

Figure 1.17 *Block diagram showing a typical closed-loop control system*

For example, the closed-loop system could be an electrical heater in a room where a desired and constant indoor temperature is required. The desired temperature is presented to the system as the electric reference signal $R(s)$ obtained from a temperature-setting knob on the wall. This signal is compared to the measured output signal $Y_m(s)$, which is the output of the temperature transducer measuring the actual temperature in the room. The measurement disturbance signal $N(s)$ represents imperfections in the temperature transducer and in this example, we assume that $N(s) = 0$. The difference between the desired and the actual temperature is the process error signal $E(s)$. The error signal is fed to the controller, which determines a proper controller output signal $U(s)$. If, for instance, the actual temperature is lower than desired, the error signal $E(s) = R(s) - Y_m(s)$ is positive. An appropriate measure of the controller would hence be to increase the output signal $U(s)$ or, in other words, increase the current to the electrical heater, increasing the heating power. The process disturbance signal $D(s)$ may represent an open window (negative sign) trying to lower the temperature in the room (we are up north now), or an electrical appliance in the room (negative sign) trying to increase the temperature.

Now, what are the characteristics of a good closed-loop control system? Let us first assume that the process disturbance is negligible, i.e. $D(s) = 0$. The **closed-loop transfer function** of the system will be

$$Y(s) = E(s)K(s)G(s) \qquad (1.77\text{a})$$

$$E(s) = R(s) - H(s)Y(s) \qquad (1.77\text{b})$$

Inserting equation (1.77b) into equation (1.77a) we get

$$
\begin{aligned}
Y(s) &= R(s)K(s)G(s) - Y(s)H(s)K(s)G(s) \\
&\Rightarrow Y(s)(1 + H(s)K(s)G(s)) = R(s)K(s)G(s) \qquad (1.78) \\
&\Rightarrow G_{\text{CL}}(s) = \frac{Y(s)}{R(s)} = \frac{K(s)G(s)}{1 + H(s)K(s)G(s)}
\end{aligned}
$$

Equation (1.78) is the transfer function for the entire system, i.e. it shows how the temperature $Y(s)$ is affected by the reference (set point) $R(s)$ signal. Some desirable properties of the system are:

- the output $Y(s)$ should follow the input $R(s)$ with good accuracy statically, i.e. the process error $E(s)$ should be small and $E(s) \to 0$ as $t \to \infty$

- the output $Y(s)$ should follow the input $R(s)$ dynamically well, i.e. a change in $R(s)$ should appear as a change $Y(s)$ in with as small a delay as possible
- the system should be stable; since the transfer function has poles, they have to be located in the left half of the s-plane (inside the unit circle for the z-plane).

Further, since the system is supposed to be linear we can set $R(s)=0$ to investigate how process disturbance $D(s)$ affects the output $Y(s)$, we get

$$Y(s) = E(s)K(s)G(s) + D(s) \tag{1.79a}$$

$$E(s) = -H(s)Y(s) \tag{1.79b}$$

$$
\begin{aligned}
Y(s) &= -Y(s)H(s)K(s)G(s) + D(s) \\
&\Rightarrow Y(s)(1 + H(s)K(s)G(s)) = D(s) \\
&\Rightarrow \frac{Y(s)}{D(s)} = \frac{1}{1 + H(s)K(s)G(s)}
\end{aligned} \tag{1.80}
$$

The transfer function (1.80) shows how the temperature $Y(s)$ is affected by the process disturbance $D(s)$. A desirable property of the system is:

- the output $Y(s)$ should be affected as little as possible by the process disturbance $D(s)$.

Now, the process $G(s)$ corresponds to an actuator (the heater) and some real world entity (the room). Commonly, the process design engineer has very limited possibilities in changing the transfer function $G(s)$. The same is true for the transducer system represented by $H(s)$. So, to be able to design a closed-loop control system having the four desired properties as above, the only thing we can do is to try to design the controller $K(s)$ in as smart way as possible. As expected, the four demands above of course counteract each other. If, for instance, a system is supposed to respond quickly it is likely to be unstable. So, in designing good controllers, we are again in a world of compromises.

Probably, the most important requirement of a closed-loop system is that it should be stable. From both equations (1.78) and (1.80) we realize that stability requires the denominator not to be zero, i.e.

$$1 + H(s)K(s)G(s) \neq 0$$

The term $H(s)K(s)G(s) = G_O(s)$ is denoted the **open-loop transfer function** of the system. If we study the open-loop transfer function in the frequency domain $s = j\omega$ we can conclude that to ensure stability the following conditions must be met

$$|G_O(\omega_i)| < 1 \quad \text{where} \quad \angle G_O(\omega_i) = -180° \tag{1.81}$$

That is, for the frequency ω_i where the phase shift of the open-loop transfer function is $-180°$, the magnitude has to be smaller than 1. This rule is often referred to as the simplified Nyquist criteria. Presenting the open-loop transfer function as a Bode plot, it is easy to find out if the system is stable and it is also easy to see how large the margins are. (Most real world open-loop transfer functions are of low-pass filter type.) There are a number of other methods to check for stability.

So far, we have only discussed "analog" systems, but the main problems are the same for digital control systems. An advantage of digital systems is that more sophisticated controllers can be designed rather than using classical, analog electronic components. The transfer functions and signal will of course be expressed using the z-transform instead of Laplace, i.e. we have: $H(z)$, $K(z)$, $G(z)$, $R(z)$ and so on. One detail that may cause problems is to get hold of the digital process function $G(z)$, since the real world processes are commonly "analog" by their nature.

Below, some controllers are discussed.

1.4.1 Proportional–integral–derivate controllers

The **proportional–integral–derivate (PID)** controller is a classical algorithm that can be found not only in numerous industrial applications, but also in diverse areas of everyday life. The analog version of the controller can be written as

$$u(t) = K \left(e(t) + \frac{1}{T_i} \int_0^t e(\tau)\, d\tau + T_d\, \frac{d}{dt} e(t) \right) \tag{1.82a}$$

or, using Laplace

$$K(s) = \frac{U(s)}{E(s)} = K \left(1 + \frac{1}{sT_i} + sT_d \right) \tag{1.82b}$$

Finding proper values of the three constants: K, proportionality; T_i, integration time and T_d, derivation time, to optimize the performance of the control system is a classical problem. Many different methods have been invented, but since the setting of these parameters is highly dependent on the properties of the plant, heuristics are common. The digital version of the PID controller can be found as follows. Analog integration can be approximated by digital summation

$$y(t) = \int_0^t x(\tau)\, d\tau \rightarrow y(k) = \sum_{i=1}^k x(i) = x(k) + y(k-1) \tag{1.83a}$$

having the z-transform

$$y(k) = x(k) + y(k-1) \Rightarrow \frac{Y(z)}{X(z)} = \frac{1}{1-z^{-1}} = \frac{z}{z-1} \tag{1.83b}$$

and analog derivation can be approximated by digital differentiation

$$y(t) = \frac{d}{dt} x(t) \rightarrow y(k) = x(k) - x(k-1) \tag{1.84a}$$

with the z-transform

$$y(k) = x(k) + x(k-1) \Rightarrow \frac{Y(z)}{X(z)} = 1 - z^{-1} \tag{1.84b}$$

Using the above, the digital version of the PID controller is

$$u(k) = K\left(e(k) + \frac{1}{f_s T_i}\sum_{i=1}^{k} e(i) + f_s T_d(e(k) - e(k-1))\right) \tag{1.85a}$$

where f_s is the sampling frequency, an alternative algorithm is

$$v(k) = v(k-1) + e(k)$$

$$u(k) = K\left(e(k) + \frac{1}{f_s T_i}v(k) + f_s T_d(e(k) - e(k-1))\right) \tag{1.85b}$$

1.4.2 Advanced controllers

The PID algorithm above is very common and sufficient for everyday applications. If, however, peak performance is required in, for instance, air and space applications, other more elaborate control algorithms must be employed. Some examples are given below.

1.4.2.1 Direct synthesis controller

A controller designed using the direct synthesis method is obtained in the following way. We are assumed to have good knowledge of the transfer function of the process $G(z)$ and the transducer system $H(z)$. Further, we have a good idea of what we would like the closed-loop transfer function to be like $G_{CL}(z)$. Starting out with the expression (1.78) modified for the z-plane, we can now solve for the transfer function $K(z)$ required obtaining the desired closed-loop transfer function

$$G_{CL}(z) = \frac{Y(z)}{R(z)} = \frac{K(z)G(z)}{1 + H(z)K(z)G(z)}$$

$$\Rightarrow K(z) = \frac{G_{CL}(z)}{G(z)(1 - G_{CL}(z)H(z))} \tag{1.86}$$

However, it is not possible to obtain a stable controller for some cases. If for instance the process has zeros outside the unit circle, they will show up as poles in the transfer function of the controller, thus resulting in an unstable controller. Another problem arises if there is a delay in the process. In such a case, the transfer function of the controller turns non-causal, i.e. needs to be able to look into the future. There are, however, other design methods for controllers used with processes having delay.

1.4.2.2 Pole placement controller

A controller designed using the pole placement method has a somewhat different structure than other controllers discussed so far (see Figure 1.18). This type of controller commonly performs very well and is capable of handling processes with delay.

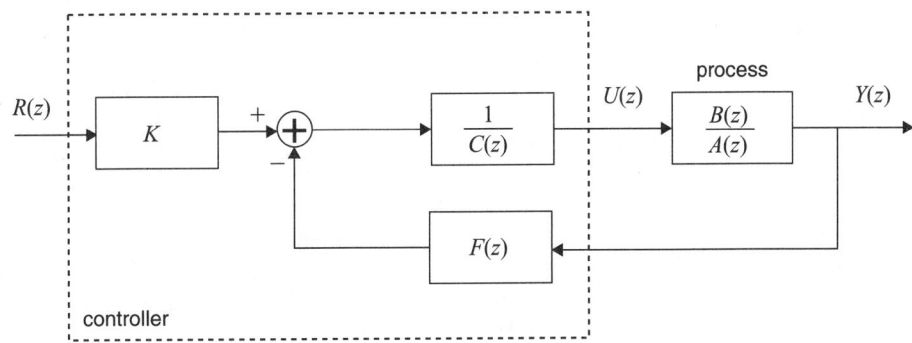

Figure 1.18 *A simplified closed-loop control system with a pole placement controller*

In Figure 1.18, we have assumed $H(z) = 1$ and that the process transfer function can be expressed as a quota between two polynomials, i.e. $G(z) = B(z)/A(z)$. The closed-loop transfer function of the system depicted in Figure 1.18 will be

$$G_{\mathrm{CL}}(z) = \frac{Y(z)}{R(z)} = \frac{KB(z)}{A(z)C(z) + B(z)F(z)} \tag{1.87}$$

The first thing to do is to determine the placement of the poles of the closed-loop transfer function. This is, of course, a delicate question well beyond the scope of this book. However, as a rule of thumb, the poles are commonly placed inside the unit circle (of course), not too close to the unit circle and in a sector ranging approximately $\Omega = \pm\pi/4$ in the z-plane. Once the location of the poles is determined, the pole placement polynomial $P(z)$ is formed having roots equal to the poles. The degree of this polynomial, i.e. the number of poles, is given by the degree of polynomial $A(z)$ plus the degree of polynomial $B(z)$ minus 1. Now, the pole placement equation is solved

$$P(z) = A(z)C(z) + B(z)F(z) \tag{1.88}$$

and the polynomials $C(z)$ and $F(z)$ are obtained. Finally, the constant K is determined using

$$K = \frac{P(1)}{B(1)} \tag{1.89}$$

The controller is complete.

1.4.2.3 Dead-beat controller

A **dead-beat controller** is a pole placement controller designed as in Section 1.4.2.2 above, using a pole placement polynomial placing all the poles at the origin

$$P(z) = \left(1 - 0 \cdot z^{-1}\right)^n = 1 \tag{1.90}$$

The closed-loop transfer function of a system with a dead-beat controller will be (compare to equation (1.87))

$$G_{\mathrm{CL}}(z) = \frac{Y(z)}{R(z)} = \frac{KB(z)}{1} = KB(z) \tag{1.91}$$

An interesting feature of such a controller is that it is converges very fast to the correct output signal $Y(z)$ after an abrupt change in the reference signal $R(z)$. The number of samples required to perform a perfect convergence is equal to the degree of the polynomial $B(z)$. A drawback with the controller is that such a fast control may require very large controller output signals $U(z)$, which may translate to many thousands of horse powers if the system controls the rudder of an oil tanker.

There are a large number of highly advanced and interesting control algorithms. In Chapter 6, for instance, the Kalman filter is discussed. More control algorithms can be found, e.g. Åström and Wittenmark (1984) and Wilkie *et al.* (2002).

Summary

In this chapter the following main topics have been addressed:

- Signals, discrete and continuous in amplitude and time
- Sampling, aliasing, the Nyquist frequency
- Quantization, resolution, dynamic range and quantization noise
- Linearity, the principle of superposition, LTI systems, causality
- Difference equations and state–space models
- Impulse response and convolution
- Transfer functions in the z-plane
- The frequency response, the gain function and the phase shift function
- Some filter architectures: non-recursive, recursive and lattice filters, FIR and IIR
- The impossibility of designing the perfect filter
- The Butterworth, Chebyshev, Cauer and Bessel approximations
- Indirect and direct filter synthesis methods
- Impulse invariance, step invariance and ramp invariance
- The bilinear transform and pre-warping, Euler's method
- The Fourier method, frequency sampling, simulation and McClellan–Parks/ Remez exchange algorithm
- Digital control, closed- and open-loop transfer functions, stability
- PID, direct synthesis, pole placement and dead-beat controllers.

Review questions

R1-1 Explain the aliasing effect in a sampled system. What is the Nyquist frequency? How can aliasing be avoided?

R1-2 Explain how quantization noise, resolution and dynamic range are related to the word length.

R1-3 Why is linearity so important? What are the requirements of a function being linear?

R1-4 In what way is the impulse response of a linear system related to the transfer function of the system?

R1-5 Why is an FIR filter always stable?

R1-6 Why is it that a "brick-wall" type filter cannot be implemented in practice?

R1-7 Why is linear phase a desirable feature in a filter?

R1-8 Give pros and cons of the filter approximations Butterworth, Chebyshev and Bessel.

R1-9 When designing filters using the bilinear transform, pre-warping may be needed. Why is it needed and under what circumstances?

R1-10 Which four demands are commonly placed on a closed-loop control system?

R1-11 Which three parameters are there to be set in a PID controller?

R1-12 What are the pros and cons of a dead-beat controller?

Solved problems

P1-1 An analog signal is converted to a digital bitstream. If the maximum frequency of the signal is 10 kHz, and we need 58 dB dynamic range, what is the data rate of the bitstream in bits/s, assuming a perfect "brick-wall" anti-aliasing filter?

P1-2 A digital controller has a zero at $z = a$ and poles at $z = b$ and $z = c$, where a, b, c are real constants. Determine the transfer function, the frequency function, the gain function and the phase shift function. Also, derive the corresponding difference equation in such a form that it could readily be implemented as computer program code.

P1-3 An FIR filter with an odd number of taps will have a linear-phase shift function if the tap weights are symmetrical, i.e. $b_n = b_{M-n}$. Show this.

P1-4 Make an Euler transform of the analog transfer function (1.73). Compare the resulting digital transfer function to equation (1.74) obtained using the bilinear transform.

P1-5 Write a MATLAB™ program to design a digital Butterworth filter and a Chebyshev filter. The filters should be bandpass filters of order eight with lower cut-off frequency 8 kHz and upper cut-off frequency 12 kHz. Maximum ripple in the passband is 1 dB. The sampling frequency is 40 kHz. The program should make Bode plots, pole–zero plots and print the filter coefficients.

2 The analog–digital interface

Background   In most systems, whether electronic, financial or social, the majority of problems arise in the interface between different sub-parts. This is of course also true for digital signal processing (DSP) systems. Most signals in real life are continuous in amplitude and time, i.e. "analog", but our digital system is working with amplitude and time discrete signals, so-called "digital" signals. Hence, the input signals entering our system need to be converted from analog-to-digital (A/D) form before the actual signal processing may take place.

For the same reason, the output signals from our DSP device need to be reconverted back from digital-to-analog (D/A) form, to be used in for instance hydraulic valves or loudspeakers or other analog actuators. These conversion processes, between the analog and digital world add some problems to our system. These matters will be addressed in this chapter, together with a brief presentation of some common techniques to perform the actual conversion processes.

Objectives   In this chapter we will cover:

- Encoding and modulation schemes, pulse code modulation, pulse amplitude modulation, pulse position modulation, pulse number modulation, pulse width modulation and pulse density modulation
- Number representation, fixed-point 2's complement, offset binary, sign and magnitude, and floating point
- Companding systems
- Multiplying, integrating and bitstream D/A converters
- Oversampling, interpolators and truncators
- Sample-and-hold, reconstruction filters and anti-aliasing filters
- Flash, successive approximation, counting and integrating A/D converters
- Dither
- Sigma–delta and bitstream A/D converters, decimation filters and comb filters.

2.1 System considerations

2.1.1 Encoding and modulation

Assuming we have now converted our analog signals to numbers in the digital world, there are many ways to **encode** the digital information into the shape of electrical signals. This process is called **modulation** (sometimes

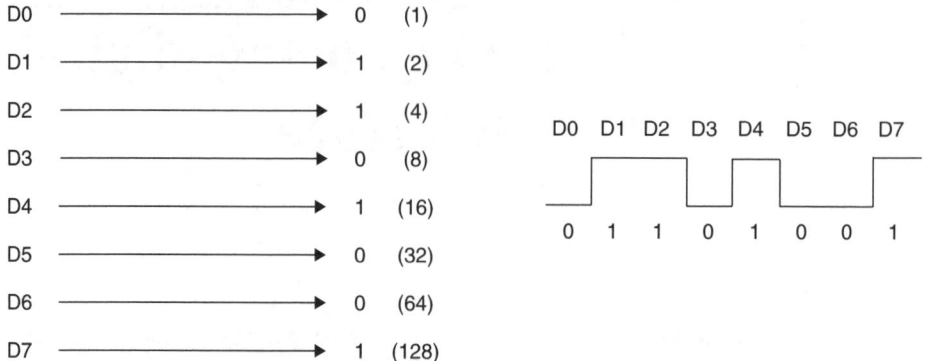

Figure 2.1 *Example, a byte (96H) encoded (weights in parenthesis) using PCM in parallel mode (parallel bus, 8 bits, eight wires) and in serial mode as an 8-bit pulse train (over one wire)*

"line modulation"). The most common method is probably **pulse code modulation (PCM)**. There are two common ways of transmitting PCM and they are **parallel** and **serial** mode. In an example of the parallel case, the information is encoded as voltage levels on a number of wires, called a parallel **bus**. We are using binary signals, which means that only two voltage levels are used, $+5\,$V corresponding to a binary "1" (or "true"), and $0\,$V meaning a binary "0" (or "false"). Hence, every wire carrying 0 or $+5\,$V contributes a binary digit ("bit"). A parallel bus consisting of eight wires will hence carry 8 bits, a byte consisting of bits D0, D1–D7 (Figure 2.1). Parallel buses are able to transfer high information data rates, since an entire data word, i.e. a sampled value, is being transferred at a time. This transmission can take place between, for instance, an analog-to-digital converter (ADC) and a digital signal processor (DSP). One drawback with parallel buses is that they require a number of wires, i.e. board space on a printed circuit board. Another problem is that we may experience skew problems, i.e. different time delays on different wires, meaning that all bits will not arrive at the same time in the receiver end of the bus, and data words will be messed up. Since this is especially true for long, high-speed parallel buses, this kind of bus is only suited for comparatively short transmission distances. Protecting long parallel buses from picking up wireless interference or to radiate interference may also be a formidable problem.

The alternative way of dealing with PCM signals is to use the **serial** transfer mode. In this case, the bits are not transferred on different wires in parallel, but in sequence on a single wire (see Figure 2.1). First bit D0 is transmitted, then D1, etc. This means of course that the transmission of, for instance, a byte requires a longer time than in the parallel case. On the other hand, only **one** wire is needed. Board space and skew problems will be eliminated and the interference problem can be easier to solve.

There are many possible modulation schemes, such as pulse amplitude modulation (PAM), pulse position modulation (PPM), pulse number modulation (PNM), pulse width modulation (PWM) and pulse density modulation (PDM). All these modulation types are used in serial transfer mode (see Figure 2.2).

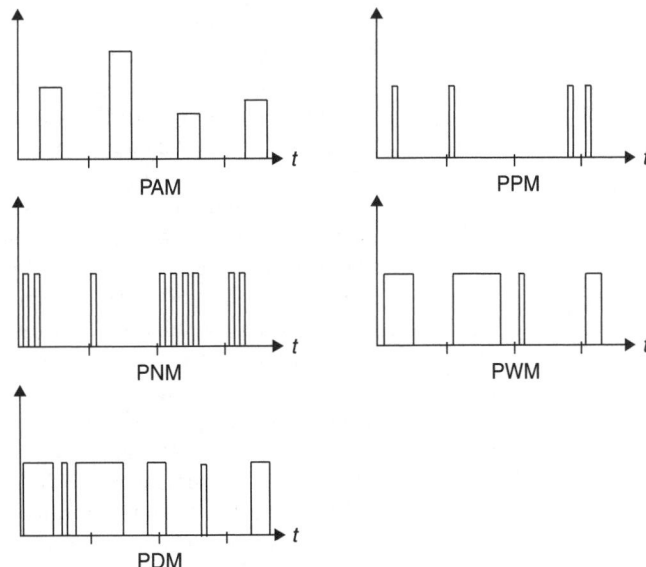

Figure 2.2 *Different modulation schemes for serial mode data communication, PAM, PPM, PNM, PWM and PDM*

- **Pulse amplitude modulation (PAM)** The actual amplitude of the pulse represents the number being transmitted. Hence, PAM is continuous in amplitude but discrete in time. The output of a sampling circuit with a zero-order hold (ZOH) is one example of a PAM signal.
- **Pulse position modulation (PPM)** A pulse of fixed width and amplitude is used to transmit the information. The actual number is represented by the position in time where the pulse appears in a given time slot.
- **Pulse number modulation (PNM)** Related to PPM in the sense that we are using pulses with fixed amplitude and width. In this modulation scheme, however, many pulses are transmitted in every time slot, and the number of pulses present in the slot represents the number being transmitted.
- **Pulse width modulation (PWM)** Quite common modulation scheme, especially in power control and power amplifier contexts. In this case, the width (duration) T_1 of a pulse in a given time slot T represents the number being transmitted. If the pulse has the amplitude A_1, the transmitted number is represented by

$$A_1 \frac{T_1}{T} \tag{2.1}$$

In most applications, the amplitude A_1 of the pulse is fixed and uninteresting. Only the time ratio is used in the transmission process. If, however, the amplitude of the pulse is also used to represent a second signal, we are using a combination of PAM and PWM. In some applications, this is a simple way of achieving a multiplication of two signals.

- **Pulse density modulation (PDM)** May be viewed as a type of degenerated PWM, in the sense that not only the width of the pulses changes, but also the periodicity (frequency). The number being transmitted is represented by the density or "average" of the pulses.

A class of variations of PDM called **stochastic representation of variables** was tried in the 1960s and 1970s. The idea was to make a "stochastic computer", replacing the analog computer consisting of operational amplifiers and integrators, working directly with the analog signals. The stochastic representation has two nice features. Firstly, the resolution can be traded for time, which implies that resolution can be improved by transmitting more pulses (longer time needed). Secondly, the calculations can be performed easily using only standard combinatorial circuits. The idea of stochastic representation has experienced a renaissance in the 1980s in some forms of neural network applications.

So far we have talked about "transmission" of digital information using different types of modulation. This discussion is of course also relevant for **storing** digital information. When it comes to optical compact disc (CD) or magnetic media, there are a number of special modulation methods (Pohlmann, 1989; Miller and Beasley, 2002) used, which will not be treated here.

As will be seen later in this chapter, some signal converting and processing chips and subsystems may use different modulation methods to communicate. This may be due to standardization or due to the way the actual circuit works. One example is the so-called **CODEC** (coder–decoder). For instance, this is a chip used in telephone systems, containing both an analog-to-digital converter (ADC) and a digital-to-analog converter (DAC) and other necessary functions to implement a full two-way analog–digital interface for voice signals. Many such chips use a serial PCM interface. Switching devices and digital signal processors commonly have built-in interfaces to handle these types of signals.

2.1.2 Number representation and companding systems

When the analog signal is quantized, it is commonly represented by binary numbers in the following processing steps. There are many possible representations of quantized amplitude values. One way is to use **fixed-point** formats like **2's complement**, **offset binary** or **sign and magnitude** (Pires, 1989). Another way is to use some kind of **floating-point** format. The difference between the fixed-point formats can be seen in Table 2.1.

The most common fixed-point representation is 2's complement. In the digital signal processing community, we often interpret the numbers as **fractions** rather than integers. This will be discussed in subsequent chapters. Other codes (Pires, 1989) are gray code and binary-coded decimal (BCD).

There are a number of floating-point formats around. They all rely on the principle of representing a number in three parts: a sign bit, an exponent and a mantissa. One such common format is the Institute of Electrical and Electronics Engineers **(IEEE) Standard 754.1985** single precision 32-bit format, where the floating-point number is represented by one sign bit, an 8-bit exponent and a 23-bit mantissa. Using this method numbers between $\pm 3.37 \cdot 10^{38}$ and $\pm 8.4 \cdot 10^{-37}$ can be represented using only 32 bits. Note however that the use of floating-point representation only expands the **dynamic range** on the expense of the resolution and system complexity. For instance, a 32-bit fixed-point system may have better resolution than a 32-bit floating-point system, since in the floating-point case, the resolution is determined by the word length of the mantissa being only 23 bits. Another problem using floating-point systems is the **signal-to-noise ratio (SNR)**. Since the size of the quantization steps will

Table 2.1 *Some fixed-point binary number formats*

Integer	2's complement	Offset binary	Sign and magnitude
7	0111	1111	0111
6	0110	1110	0110
5	0101	1101	0101
4	0100	1100	0100
3	0011	1011	0011
2	0010	1010	0010
1	0001	1001	0001
0	0000	1000	0000
−1	1111	0111	1000
−2	1110	0110	1001
−3	1101	0101	1010
−4	1100	0100	1011
−5	1011	0011	1100
−6	1010	0010	1101
−7	1001	0001	1110
−8	1000	0000	1111

change as the exponent changes, so will the quantization noise. Hence, there will be discontinuous changes in SNR at specific signal levels. In an audio system, audible distortion (Pohlmann, 1989) may result from the modulation and quantization noise created by barely audible low-frequency signals causing numerous exponent switches.

From the above, we realize that fixed-point (linear) systems yield **uniform** quantization of the signal. Meanwhile floating-point systems, due to the range changing, provide a **non-uniform** quantization. Non-uniform quantization is often used in systems where a compromise between word length, dynamic range and distortion at low signal levels has to be found. By using larger quantization steps for larger signal levels and smaller steps for weak signals, a good dynamic range can be obtained without causing serious distortion at low signal levels or requiring unreasonable word lengths (number of quantization steps). A digital telephone system may serve as an example where small signal levels are the most probable ones, thus causing the need for good resolution at low levels to keep distortion low. On the other hand, sometimes stronger signals will be present, and distortion due to saturation is not desirable. Due to the large number of connections in a digital telephone switch, the word length must be kept low, commonly not more than 8 bits.

Another way of accomplishing non-uniform quantization is **companding** (Pohlmann, 1989; Miller and Beasley, 2002), a method which is not "perfect", but easy to implement. The underlying idea is to use a system utilizing common uniform quantization and reconstruction. At the input of the system a **compressor** is connected and at the output of the system an **expander** (hence, **comp**ressor–exp**ander**: "**compander**") is added (see Figure 2.3).

The compressor is mainly a non-linear amplifier, often logarithmic, having a lower gain for stronger signals than for weaker ones. In this way, the dynamic

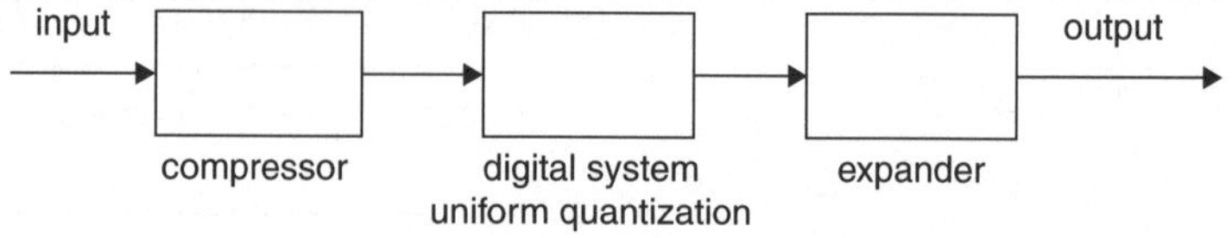

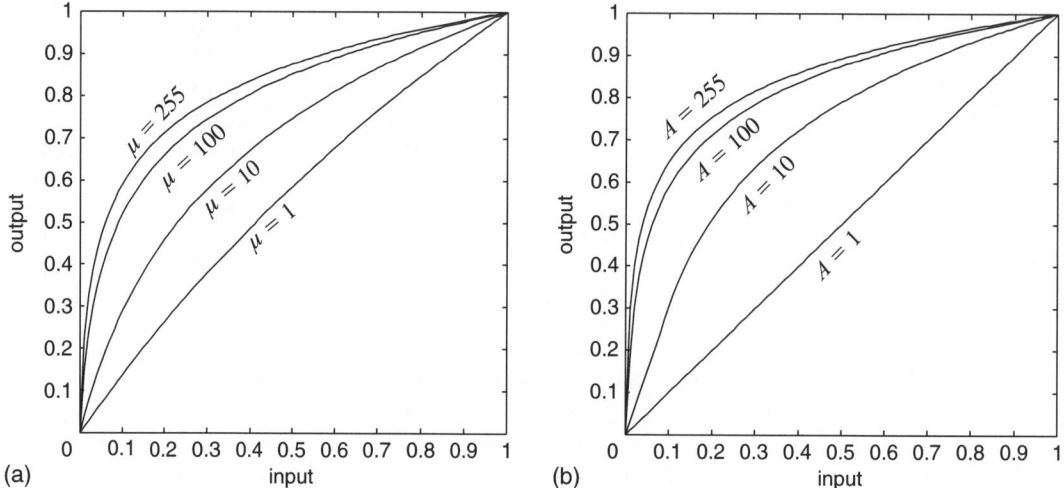

Figure 2.3 *A companding system consisting of a **compressor**, a common digital system using uniform quantization intervals and an **expander**. Note! The signal processing algorithms used must take compression and expansion into account*

Figure 2.4 *The (a) μ-law and (b) A-law transfer functions for different parameter values*

range of the input signal is compressed. The expander is another non-linear amplifier having the function of being the inverse of the compressor. Hence, the expander "restores" the dynamic range of the signal at the output of the system. The total system will now act as a system using non-uniform quantization. Note: the signal processing algorithms in the system must take the non-linearity of the compressor and expander into account.

For speech applications, typically found in telephone systems, there are two common non-linearities (Pohlmann, 1989) used: the **μ-law** and the **A-law**. The μ-law is preferred in the US and the logarithmic compression characteristic has the form

$$y = \frac{\log(1 + \mu x)}{\log(1 + \mu)} \quad \text{for } x \geq 0 \tag{2.2}$$

where y is the magnitude of the output, x the magnitude of the input and μ a positive parameter defined to yield the desired compression characteristic. A parameter value of 0 corresponds to linear amplification, i.e. no compression and uniform quantization. A value of $\mu = 255$ is often used to encode speech signals (see Figure 2.4). In such a system, an 8-bit implementation can achieve a good SNR and a dynamic range equivalent to that of a 12-bit system using uniform quantization. The inverse function is used for expansion.

The A-law is primarily used in Europe. Its compression characteristic has the form

$$y = \begin{cases} \dfrac{Ax}{1 + \log(A)} & \text{for } 0 \le x < \dfrac{1}{A} \\[3ex] \dfrac{1 + \log(Ax)}{1 + \log(A)} & \text{for } \dfrac{1}{A} \le x < 1 \end{cases} \tag{2.3}$$

where y is the magnitude of the output, x the magnitude of the input and A is a positive parameter defined to yield the desired compression characteristic.

Figure 2.4 shows μ-law and A-law transfer functions for some values of μ and A. Companding techniques are also used in miscellaneous noise reduction systems, for instance the DOLBY^{TM} systems.

2.2 Digital-to-analog conversion

From now on we will only discuss systems using uniform quantization intervals. The task of the digital-to-analog converter (**DAC**) is to convert a numerical, commonly binary so-called "digital" value into an "analog" output signal. The DAC is subject to many requirements, such as offset, gain, linearity, monotonicity and settling time. Assume that the output voltage of an n bit perfect DAC with digital input value m (shown as a dotted line in Figure 2.5) can be written as

$$\frac{V_{\text{FS}}}{2^n} m = Am$$

where V_{FS} is the full-scale (FS) output voltage of the DAC and the constant A corresponds to the output voltage step size for one least significant bit (LSB). In this case, the output voltage $v(m)$ of a real world DAC can be expressed as

$$v(m) = Am + \varepsilon_{\text{O}} + Am\varepsilon_{\text{G}} + \varepsilon_N(m)$$

which consists of the true output voltage plus three error terms defined below.

Offset is the analog output when the digital input calls for a zero output. This should of course ideally be zero. The offset error ε_{O} affects all output signals with the same additive amount and in most cases it can be sufficiently compensated for by external circuits or by **trimming** the DAC.

Gain or scale factor is the slope of the transfer curve from digital numbers to analog levels. Hence, the gain error ε_{G} is the error in the slope of the transfer curve. This error affects all output signals by the same percentage amount, and can normally be (almost) eliminated by trimming the DAC or by means of external circuitry.

Linearity can be sub-divided into integral linearity (relative accuracy) and differential linearity. **Integral linearity** error is the deviation of the transfer curve from a straight line (the output of a perfect DAC) and is commonly expressed in number of LSBs $(\varepsilon_N(m))/A$. This error is not possible to adjust or compensate for easily, since it may change more or less randomly as a function of the input digital code *m*.

Differential linearity measures the difference between any two adjacent output levels minus the step size for one LSB, or in other words, $(\varepsilon_N(m) - \varepsilon_N(m - 1))/A$ expressed in LSBs. If the output level for one step differs from the previous step by exactly the value corresponding to one least significant bit (LSB) of the digital value, the differential non-linearity is zero. Differential linearity errors cannot be eliminated easily.

Monotonicity implies that the analog output must increase as the digital input increases, and decrease as the input decreases for all values over the specified signal range. Non-monotonicity is a result of excess differential

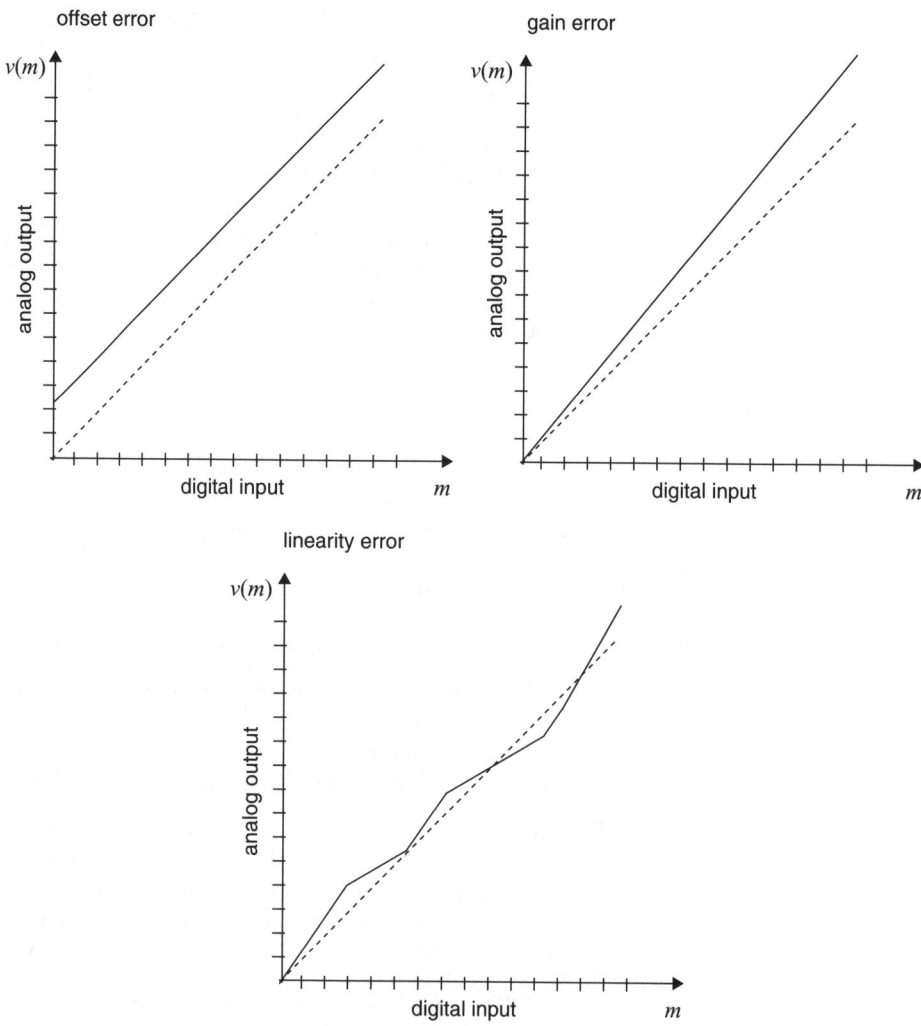

Figure 2.5 *Errors of a DAC, the dotted line shows the performance of a perfect converter*

non-linearity (≥ 1 LSB). This implies that a DAC which has a differential non-linearity specification of maximum ± 0.5 LSB is more tightly specified than one for which only monotonicity is guaranteed. Monotonicity is essential in many control applications to maintain precision and to avoid instabilities in feedback loops.

Absolute accuracy error is the difference between the measured analog output from a DAC compared to the expected output for a given digital input. The absolute accuracy is the compound effect of the offset error, gain error and linearity errors described above.

Settling time of a DAC is the time required for the output to approach a final value within the limits of an allowed error band for a step change in the digital input. Measuring the settling time may be difficult in practice, since some DACs produce glitches when switching from one level to another. These glitches, being considerably larger than the fraction of the 1 LSB step of interest, may saturate, for instance, an oscilloscope input amplifier, thereby causing significant measuring errors. DAC settling time is a parameter of importance mainly in high sampling rate applications.

One important thing to remember is that the parameters above may be affected by supply voltage and temperature. In DAC data sheets, the parameters are only specified for certain temperatures and supply voltages, e.g. normal room temperature $+25°$C and nominal supply voltage. Considerable deviations from the specified figures may occur in a practical system.

2.2.1 Multiplying digital-to-analog converters

This is the most common form of DAC. The output is the product of an input current or reference voltage and an input digital code. The digital information is assumed to be in PCM parallel format. There are also DACs with a built-in shift register circuit, converting serial PCM to parallel. Hence, there are multiplying DACs for both parallel and serial transfer mode PCM available. Multiplying DACs have the advantage of being fast.

In Figure 2.6 a generic current source multiplying DAC is shown. The bits in the input digital code are used to turn on a selection of current sources, which

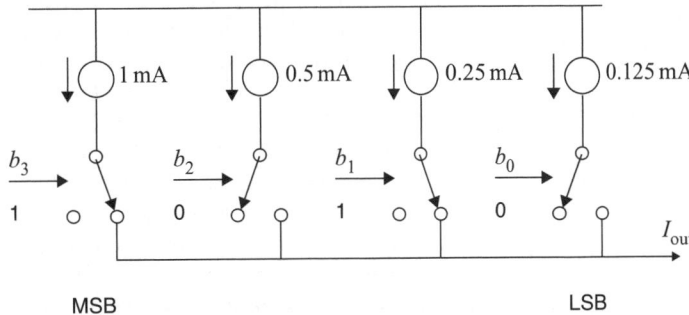

Figure 2.6 *A generic multiplying DAC using current sources controlled by the bits in the digital input code*

are then summed to obtain the output current. The output current can easily be converted into an output voltage using an operational amplifier.

Another way of achieving the different current sources in Figure 2.6 would be to use a constant input reference voltage U_{ref} and a set of resistors. In this way, the currents for the different branches would simply be obtained by

$$I_i = \frac{U_{ref}}{R_i} b_i = \frac{U_{ref}}{2^{N-1-i} R_{N-1}} b_i \qquad (2.4)$$

where R_i is the resistance of the resistor in the i-th branch being controlled by the i-th bit: b_i being 1 or 0. The word length of the digital code is N and $i = 0$ is LSB. The total output current can then be expressed as the sum of the currents from the branches

$$I_{out} = \sum_{i=0}^{N-1} I_i = \frac{U_{ref}}{R_{N-1}} \sum_{i=0}^{N-1} \frac{b_i}{2^{N-1-i}} \qquad (2.5)$$

Building such a DAC in practice would however cause some problems. This is especially true as the word length increases. Assume for instance that we are to design a 14-bit DAC. If we choose the smallest resistor, i.e. the resistor R_{13} corresponding to $i = N - 1 = 13$ to be 100 ohms, the resistor R_0 will then need to be $R_0 = 2^{13} \cdot 100 = 819\,200$ ohms. Now, comparing the currents flowing in branch 0 and 13, respectively, we find: $I_0 = (U_{ref}/819\,200)$ A and $I_{13} = (U_{ref}/100)$ A. To obtain good differential linearity, the error of current I_{13} (corresponding to the most significant bit, MSB) must be smaller than the smallest current I_0 (corresponding to the LSB). Hence the resistance of R_{13} is required to be correct within ± 122 ppm (parts per million). Another problem is that different materials and processes may be required when making resistors having high or low resistance, respectively. This will result in resistors in the DAC having different aging and temperature stability properties, thus making it harder to maintain specifications under all working conditions.

It is possible to achieve the required precision by laser trimming, but there is a smarter and less expensive way to build a good DAC using an **R–$2R$ ladder** structure. In an R–$2R$ ladder, there are considerably more than N resistors, but they all have the same resistance R and are hence easier to manufacture and integrate on a single silicon chip. The R–$2R$ ladder structure uses current division. A simple 3-bit R–$2R$ ladder DAC is shown in Figure 2.7.

All resistors R1–R11 have the same resistance R. As can be seen from the figure, it is quite common that two resistors are connected in series, e.g. R1–R2, R3–R4, R5–R6 and R7–R8. Each of these pairs can of course be replaced by a single resistor having the resistance $2R$. That is why this structure is called R–$2R$ ladder, it can be built simply by using two resistance values: R and $2R$.

The switches b_2, b_1, b_0 are controlled by bits 2, 1 and 0 in the digital code. If a bit is set to one, the corresponding switch is in its right position, i.e. switched to the negative input of the operational amplifier. If the bit is set to zero, the switch is in its left position, connecting the circuit to ground. The negative input of the operational amplifier is a current summation point, and due to the feedback amplifier having its positive input connected to ground, the negative input will be held at practically zero potential, i.e. ground ("virtual ground"). Hence, the

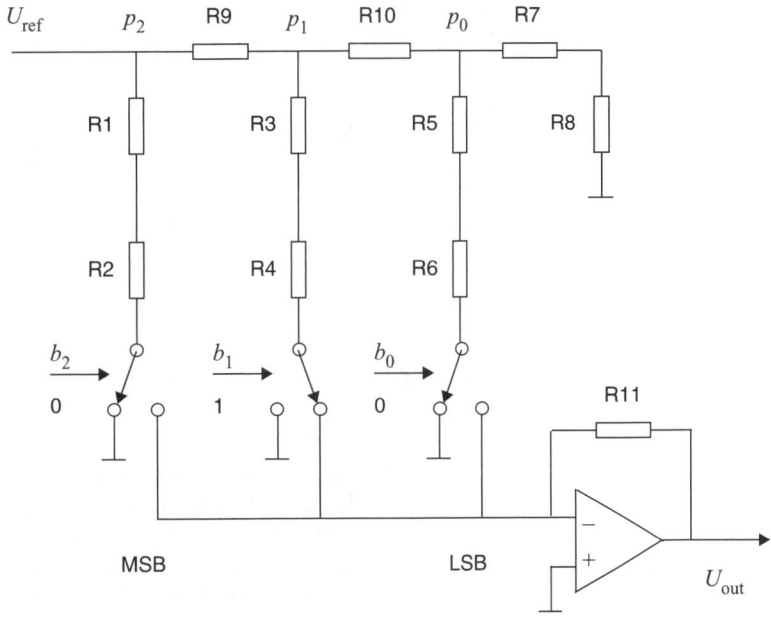

Figure 2.7 *An example 3 bit, R–2R ladder multiplying DAC, with voltage output, all resistors have the same resistance, R. Input digital code: 010*

resistor ladder circuit will be loaded in the same way and the position of the switches does not matter.

Now, let us examine the resistor ladder structure a bit more closely. Starting in junction point p_0 we conclude that the current flowing through resistor R10 will be divided into the two branches R5–R6 and R7–R8, respectively. As seen above, both branches have the same resistance, namely $2R$. This means that the current will be divided equally in the two branches, i.e. half of the current passing through R10 will pass through the switch b_0. Further, if we calculate the total resistance of the circuit R5–R6 in parallel with R7–R8, we will find it to be R. If we now move to point p_1 we will have a similar situation as in p_0. The total resistance of R5, R6, R7 and R8 is R. This combination is connected in series with R10 having resistance R ohms, hence the total resistance in the circuit passing through R10 to ground will be $2R$. Since the total resistance through R3–R4 is also $2R$, the current flowing through resistor R9 will be divided equally in the two branches as in the earlier case. One half of the current will pass through switch b_1 and one half will continue down the ladder via R10. For every step in the ladder, the same process is repeated. The current created by the reference voltage U_{ref}, will be divided by 2 for every step in the ladder, and working with binary numbers, this is exactly what we want.

Assuming the operational amplifier to be an ideal one (zero offset voltage, zero bias current and infinite gain) the output voltage from the DAC can then be written as

$$U_{\text{out}} = -\frac{U_{\text{ref}}}{2}\left(b_2 + \frac{b_1}{2} + \frac{b_0}{4}\right) = -\frac{U_{\text{ref}}}{2}\sum_{i=0}^{N-1}\frac{b_i}{2^{N-1-i}} \qquad (2.6)$$

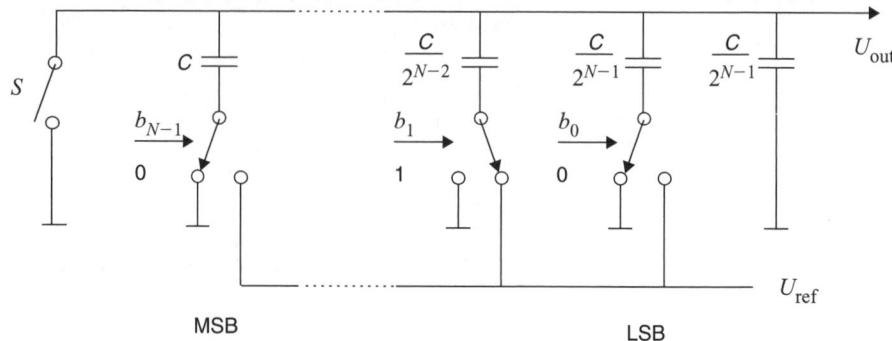

Figure 2.8 *A generic charge redistribution DAC*

Another way of building a DAC is by utilizing **charge redistribution**. This technique is quite common in complementary metal oxide semiconductor (CMOS) single-chip computers and the DAC can be implemented on a silicon chip in the same way as, for instance, switched capacitor (SC) filters. This type of DAC makes use of the fact that if a fixed voltage U_{ref} is applied to a (variable) capacitor C, the charge in the capacitor will be

$$Q = CU_{ref} \tag{2.7}$$

After charging the capacitor C, it is disconnected from the voltage source and connected to another (discharged) capacitor C_1. The electric charge Q is then redistributed between the two capacitors and the voltage over the capacitors will be

$$U = \frac{Q}{C + C_1} = U_{ref}\frac{C}{C + C_1} \tag{2.8}$$

An example of a generic, charge redistribution DAC is shown in Figure 2.8. The DAC works in two phases: in phase 1, the reset phase, all switches that are connected to ground close, discharging all capacitors; in phase 2, switch S is opened and the remaining switches are controlled by the incoming digital word. If a bit b_i is one, the corresponding capacitor $C_i = C/2^i$ is connected to U_{ref}, otherwise it is connected to ground.

It is quite straightforward to see how the DAC works. In phase 2, when the capacitors are charged and the total capacitance C_{tot} as seen from the voltage source U_{ref} is

$$C_{tot} = \frac{C_{one}\left(C_{zero} + \frac{C}{2^{N-1}}\right)}{C_{one} + \left(C_{zero} + \frac{C}{2^{N-1}}\right)} \tag{2.9}$$

where C_{one} is the total capacitance of the capacitors corresponding to the bits that are 1 in the digital word (these capacitors will appear to be connected in parallel) and C_{zero} is the total capacitance of the capacitors corresponding to the bits that are 0. Hence the total charge of the circuit will be

$$Q_{tot} = C_{tot}U_{ref} \tag{2.10}$$

Now, the output voltage U_{out} is the voltage over the capacitors connected to ground

$$U_{\text{out}} = \frac{Q_{\text{tot}}}{\left(C_{\text{zero}} + \frac{C}{2^{N-1}}\right)} = \frac{U_{\text{ref}} C_{\text{tot}}}{\left(C_{\text{zero}} + \frac{C}{2^{N-1}}\right)} \tag{2.11}$$

Inserting equation (2.9) into equation (2.11) and expressing the capacitances as sums, we finally obtain the output voltage of the DAC as a function of the bits b_i in the digital code

$$\begin{aligned} U_{\text{out}} &= \frac{U_{\text{ref}} C_{\text{one}} \left(C_{\text{zero}} + \frac{C}{2^{N-1}}\right)}{\left(C_{\text{zero}} + \frac{C}{2^{N-1}}\right) \left(C_{\text{one}} + \left(C_{\text{zero}} + \frac{C}{2^{N-1}}\right)\right)} \\ &= \frac{U_{\text{ref}} C_{\text{one}}}{C_{\text{one}} + C_{\text{zero}} + \frac{C}{2^{N-1}}} \\ &= \frac{U_{\text{ref}} C \sum_{i=0}^{N-1} \frac{b_i}{2^i}}{C \sum_{i=0}^{N-1} \frac{1}{2^i} + \frac{C}{2^{N-1}}} = U_{\text{ref}} \sum_{i=1}^{N} b_i 2^{-i} \end{aligned} \tag{2.12}$$

There are many alternative charge-based circuits around. It is, for instance, possible to design a type of C–$2C$ ladder circuit.

2.2.2 Integrating digital-to-analog converters

This class of DACs is also called **counting** DACs. These DACs are often slow compared to the previous type of converters. On the other hand, they may offer high resolution using quite simple circuit elements. No high precision resistors, etc., are needed.

The basic building blocks are: an "analog accumulator" usually called an integrator, a voltage (or current) reference source, an analog selector, a digital counter and a digital comparator. Figure 2.9 shows an example of an integrating DAC. The incoming N-bits PCM data (parallel transfer mode) is fed to one input of the digital comparator. The other input of the comparator is connected to the binary counter having N bits, counting pulses from a clock oscillator running at frequency f_c

$$f_c \geq 2^N f_s \tag{2.13}$$

where f_s is the sampling frequency of the system and N is the word length. Now, assume that the counter starts counting from 0. The output of the comparator will be zero and there will be a momentary logic one output from the comparator when the counter digital value equals the digital input PCM code. This pulse will set the bistable flip–flop circuit. The flip–flop will then be reset when the counter wraps around from 2^{N-1} to 0 and a carry signal is generated.

The output from the flip–flop controls the analog selector in such a way that when the flip–flop is reset, $-U_{\text{ref}}$ is connected to the input of the integrator, and when the flip–flop is set, $+U_{\text{ref}}$ is selected. Note that the output of the comparator is basically a PPM version of the PCM input data, and the output of the flip–flop is a PWM version of the same quantity. Hence, if we happen

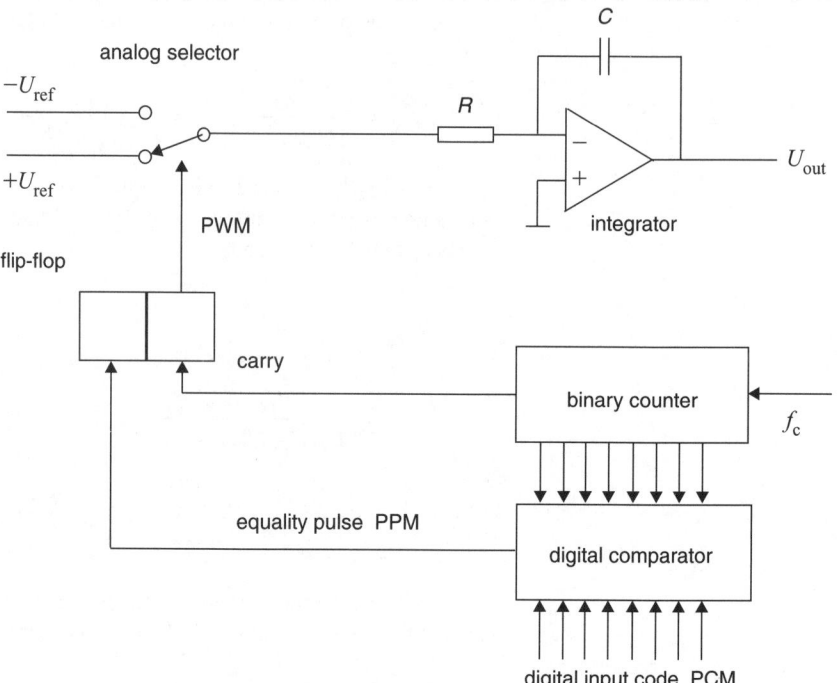

Figure 2.9 *An example integrating (counting) DAC. (In a real world implementation, additional control and synchronization circuits are needed.)*

to have the digital code available in PWM or PPM format instead of parallel PCM, the circuit can be simplified accordingly.

The integrator simply averages the PWM signal presented to the input, thus producing the output voltage U_{out}. The precision of this DAC depends on the stability of the reference voltages, the performance of the integrator and the timing precision of the digital parts, including the analog selector.

There are many variants of this basic circuit. In some types, the incoming PCM data is divided into a "high" and "low" half, controlling two separate voltage or current reference selectors. The reference voltage controlled by the "high" data bits is higher than the one controlled by the "low" data bits. This type of converter is often referred to as a **dual slope converter**.

2.2.3 Bitstream digital-to-analog converters

This type of DAC relies on the **oversampling** principle, i.e. using a considerably higher sampling rate than required by the Nyquist criteria. Using this method, sampling rate can be traded for accuracy of the analog hardware and the requirements of the analog reconstruction filter on the output can be relaxed. Oversampling reduces the problem of accurate N-bit data conversion to a rapid succession of, for instance, 1-bit D/A conversions. Since the latter operation involves only an analog switch and a reference voltage source, it can be performed with high accuracy and linearity.

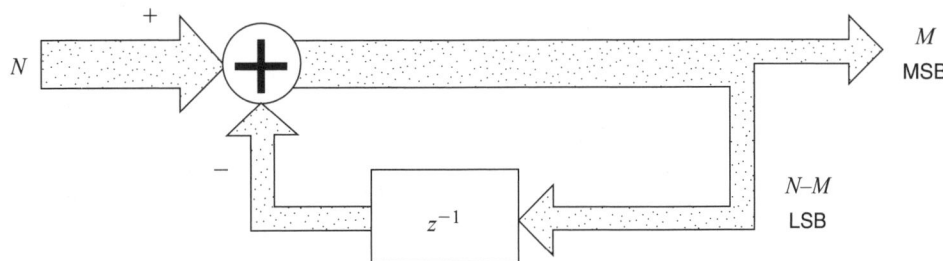

Figure 2.10 *A truncator with noise shaping feedback, input* N *bits, truncated to output* M *bits*

The concept of oversampling is to increase a fairly low sampling frequency to a higher one by a factor called the oversampling ratio (OSR). Increasing the sampling rate implies that more samples are needed than are available in the original data stream. Hence, "new" sample points in between the original ones have to be created. This is done by means of an **interpolator**, also called an **oversampling filter** (Pohlmann, 1989). The simplest form of interpolator creates new samples by making a linear interpolation between two "real" samples. In many systems, more elaborate interpolation functions are often used, implemented as a cascade of digital filters. As an example, an oversampling filter in a CD player may have 16-bit input samples at 44.1 kHz sampling frequency and an output of 28-bit samples at 176.4 kHz, i.e. an OSR of 4.

The interpolator is followed by the **truncator** or M-bit **quantizer** (Figure 2.10). The task of the truncator is to reduce the number of N bits in the incoming data stream to M bits in the outgoing data stream ($N > M$). The truncation process is simply performed by taking the M most significant bits out of the incoming N bits. This process however creates a strong quantization noise in the passband of interest. This is counteracted by means of the **noise shaping feedback loop**, consisting of a delay element and an adder. The $M–N$ least significant bits are fed back through the delay element and subtracted from the incoming data stream. Hence, the signal transfer function of the truncator equals 1 for the signal, but constitutes a high-pass filter having the transfer function

$$H(z) = 1 - z^{-1} \tag{2.14}$$

for the quantization noise. The noise level in the interesting passband is attenuated, while at higher frequencies, the noise increases. The higher-frequency noise components will however be attenuated by the analog reconstruction filter following the DAC. To conclude our example above, the CD player may then have a truncator with $N = 28$ bits and $M = 16$ bits. The sampling rate of both the input and output data streams are 176.4 kHz. In this example we still have to deal with 16-bit data in the DAC, but we have simplified the analog reconstruction filter on the output.

Oversampling could be used to derive yet one more advantage. Oversampling permits the noise shaper to transfer information in the 17th and 18th bits of the signal into a 16-bit output with duty cycle modulation of the 16th bit. In this way a 16-bit DAC can be used to convert an 18-bit digital signal. The 18-bit capability is retained because the information in the two "surplus" bits is transferred in the four times oversampled signal (OSR = 4). Or, in other words,

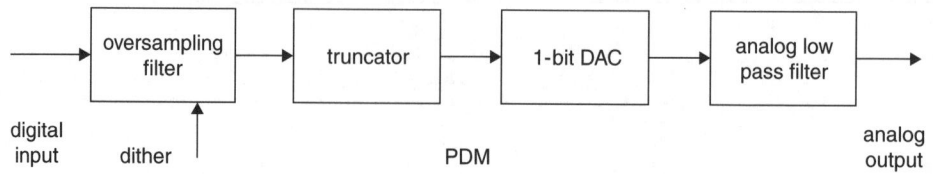

Figure 2.11 *A PDM bitstream 1-bit DAC, with (dither) pseudo-noise added*

the averaged (filtered) value of the quarter-length 16-bit samples is as accurate as that of 18-bit samples at the original sampling rate.

Now, if the truncator is made in such a way, the output data stream is 1 bit, i.e. $M = 1$; this bitstream is a PDM version of the original N-bit wide-data stream. Hence, a 1-bit DAC can be used to convert the digital signal to its analog counterpart (see Figure 2.11).

To improve the resolution, a **dither** signal is added to the LSB. This is a pseudo-random sequence, a kind of noise signal. Assume that the digital over-sampled value lies somewhere in between two quantization levels. To obtain an increased accuracy, smaller quantization steps, i.e. an increased word length (more bits) is needed. Another way of achieving improved resolution is to add random noise to the signal. If the signal lies half way between two quantization levels, the signal plus noise will be equally probable to take on the high and low quantized value, respectively. The actual level of the signal will hence be represented in a stochastic way. Due to the averaging process in the analog reconstruction low-pass filter, a better estimate of the actual signal level will be obtained than without dither. If the signal lies closer to one quantization level than the other, the former quantized value will be more probable and hence, the average will approach this quantized level accordingly.

Dither signals are also used in control systems including actuators like electrohydraulic valves. In this case, the dither signal may be a single sinus tone, superimposed on the actuator control signal. The goal is to prohibit the mechanics of the valve to "hang up" or stick.

2.2.4 Sample-and-hold and reconstruction filters

The output from a DAC can be regarded as a PAM representation of the digital signal at the sampling rate. An ideal sample represents the value of the corresponding analog signal in a single **point** in time. Hence, in an ideal case, the output of a DAC is a train of impulses, each having an infinitesimal width, thus eliminating the **aperture error**. The aperture error is caused by the fact that a sample in a practical case does occupy a certain interval of time. The narrower the pulse width of the sample, the less the error. Of course, ideal DACs cannot be built in practice.

Another problem with real world DACs is that during the transition from one sample value to another, glitches, ringing and other types of interference may occur. To counteract this, a **sample-and-hold** device (S&H or S/H) is used. The most common type is the **zero-order hold (ZOH)** presented in Chapter 1. This device keeps the output constant until the DAC has settled on the next sample value. Hence, the output of the S/H is a staircase waveform approximation of the sampled analog signal. In many cases, the S/H is built into the DAC itself. Now, the S/H having a sample pulse impulse response has a corresponding transfer function of the form

$$H(f) = \frac{\sin(\pi f T)}{\pi f T} = \mathrm{sinc}(f T) \tag{2.15}$$

where T is the sampling rate, i.e. the holding time of the S/H. The function (2.15) represents a low-pass filter with quite mediocre passband properties. In Chapter 1 we concluded that the function required to ideally reconstruct the analog signal would be an ideal low-pass filter, having a completely flat passband and an extremely sharp cut-off at the Nyquist frequency. Obviously, the transfer function of the S/H is far from ideal. In many cases an analog **reconstruction filter** or **smoothing filter** (or anti-image filter) is needed in the signal path after the S/H to achieve a good enough reconstruction of the analog signal. Since the filter must be implemented using analog components, it tends to be bulky and expensive and it is preferably kept simple, of the order of 3 or lower. A good way of relaxing the requirements of the filter is to use oversampling as described above. There are also additional requirements on the reconstruction filter depending on the application. In a high-quality audio system, there may be requirements regarding linear-phase shift and transient response, while in a feedback control system time delay parameters may be crucial.

In most cases a reconstruction filter is necessary. Even if a poorly filtered output signal has an acceptable quality, the presence of imaged high-frequency signal components may cause problems further down the signal path. For example, assume that an audio system has the sampling frequency 44.1 kHz and the sampled analog sinusoidal signal has a frequency of $f_1 = 12$ kHz, i.e. well below the Nyquist frequency. Further, the system has a poor reconstruction filter at the output and the first high-frequency image signal component $f_2 = 44.1 - f_1 = 44.1 - 12 = 32.1$ kHz leaks through. Now, the 32.1 kHz signal is not audible and does not primarily cause any problems. If, however, there is a non-linearity in the signal path, for instance an analog amplifier approaching saturation, the signal components will be mixed. New signal components at lower frequencies may be created, thus causing audible distortion. For instance, a third-order non-linearity, quite common in bipolar semiconductor devices, will create the following "new" signal components due to intermodulation (mixing)

$$f_3 = 8.1\,\text{kHz} \quad f_4 = 36.0\,\text{kHz} \quad f_5 = 52.2\,\text{kHz}$$
$$f_6 = 56.1\,\text{kHz} \quad f_7 = 76.2\,\text{kHz} \quad f_8 = 96.3\,\text{kHz}$$

Note that the first frequency 8.1 kHz is certainly audible and will cause distortion. The high-frequency components can of course also interfere with e.g. bias oscillators in analog tape recorders and miscellaneous radio frequency equipment, causing additional problems.

Finally, again we find an advantage using oversampling techniques. If the sampling frequency is considerably higher than twice the Nyquist frequency, the distance to the first mirrored signal spectra will be comparatively large (see Chapter 1). The analog filter now needs to cut-off at a higher frequency, hence the gain can drop-off slower with frequency and a simpler filter will be satisfactory.

2.3 Analog-to-digital conversion

The task of the analog-to-digital converter (ADC) is the "inverse" of the digital-to-analog converter (DAC), i.e. to convert an "analog" input signal into a numerical, commonly binary so-called "digital" value. The specifications for an ADC is similar to those for a DAC, i.e. offset, gain, linearity, missing codes, conversion time and so on. Since a DAC has a finite number of digital input codes, there are also a finite number of analog output levels. The ADC, on the other hand, has an infinite number of possible analog input levels, but only a finite number of digital output codes. For this reason, ADC errors are commonly assumed to be the difference between the actual and the ideal analog input levels, causing a transition between two adjacent digital output codes. For a perfect n-bit ADC the width of a quantization band is

$$A = \frac{V_{FS}}{2^n}$$

where V_{FS} is the full-scale input voltage of the ADC. Ideally, the transition level for each code should be situated in the center of the respective quantization band. These points are shown as a dotted line in Figure 2.12. Hence, the digital output

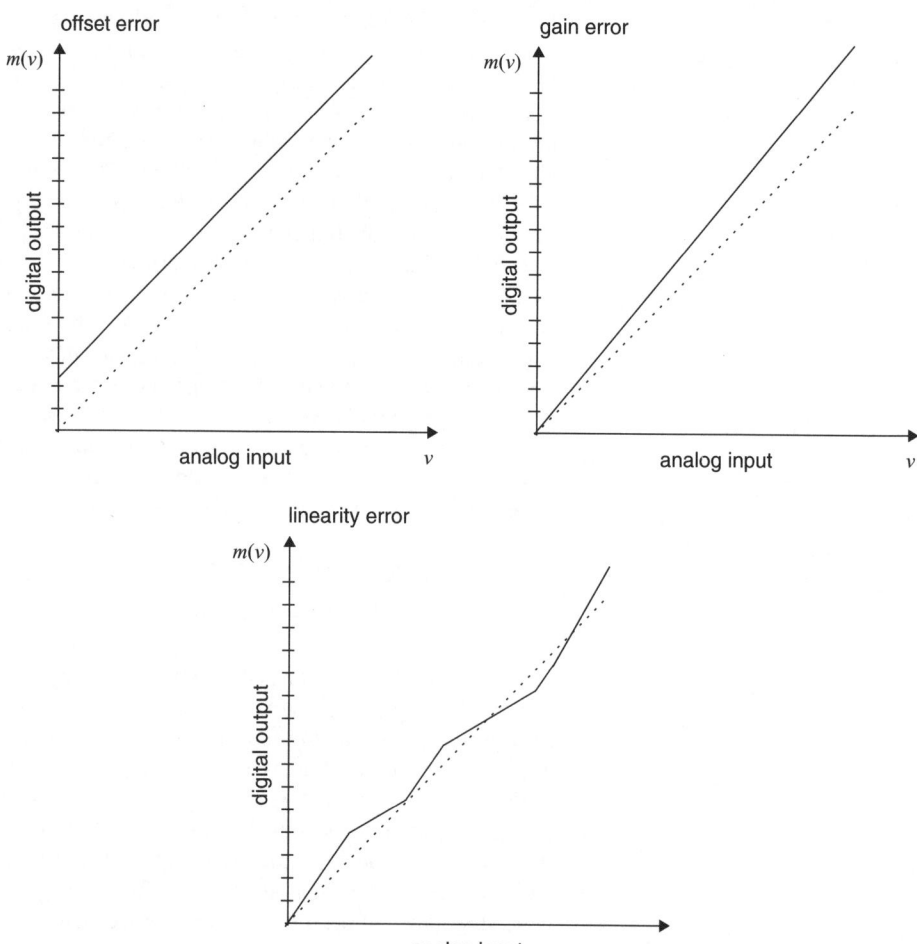

Figure 2.12 *Errors of an ADC, the dotted line shows the ideal placement of the transition levels*

code $m(v)$ as a function of the analog input voltage v for a perfect ADC can be written as

$$m(v) = \left\lfloor \frac{v}{A} + \frac{1}{2} \right\rfloor$$

where $\lfloor \ \rfloor$ is the "floor" operator, that is the first integer less than or equal to the argument. Taking the errors into account, the output of a real world ADC can be expressed as

$$m(v) = \left\lfloor \frac{v}{A} + \frac{1}{2} + \varepsilon_O + \frac{v}{A}\varepsilon_G + \varepsilon_N(v) \right\rfloor$$

The error terms will be explained below.

Offset error ε_O is the difference between the analog input level which causes a first bit transition to occur and the level corresponding to 1/2 LSB. This should of course ideally be zero, i.e. the first bit transition should take place at a level representing exactly 1/2 LSB. The offset error affects all output codes with the same additive amount and can in most cases be sufficiently compensated for by adding an analog DC level to the input signal and/or by adding a fixed constant to the digital output.

Gain or scale factor is the slope of the transfer curve from analog levels to digital numbers. Hence, the gain error ε_G is the error in the slope of the transfer curve. It affects all output codes by the same percentage amount, and can normally be counteracted by amplification or attenuation of the analog input signal. Compensation can also be done by multiplying the digital number with a fixed gain calibration constant.

As pointed out above, offset and gain errors can, to a large extent, be compensated for by preconditioning the analog input signal or by processing the digital output code. The first method requires extra analog hardware, but has the advantage of utilizing the ADC at its best. The digital compensation is often easier to implement in software, but cannot fully eliminate the occurrence of unused quantization levels and/or overloading the ADC.

Linearity can be sub-divided into integral linearity (relative accuracy) and differential linearity.

Integral linearity error is the deviation of code mid-points of the transfer curve from a straight line, and is defined as $(\varepsilon_N(v))/A$ expressed in LSBs. This error is not possible to adjust or compensate for easily.

Differential linearity measures the difference between input levels corresponding to any two adjacent digital codes, $(\varepsilon_N(v_k) - \varepsilon_N(v_{k-1}))/A$, where $m(v_k) = m(v_{k-1}) + 1$. If the input level for one step differs from the previous step by exactly the value corresponding to one least significant bit (LSB), the differential non-linearity is zero. Differential linearity errors cannot be eliminated easily.

Monotonicity implies that increasing the analog input level never results in a decrease of the digital output code. Non-monotonicity may cause stability problems in feedback controls systems.

Missing codes in an ADC means that some digital codes can never be generated. It indicates that differential non-linearity is larger than 1 LSB. The problem of missing codes is generally caused by a non-monotonic behavior of the internal DAC. As will be shown in later sections, some types of ADCs use a built-in DAC.

Absolute accuracy error is the difference between the actual analog input to an ADC compared to the expected input level for a given digital output. The absolute accuracy is the compound effect of the offset error, gain error and linearity errors described above.

Conversion time of an ADC is the time required by the ADC to perform a complete conversion process. The conversion is commonly started by a "strobe" or synchronization signal, controlling the sampling rate.

As for DACs, it is important to remember that the parameters above may be affected by supply voltage and temperature. Data sheets only specify the parameters for certain temperatures and supply voltages. Significant deviations from the specified figures may hence occur in a practical system.

2.3.1 Anti-aliasing filters and sample-and-hold

As pointed out in Chapter 1, the process of sampling the analog time-continuous signal requires an efficient **anti-aliasing filter** to obtain an unambiguous digital, time-discrete representation of the signal. Ideally, a low-pass filter having a flat passband and extremely sharp cut-off at the Nyquist frequency is required. Of course, building such a filter in practice is impossible and approximations and compromises thus have to be made. The problem is quite similar to building the "perfect" reconstruction filter.

The first signal processing block of a digital signal processing system is likely to be an analog low-pass anti-aliasing filter. Depending on the application, different requirements of the filter may be stated. In audio systems, linear-phase response may, for instance, be an important parameter, while in a digital DC-voltmeter instrument, a low offset voltage may be imperative. Designing proper anti-aliasing filters is generally not a trivial task, especially if practical limitations, such as circuit board space and cost, also have to be taken into account.

Anti-aliasing filters are commonly implemented as active filters using feedback operational amplifiers or as SC filters. One way to relax the requirements of the analog anti-aliasing filter is to use **oversampling** techniques (see also reconstruction filters for DAC above). In this case, the signal is sampled with a considerably higher rate than required to fulfill the Nyquist criteria. Hence, the distance to the first mirrored signal spectra on the frequency axis will be much longer than if sampling were performed at only twice the Nyquist frequency. Figure 2.13 shows an example of an oversampled system. This could for instance be an audio system, where the highest audio frequency of interest would be about 20 kHz. The sampling rate for such a system could be $f_s = 48$ kHz. Without oversampling, the analog anti-aliasing filter should hence have a gain of about 1 at 20 kHz and an attenuation of, say, 30 dB at 24 kHz. This corresponds to a slope of roughly 150 dB/octave, a quite impressive filter and very hard to implement using analog components.

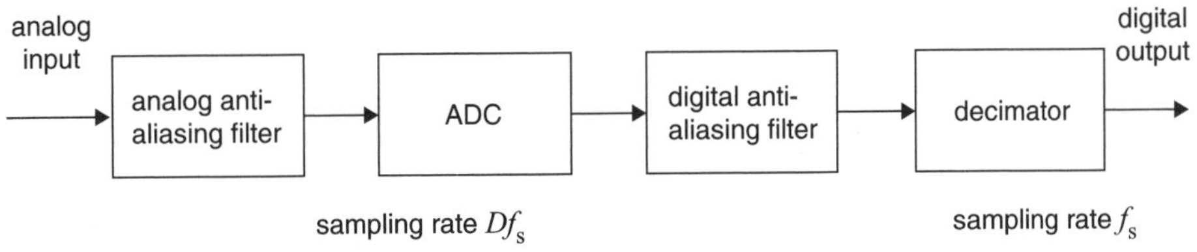

Figure 2.13 *D times oversampling ADC relaxing demands on the analog anti-aliasing filter*

Now, assume that we sample the analog input signal using $Df_s = 16 \cdot 48 = 768$ kHz, i.e. using an OSR of $D = 16$. To avoid aliasing distortion in this case, the analog anti-aliasing filter has quite relaxed requirements: gain of 1 at 20 kHz (as before) and attenuation of 30 dB at 384 kHz, which corresponds to a slope of 1.6 dB/octave. Such a filter is very easy to realize using a simple passive RC network.

Following the ADC, there is another anti-aliasing filter having the same tough requirements as stated at first. On the other hand, this filter is a digital one, being implemented using digital hardware or as software in a DSP. Designing a digital filter having the required passband characteristics is not very difficult. Preferably, a finite impulse response (FIR) filter structure may be used, having a linear-phase response.

At last, following the digital anti-aliasing filter is a **decimator** or **downsampler**. To perform the downsampling (Pohlmann, 1989), the device only passes every D-th sample from input to output and ignores the others, hence

$$y(n) = x(Dn) \tag{2.16}$$

In our example the decimator only takes every 16th sample and passes it on, i.e. the sampling rate is now brought back to $f_s = 48$ kHz.

Another detail that needs to be taken care of when performing analog-to-digital conversion is the actual sampling of the analog signal. An ideal sampling implies measuring the analog signal level during an infinitely short period of time (the "aperture"). Further, the ADC requires a certain time to perform the conversion and during this process the analog level must not change or conversion errors may occur. This problem is solved by feeding the analog signal to a sample and hold (S/H) device before it reaches the ADC. The S/H will take a quick snapshot sample and hold it constant during the conversion process. Many ADCs have a built-in S/H device.

2.3.2 Flash analog-to-digital converters

Flash type (or parallel) ADCs are the fastest due to the short conversion time and can hence be used for high sampling rates. Hundreds of megahertz is common today. On the other hand, these converters are quite complex, they have limited word length and hence resolution (10 bits or less), they are quite expensive and often suffer from considerable power dissipation.

The block diagram of a simple 2-bit flash ADC is shown in Figure 2.14. The analog input is passed to a number of analog level comparators in parallel (i.e. a bank of fast operational amplifiers with high gain and low offset).

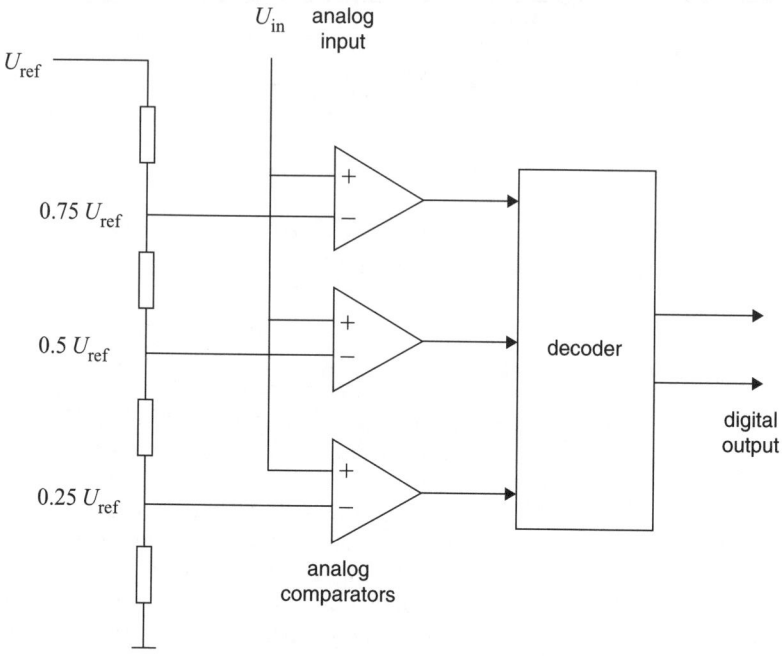

Figure 2.14 *An example 2-bit flash ADC*

If the analog input level U_{in} on the positive input of a comparator is greater than the level of the negative input, the output will be a digital "one". Otherwise, the comparator outputs a digital "zero". Now, a reference voltage U_{ref} is fed to the voltage divider chain, thus obtaining a number of reference levels

$$U_k = \frac{k}{2^N} U_{ref} \qquad (2.17)$$

where k is the quantization threshold number and N is the word length of the ADC. The analog input voltage will hence be compared to all possible quantization levels at the same time, rendering a "thermometer" output of digital ones and zeros from the comparators. These ones and zeros are then used by a digital decoder circuit to generate digital parallel PCM data on the output of the ADC.

As pointed out above, this type of ADC is fast, but is difficult to build for large word lengths. The resistors in the voltage divider chain have to be manufactured with high precision and the number of comparators and the complexity of the decoder circuit grows fast as the number of bits is increased.

2.3.3 Successive approximation analog-to-digital converters

These ADCs, also called successive approximation register (SAR), converters are the most common ones today. They are quite fast, but not as fast as flash converters. On the other hand, they are easy to build and inexpensive, even for larger word lengths.

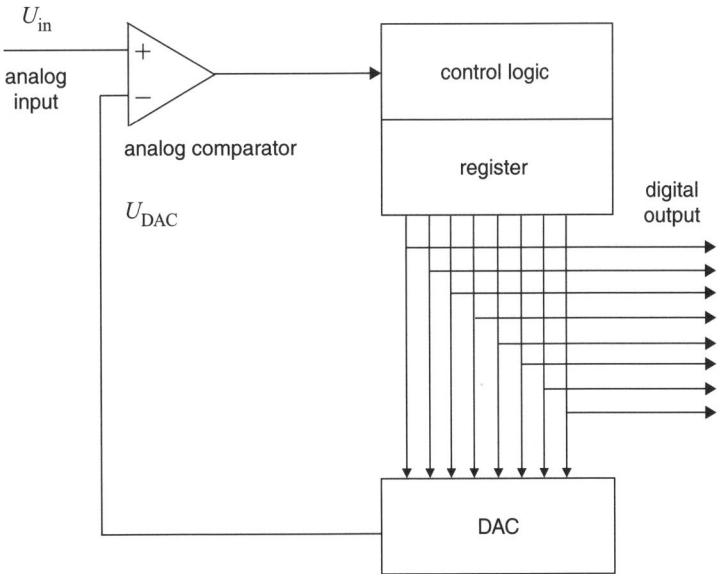

Figure 2.15 *An example SAR ADC (simplified block diagram)*

The main parts of the ADC are: an analog comparator, a digital register, a DAC and some digital control logic (see Figure 2.15). Using the analog comparator, the unknown input voltage U_{in} is compared to a voltage U_{DAC} created by a DAC, being a part of the ADC. If the input voltage is greater than the voltage coming from the DAC, the output of the comparator is a logic "one", otherwise a logic "zero". The DAC is fed an input digital code from the register, which is in turn controlled by the control logic. Now, the principle of successive approximation works as follows.

Assume that the register contains all zeros to start with, hence, the output of the DAC is $U_{DAC} = 0$. Now, the control logic will start to toggle the MSB to a one, and the analog voltage coming from the DAC will be half of the maximum possible output voltage. The control logic circuitry samples the signal coming from the comparator. If this is a one, the control logic knows that the input voltage is still larger than the voltage coming from the DAC and the "one" in the MSB will be left as is. If, on the other hand, the output of the comparator has turned zero, the output from the DAC is larger than the input voltage. Obviously, toggling the MSB to a one was just too much, and the bit is toggled back to zero. Now, the process is repeated for the second most significant bit and so on until all bits in the register have been toggled and set to a one or zero, see Figure 2.16.

Hence, the SAR ADC always has a constant conversion time. It requires n approximation cycles, where n is the word length, i.e. the number of bits in the digital code. SAR-type converters of today may be used for sampling rates up to some megahertz.

An alternative way of looking at the converter is to see it as a DAC + register put in a control feedback loop. We try to "tune" the register to match the analog input signal by observing the error signal from the comparator. Note that the DAC can of course be built in a variety of ways (see previous sections).

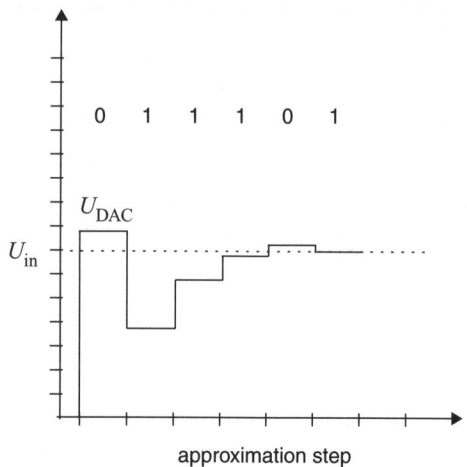

Figure 2.16 *Operation of a SAR converter*

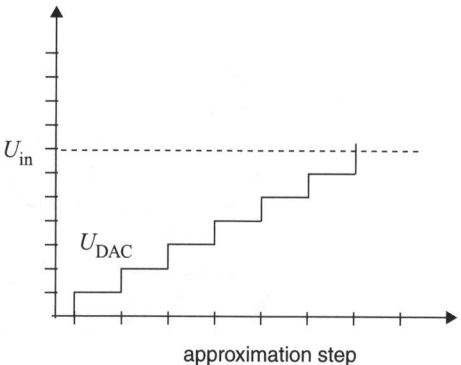

Figure 2.17 *Operation of a counting ADC*

Today, charge redistribution-based devices are quite common, since they are straightforward to implement using CMOS technology.

2.3.4 Counting analog-to-digital converters

An alternative, somewhat simpler ADC type is the **counting** ADC. The converter is mainly built in the same way as the SAR converter (Figure 2.15), but the control logic and register is simply a binary counter. The counter is reset to all zeros at the start of a conversion cycle, and is then incremented step by step. Hence, the output of the DAC is a staircase ramp function. The counting maintains until the comparator output switches to a zero and the counting is stopped (see Figure 2.17).

The conversion time of this type of ADC depends on the input voltage, the higher the level, the longer the conversion time (counting is assumed to take place at a constant rate). The interesting thing about this converter is that the output signal of the comparator is a PWM representation of the analog input signal. Further, by connecting an edge-triggered, monostable flip-flop to the comparator output, a PPM representation can also be obtained.

This type of converter is not very common. Using only a DAC and a comparator, the digital part can easily be implemented as software, such as a microcontroller. SAR-type converters can of course also be implemented in the same way as well as "tracking" type converters.

Tracking type converters can be seen as a special case of the generic counting converter. The tracking converter assumes that the change in the input signal level between consecutive samples is small. The counter is not restarted from zero at every conversion cycle, but starts from the previous state. If the output from the comparator is one, the counter is incremented until the comparator output toggles. If the output is zero when conversion is initiated, the counter is decremented until the output of the comparator toggles.

2.3.5 Integrating analog-to-digital converters

Integrating ADCs (sometimes also called counting converters) are often quite slow, but inexpensive and accurate. A common application is digital multimeters and similar equipment, in which precision and cost are more important than speed.

There are many different variations of the integrating ADC, but the main idea is that the unknown analog voltage (or current) is fed to the input of an analog integrator with a well-known integration time constant $\tau = RC$. The slope of the ramp on the output of the integrator is measured by taking the time between the output level passing two or more fixed reference threshold levels. The time needed for the ramp to go from one threshold to the other is measured by starting and stopping a binary counter running at a constant speed. The output of the counter is hence a measure of the slope of the integrator output, which in turn is proportional to the analog input signal level. Since this type of ADC commonly has a quite long conversion time, i.e. integration time, the input signal is required to be stable or only slowly varying. On the other hand, the integration process will act as a low-pass filter, averaging the input signal and hence suppressing interference superimposed on the analog input signal to a certain extent.

Figure 2.18 shows a diagram of a simplified integrating ADC. The basic function (ramp method) works as follows. Initially, the input switch is connected to ground in order to reset the analog integrator. The switch then connects the unknown analog input voltage $-U_{\text{in}}$ to the input of the integrator. The output level of the integrator will then start to change as

$$U(t) = U_0 + U_{\text{in}}\frac{t - t_0}{RC} + \varepsilon\frac{t - t_0}{RC} \tag{2.18}$$

where U_0 is the initial output signal from the integrator at time t_0 and ε is an error term representing the effect of offset voltage, leak currents and other shortcomings of a practical integrator. In practice, this term causes the output level of the integrator to drift slowly away from the starting level, even if the input voltage is held at zero.

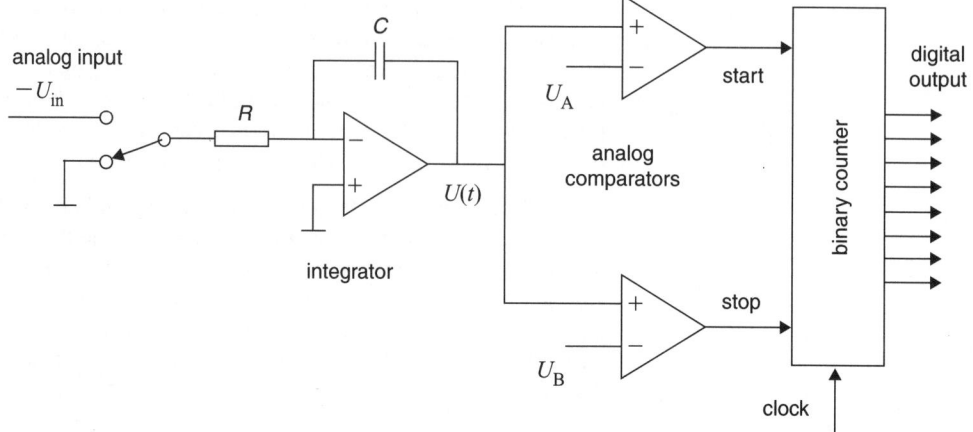

Figure 2.18 *A simplified integrating ADC (assuming $U_0 = 0$)*

Now, when the input signal is applied, the output level of the integrator starts to increase, and at time t_1 it is larger than the fixed threshold level U_A and a "start" signal will be generated. The binary counter (that has been reset to zero) starts counting at a fixed rate, determined by the stable clock pulse frequency. The output level of the integrator continues to increase until, at time t_2, it is larger than the upper threshold U_B, and a "stop" signal results. The binary counter stops and the binary PCM counter value now represents the time difference $t_2 - t_1$ and hence the analog input voltage

$$U(t_1) = U_A \quad \text{start signal at time } t_1$$

$$U(t_2) = U_B \quad \text{stop signal at time } t_2$$

$$U_A = U(t_1) = U_0 + U_{in}\frac{t_1 - t_0}{RC} + \varepsilon\frac{t_1 - t_0}{RC}$$

$$= U_0 + (U_{in} + \varepsilon)\frac{t_1 - t_0}{RC} \tag{2.19}$$

which can be rewritten as (assuming $U_0 = 0$)

$$t_1 - t_0 = U_A\frac{RC}{U_{in} + \varepsilon} \tag{2.20}$$

and in a similar way, we obtain

$$t_2 - t_0 = U_B\frac{RC}{U_{in} + \varepsilon} \tag{2.21}$$

Now, expressing the time difference measured by the counter we get

$$t_2 - t_1 = t_2 - t_0 - t_1 + t_0 = (U_B - U_A)\frac{RC}{U_{in} + \varepsilon} \tag{2.22}$$

Rearranging equation (2.22) yields

$$U_{in} + \varepsilon = (U_B - U_A)\frac{RC}{t_2 - t_1} = \frac{K}{t_2 - t_1} \tag{2.23}$$

As can be seen from equation (2.23) the unknown analog input voltage can easily be determined from the time difference recorded by the binary counter. Unfortunately, the error term will also be present. To obtain good precision in the ADC, we hence need to design the circuitry carefully to reduce leak current, etc. to a minimum. This is a problem associated with this type of ADC, and is one of the reasons this basic conversion method is seldom used.

One way to eliminate the error term is to use a **dual slope** method. The term "dual slope" refers to various methods in the literature, however, here only one method will be presented. Using this method, the timing measurement is performed in two phases. Phase 1 works as the simple ramp method described above. As soon as the output level of the integrator reaches the upper threshold, i.e. $U(t_2) = U_B$, phase 2 is initiated. During this phase, the polarity of the input signal is reversed and $+U_{in}$ is fed to the input of the integrator. The output level of the integrator will now start to decrease and at time t_3, the lower threshold is reached again, i.e. $U(t_3) = U_A$ which completes the conversion process. We now have two time differences: $t_2 - t_1$ and $t_3 - t_2$ which can be used to express the analog input voltage with the error term eliminated. At time t_3 we have

$$\begin{aligned}
U_A = U(t_3) &= U(t_2) - U_{in}\frac{t_3 - t_2}{RC} + \varepsilon\frac{t_3 - t_2}{RC} \\
&= U_B - (U_{in} - \varepsilon)\frac{t_3 - t_2}{RC}
\end{aligned} \tag{2.24}$$

Rewriting equation (2.24) we get

$$-U_{in} + \varepsilon = -(U_B - U_A)\frac{RC}{t_3 - t_2} = -\frac{K}{t_3 - t_2} \tag{2.25}$$

Subtracting equation (2.25) from equation (2.23) and dividing by 2 we obtain an expression for the analog input voltage without the error term

$$\begin{aligned}
U_{in} &= \frac{U_{in} + \varepsilon - (-U_{in} + \varepsilon)}{2} \\
&= \frac{1}{2}\left((U_B - U_A)\frac{RC}{t_2 - t_1} + (U_B - U_A)\frac{RC}{t_3 - t_2}\right) \\
&= \frac{K}{2}\left(\frac{1}{t_2 - t_1} + \frac{1}{t_3 - t_2}\right) = \frac{K}{2}\frac{t_3 - t_1}{(t_2 - t_1)(t_3 - t_2)}
\end{aligned} \tag{2.26}$$

An extension of the circuit in Figure 2.18 can be made in such a way that the system is astable and continuously generates a square wave output having the period $t_3 - t_1$. During $t_2 - t_1$ the square wave is low and during $t_3 - t_2$ it is high. A quite simple microcontroller can be used to calculate the value of the analog input voltage according to expression (2.26) above. An impressive resolution can be obtained using low-cost analog components.

2.3.6 Dither

Earlier we discussed the use of dither techniques and oversampling to improve resolution of a DAC. Dither can be used in a similar way to improve the resolution of an ADC and hence reduce the effect of quantization distortion.

In the case of an ADC, analog noise with amplitude of typically 1/3 LSB is added to the input signal before quantization takes place. If the analog input signal level is between two quantization thresholds, the added noise will make the compound input signal cross the closest quantization level now and then. The closer to the threshold the original signal level, the more frequent the threshold will be crossed. Hence, there will be a stochastic modulation of the binary PCM code from the ADC containing additional information about the input signal. Averaging the PCM samples, resolution below the LSB can be achieved.

Dither is common in high-quality digital audio systems (Pohlmann, 1989) nowadays, but has been used in video applications since 1950 and before that, to counteract gear backlash problems in early radar servo mechanisms.

2.3.7 Sigma–delta analog-to-digital converters

The **sigma–delta** ADC, sometimes also called **bitstream** ADC utilizes the technique of oversampling, discussed earlier. The sigma–delta modulator was first introduced in 1962, but until recent developments in digital very large scale integration (VLSI) technology it was difficult to manufacture with high resolution and good noise characteristics at competitive prices.

One of the major advantages of the sigma–delta ADC using oversampling is that it is able to use digital filtering and relaxes the demands on the analog anti-aliasing filter. This also implies that about 90% of the die area is purely digital, cutting production costs. Another advantage of using oversampling is that the quantization noise power is spread evenly over a larger frequency spectrum than the frequency band of interest. Hence, the quantization noise power in the signal band is lower than in the case of traditional sampling based on the Nyquist criteria.

Now, let us take a look at a simple 1-bit sigma–delta ADC. The converter uses a method that was derived from the delta modulation technique. This is based on quantizing the **difference** between successive samples, rather the quantizing the absolute value of the samples themselves. Figure 2.19 shows a **delta modulator** and demodulator with the modulator working as follows. From the analog input signal $x(t)$ a locally generated estimate $\tilde{x}(t)$ is subtracted. The difference $\varepsilon(t)$ between the two is fed to a 1-bit quantizer. In this simplified case, the quantizer may simply be the sign function, i.e. when $\varepsilon(t) \geq 0$ $y(n) = 1$, else $y(n) = 0$. The quantizer is working at the oversampling frequency, i.e. considerably faster than required by the signal bandwidth. Hence, the 1-bit digital output $y(n)$ can be interpreted as a kind of digital error signal:

$y(n) = 1$: estimated input signal level too small, increase level

$y(n) = 0$: estimated input signal level too large, decrease level

Now, the analog integrator situated in the feedback loop of the delta modulator (DM) is designed to function in exactly this way. Hence, if the analog

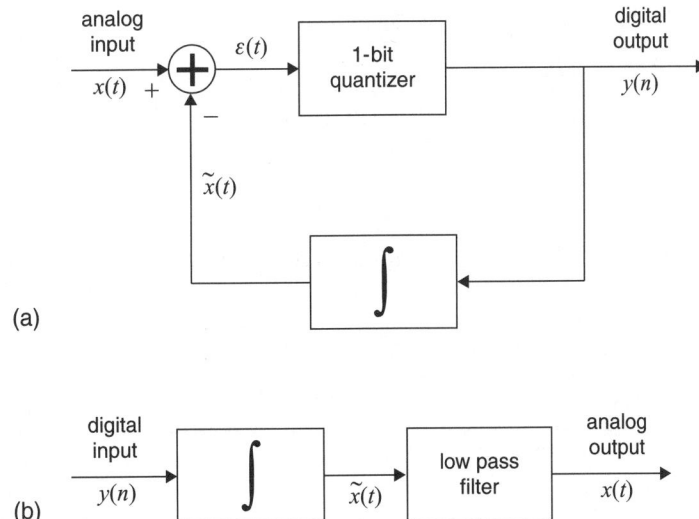

Figure 2.19 *A simplified (a) delta modulator and (b) demodulator*

input signal $x(t)$ is held at a constant level, the digital output $y(n)$ will (after convergence) be a symmetrical square wave (0 1 0 1 …), i.e. decrease, increase, decrease, increase … a kind of stable limit oscillation.

The delta demodulator is shown in the lower portion of Figure 2.19. The function is straightforward. Using the digital 1-bit "increase/decrease" signal, the estimated input level $\tilde{x}(t)$ can be created using an analog integrator of the same type as in the modulator. The output low-pass filter will suppress the ripple caused by the increase/decrease process.

Since integration is a linear process, the integrator in the demodulator can be moved to the input of the modulator. Hence, the demodulator will now only consist of the low-pass filter. We now have similar integrators on both inputs of the summation point in the modulator. For linearity reasons, these two integrators can be replaced by one integrator having $\varepsilon(t)$ connected to its input, and the output connected to the input of the 1-bit quantizer. The delta modulator has now become a sigma–delta modulator. The name is derived from the summation point (sigma) followed by the delta modulator.

If we now combine the oversampled 1-bit sigma–delta modulator with a digital **decimation filter** (rate reduction filter) we obtain a basic sigma–delta ADC (see Figure 2.20). The task of the decimation filter is threefold: to reduce the sampling frequency, to increase the word length from 1 bit to N bits and to reduce any noise pushed back into the frequency range of interest by the crude 1-bit modulator. A simple illustration of a decimation filter, decimating by a factor 5, would be an averaging process as shown in Table 2.2.

The decimation filter is commonly built using a decimator and a **comb filter** (Marven and Ewers, 1993; Lynn and Fuerst, 1998). The comb filter belongs to the class of frequency sampling filters, and has the advantage of being easy to implement in silicon or in a DSP, using only additions and subtractions. A comb filter typically has $2k$ zeros on the unit circle in the z-plane, resulting in zero

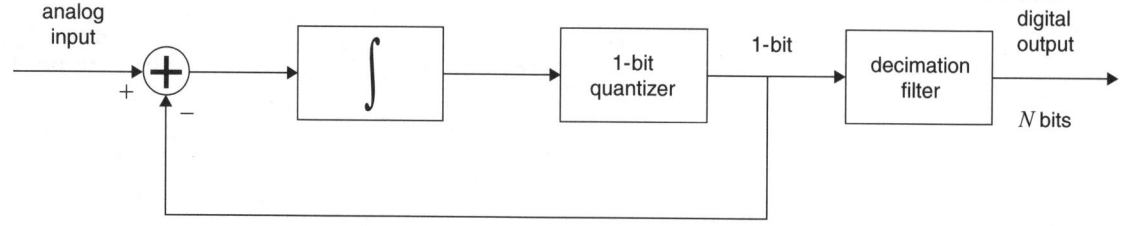

Figure 2.20 *A simplified, oversampled bitstream sigma–delta ADC*

Table 2.2 *Decimation filter example*

Input	0 0 1 1 0	0 1 1 1 0	1 1 1 0 1	1 0 1 0 1	0 0 0 1 1
Averaging process	$3 \times 0; 2 \times 1$	$2 \times 0; 3 \times 1$	$1 \times 0; 4 \times 1$	$2 \times 0; 3 \times 1$	$3 \times 0; 2 \times 1$
Output	0	1	1	1	0

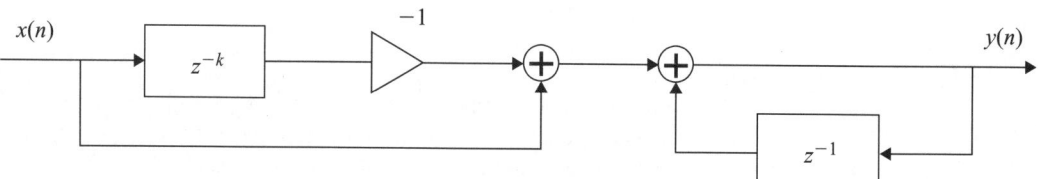

Figure 2.21 *Block diagram of an example comb filter. Note! Only additions, subtractions and delays are used*

gain at the frequencies $\pm(f_s/k), \pm 2(f_s/k), \pm 3(f_s/k), \ldots, (f_s/2)$. The transfer function of a common type of comb filter can be written as

$$H(z) = \frac{1 - z^{-k}}{1 - z^{-1}} \tag{2.27}$$

The corresponding block diagram is shown in Figure 2.21. Note that the delay z^{-k} can, of course, be implemented as a cascade of k standard z^{-1} delays.

Summary

In this chapter the following topics have been treated:

- PCM, PAM, PPM, PNM, PWM and PDM
- Fixed-point 2's complement, offset binary, sign and magnitude and floating-point
- Companding systems
- Multiplying, Integrating and Bitstream D/A converters
- Oversampling, interpolators, truncators and dither
- Sample and hold reconstruction filters and anti-aliasing filters
- Flash, Successive approximation, Counting and Integrating A/D converters
- Sigma–delta and Bitstream A/D converters, Decimation filters, Comb filters.

Review questions

R2-1 What is PCM, PAM, PPM, PNM, PWM and PDM?

R2-2 Explain the following terms: fixed-point 2's complement and floating point.

R2-3 Why are companders used? What are the pros and cons?

R2-4 Draw a block diagram of a R–$2R$ multiplying DAC and explain how it works. What is the advantage of using only R and $2R$ resistors?

R2-5 Draw a block diagram of a flash ADC and explain how it works. What is the great advantage of a flash converter? Are there any disadvantages?

R2-6 Draw a block diagram of a SAR ADC and explain how it works.

R2-7 Why is oversampling used in DACs? Why is it used in ADCs?

Solved problems

P2-1 Show that the μ-law turns into a linear function if the parameter is chosen as $\mu = 0$.

P2-2 A 14-bit ADC has a full-scale input voltage of 5 V. Disregarding all errors except the offset error, we get an erroneous reading from the ADC by 3 LSB. What is the approximate offset error expressed in mV?

P2-3 Write a MATLAB™ program to obtain a plot of the poles and zeros of the comb filter in equation (2.27), assume $k = 9$. Are there any numeric risks?

P2-4 Write a MATLAB™ program to obtain a Bode plot of the comb filter in equation (2.27), assume $k = 9$.

3 Adaptive digital systems

Background In Chapter 1, digital filters were presented. In the simple case, we assumed a good knowledge of the input signal (for instance, a desired signal plus interference) and the output signal (only the desired signal). Hence, a filter specification was easy to formulate, from which filter structure and filter parameters could be obtained using proper filter synthesis methods.

In some cases, the situation may, however, be harder. We may, for instance, not have a good knowledge of the properties of the input and output signal, or the signal properties may change randomly over time. Another problem might be that there are no proper synthesis methods available. In such situations, an **adaptive filter** may be the solution. Given a performance measure, an adaptive filter is able to find good filter parameters by itself in an iterative way. The filter can also "track" and readjust if the properties of the input signal changes.

In this chapter, the theory of adaptive filters and related systems is discussed. Example applications are also included like interference canceling, equalizers and beam-forming systems. Adaptive filters are common in telecommunication applications like high-speed modems and cell phones.

Objectives In this chapter we will discuss:

- The structure of a generic closed-loop adaptive system
- The linear combiner and variations
- The mean square error (MSE) performance function of an adaptive system
- Some common adaptation algorithms, steepest descent, Newton and least mean square (LMS)
- Example applications, adaptive interference canceling, equalizers and beamforming.

3.1 Introduction

An adaptive signal processing system is a system which has the ability to change its processing behavior in a way to maximize a given performance measure. An **adaptive system** is self-adjusting and is, by its nature, a time varying and non-linear system. Hence, when using classical mathematical models and tools assuming linearity for the analysis of adaptive systems, care must be exercised.

A simple example of an adaptive system is the automatic gain control (AGC) in a radio receiver (RX). When the input antenna signal is strong, the AGC circuitry reduces the amplification in the RX to avoid distortion caused by saturation of the amplifying circuits. At weak input signals, the amplification is increased to make the signal readable.

Adaptive systems should, however, not be confused with pure feedback control systems. An electric heater controlled by a thermostat is, for instance, a feedback control system and not an adaptive system, since the control function is not changed (e.g. the thermostat always switches off the heater at 22°C).

The idea of self-adjusting systems (partly "self-designing" systems) is not new. Building such systems using "analog" signals and components is, however, very hard, except in some simple cases. The advent of very large scale integration (VLSI) digital signal processing (DSP) and computing devices has made digital adaptive systems possible in practice.

In this chapter, we will discuss a class of adaptive digital systems based on closed-loop (feedback) adaptation. Some common systems and algorithms will be addressed. There are, however, many variations possible (Widrow and Stearns, 1985). The theory of adaptive systems is on the border line of optimization theory and neural network technology (treated in Chapter 4).

3.1.1 System structure

Figure 3.1 shows a generic adaptive signal processing system. The system consists of three parts: the processor, the performance function and the adaptation algorithm.

The **processor** is the part of the system that is responsible for the actual processing of the input signal x, thus generating the output signal y. The processor can, for instance, be a digital finite impulse response (FIR) filter.

The **performance function** takes the signals x, y as inputs as well as "other data" d, that may affect the performance of the entire system from x to y. The performance function is a quality measure of the adaptive system. In optimization theory, this function corresponds to the "objective function" and in control theory it corresponds to the "cost function". The output ε from the performance function is a "quality signal" illustrating the processor at its present state and indicating whether it is performing well, taking into account the input signals, output signals and other relevant parameters.

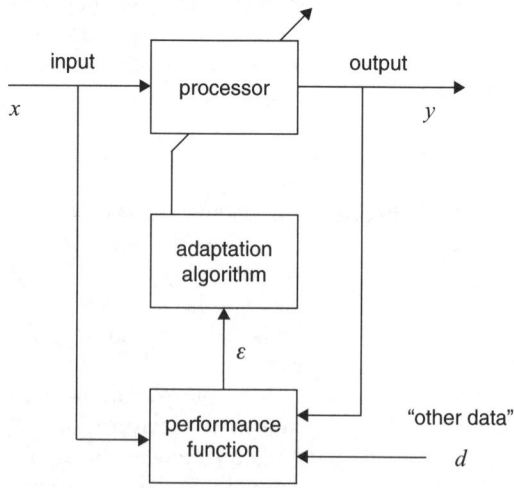

Figure 3.1 *A generic closed-loop adaptive system*

The quality signal ε is finally fed to the **adaptation algorithm**. The task of the adaptation algorithm is to change the parameters of the processor in such a way that the performance is maximized. In the example of an FIR filter, this would imply changing the tap weights.

Closed-loop adaptation has the advantage of being usable in many situations in which no analytic synthesis procedure either exists or is known and/or when the characteristics of the input signal vary considerably. Further, in cases where physical system component values are variable or inaccurately known, closed-loop adaptation will find the best choice of parameters. In the event of partial system failure, the adaptation mechanism may even be able to readjust the processor in such a way that the system will still achieve an acceptable performance. System reliability can often be improved using performance feedback.

There are, however, some inherent problems as well. The processor model and the way the adjustable parameters of the processor model is chosen affect the possibilities of building a good adaptive system. We must be sure that the desired response of the processor can really be achieved by adjusting the parameters within the allowed ranges. Using a too complicated processor model would, on the other hand, make analytic analysis of the system cumbersome or even impossible.

The performance function also has to be chosen carefully. If the performance function does not have a unique extremum, the behavior of the adaptation process may be uncertain. Further, in many cases it is desirable (but not necessary) that the performance function is differentiable. Choosing "other data" carefully may also affect the usefulness of the performance function of the system.

Finally, the adaptation algorithm must be able to adjust the parameters of the processor in such a way as to improve the performance as fast as possible and eventually to converge to the optimum solution (in the sense of the performance function). A too "fast" adaptation algorithm may, however, result in instability or undesired oscillatory behavior. These problems are well known from optimization and control theory.

3.2 The processor and the performance function

In this section, a common processor model, the **adaptive linear combiner**, and a common performance function, the **mean square error (MSE)** will be presented. Throughout this text, vector notation will be used, where all vectors are assumed to be column vectors and are usually indicated as a transpose of a row vector, unless otherwise stated.

3.2.1 The adaptive linear combiner

One of the most common processor models is the so-called adaptive linear combiner (Widrow and Stearns, 1985), shown in Figure 3.2. In the most general form, the combiner is assumed to have $L + 1$ inputs denoted

$$\mathbf{X}_k = [x_{0k} \quad x_{1k} \quad \cdots \quad x_{Lk}]^{\mathrm{T}} \tag{3.1}$$

where the subscript k is used as the time index. Similarly, the gains or weights of the different branches are denoted

$$\mathbf{W}_k = [w_{0k} \quad w_{1k} \quad \cdots \quad w_{Lk}]^{\mathrm{T}} \tag{3.2}$$

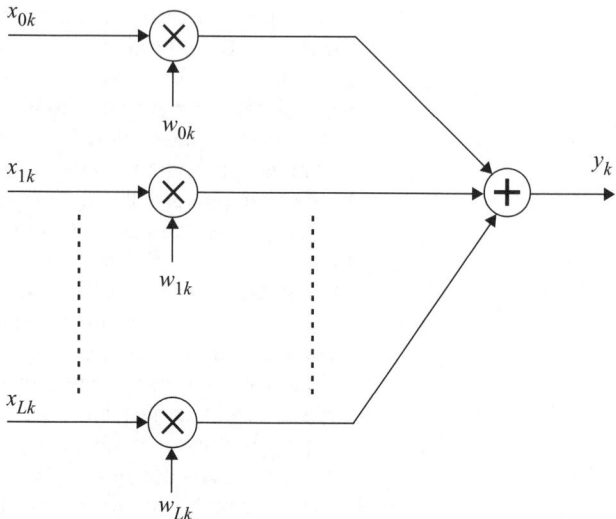

Figure 3.2 *The general form of adaptive linear combiner with multiple inputs*

Since we are now dealing with an adaptive system, the weights will also vary in time and hence have a time subscript k. The output y of the adaptive multiple-input linear combiner can be expressed as

$$y_k = \sum_{l=0}^{L} w_{lk} x_{lk} \tag{3.3}$$

As can be seen from equation (3.3) this is nothing else but a dot product between the input vector and the weight vector, hence

$$y_k = \mathbf{X}_k^{\mathrm{T}} \mathbf{W}_k = \mathbf{W}_k^{\mathrm{T}} \mathbf{X}_k \tag{3.4}$$

The adaptive linear combiner can also take another form working in the temporal domain, with only a single input (see Figure 3.3). This is the general form of the adaptive linear combiner shown in Figure 3.2, extended with a delay line. In this case, the output is given by the convolution sum

$$y_k = \sum_{l=0}^{L} w_{lk} x_{k-l} \tag{3.5}$$

which is the expression for the well-known FIR filter or transversal filter. Now, if the input vector is defined as

$$\mathbf{X}_k = [x_k \quad x_{k-1} \quad \cdots \quad x_{k-L}]^{\mathrm{T}} \tag{3.6}$$

equation (3.4) still holds.

It is possible to use other types of structure, e.g. infinite impulse response (IIR) filters or lattice filters as a processor in an adaptive system. The adaptive linear combiner is, however, by far the most common since it is quite straightforward to analyze and design. It is much harder to find good adaptation

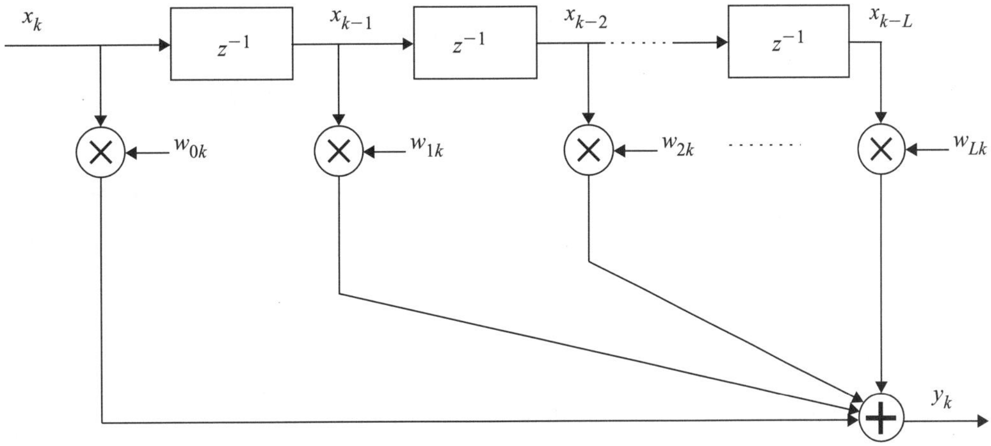

Figure 3.3 *A temporal version of an adaptive linear combiner with a single input – an adaptive transversal filter (FIR filter or tapped delay-line filter)*

algorithms for IIR filters, since the weights are not permitted to vary arbitrarily or instability and oscillations may occur.

3.2.2 The performance function

The performance function can be designed in a variety of ways. In this text, we will concentrate on the quite common mean square error (**MSE**) measure. To be able to derive the following expressions, we will assume that "other data" d_k will be the desired output from our system, sometimes called "**training signal**" or "**desired response**" (see Figure 3.4). One could argue: why employ an adaptive system at all if the desired response is known in advance? However, presently we shall assume the availability of such a signal. Later, we will discuss its derivation in more detail. Considerable ingenuity is, however, needed in most cases to find suitable "training signals".

In this case, the performance function will be based on the error signal ε_k

$$\varepsilon_k = d_k - y_k = d_k - \mathbf{X}_k^{\mathrm{T}}\mathbf{W} = d_k - \mathbf{W}^{\mathrm{T}}\mathbf{X}_k \tag{3.7}$$

In this discussion, we assume that the adaptation process is slow compared to the variations in time of the input signal. Hence, the time subscript on the weight vector has been dropped. From equation (3.7), we will now derive the performance function as the mean square error (MSE). We then have to minimize the MSE (the power of the error signal) in order to maximize the performance of the corresponding adaptive system. Now, take the square of equation (3.7)

$$\varepsilon_k^2 = \left(d_k - \mathbf{W}^{\mathrm{T}}\mathbf{X}_k\right)\left(d_k - \mathbf{X}_k^{\mathrm{T}}\mathbf{W}\right) = d_k^2 + \mathbf{W}^{\mathrm{T}}\mathbf{X}_k\mathbf{X}_k^{\mathrm{T}}\mathbf{W} - 2d_k\mathbf{X}_k^{\mathrm{T}}\mathbf{W} \tag{3.8}$$

Assume that ε_k, d_k and $\mathbf{X}_k$ are statistically stationary (Papoulis and Pillai, 2001), and take the expected value of equation (3.8) over k to obtain the MSE

$$\xi = \mathrm{E}\left[\varepsilon_k^2\right] = \mathrm{E}\left[d_k^2\right] + \mathbf{W}^{\mathrm{T}}\mathrm{E}\left[\mathbf{X}_k\mathbf{X}_k^{\mathrm{T}}\right]\mathbf{W} - 2\mathrm{E}\left[d_k\mathbf{X}_k^{\mathrm{T}}\right]\mathbf{W}$$

$$= \mathrm{E}\left[d_k^2\right] + \mathbf{W}^{\mathrm{T}}\mathbf{R}\mathbf{W} - 2\mathbf{P}^{\mathrm{T}}\mathbf{W} \tag{3.9}$$

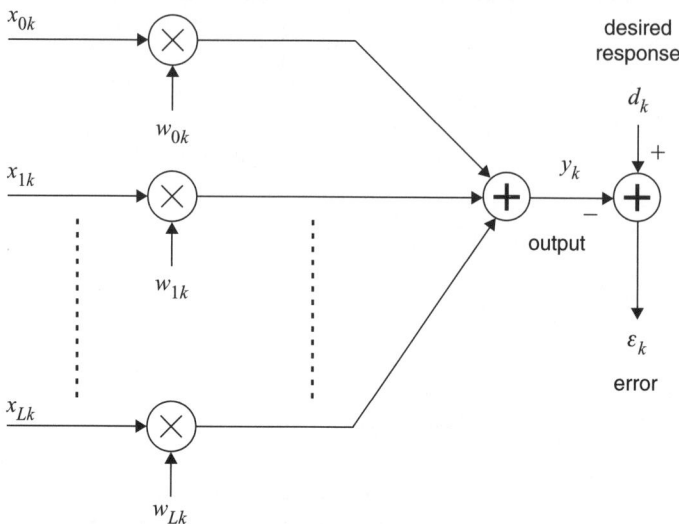

Figure 3.4 *Adaptive linear combiner with desired response ("other data") and error signals*

where the matrix $\mathbf{R}$ is the square input **correlation matrix**. The main diagonal terms are the mean squares of the input components, i.e. the variance and the power of the signals. The matrix is symmetric. If the single input form of the linear combiner is used, this matrix constitutes the **auto-correlation** matrix of the input signal. In the latter case, all elements on the diagonal will be equal.

The matrix $\mathbf{R}$ looks like

$$\mathbf{R} = \mathrm{E}\left[\mathbf{X}_k\mathbf{X}_k^{\mathrm{T}}\right] = \mathrm{E}\begin{bmatrix} x_{0k}^2 & x_{0k}x_{1k} & \cdots & x_{0k}x_{Lk} \\ x_{1k}x_{0k} & x_{1k}^2 & \cdots & x_{1k}x_{Lk} \\ \vdots & \vdots & & \vdots \\ x_{Lk}x_{0k} & x_{Lk}x_{1k} & & x_{Lk}^2 \end{bmatrix} \tag{3.10}$$

The vector $\mathbf{P}$ is the **cross-correlation** between the desired response and the input components

$$\mathbf{P} = \mathrm{E}[d_k x_{0k} \quad d_k x_{1k} \quad \cdots \quad d_k x_{Lk}]^{\mathrm{T}} \tag{3.11}$$

Since the signals d_k and x_{ik} are generally not statistically independent, the expected value of the product $d_k\mathbf{X}_k^{\mathrm{T}}$ **cannot** be rewritten as a product of expected values. If d_k and all x_{ik} are indeed statistically independent, they will also be uncorrelated and $\mathbf{P} = \mathbf{0}$. It can easily be seen from equation (3.9) that in this case, minimum MSE would occur when setting $\mathbf{W} = \mathbf{0}$, i.e. setting all weights equal to 0, or in other words switching off the processor completely. It does not matter what the adaptive system tries to do using the input signals; the MSE cannot be reduced using any input signal or combination thereof. In this case, we have probably made a poor choice of d_k or $\mathbf{X}_k$, or the processor model is not relevant.

From equation (3.9) we can see that the MSE is a quadratic function of the components of the weight vector. This implies that the surface of the

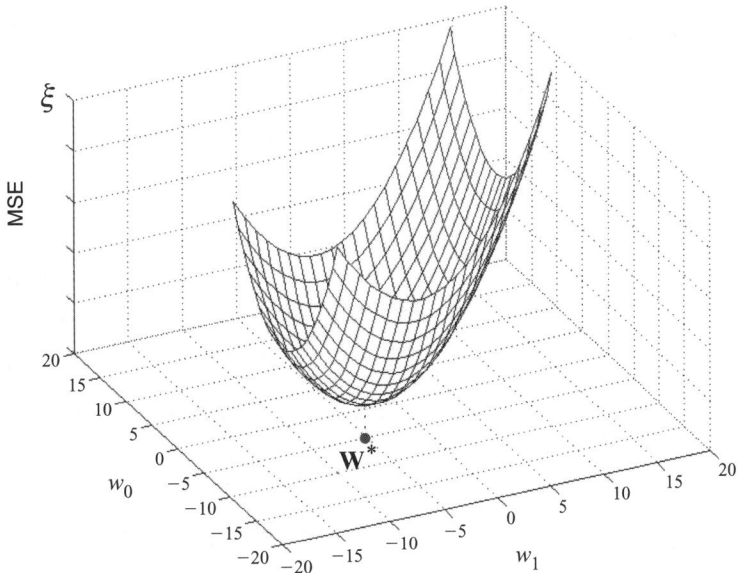

Figure 3.5 *Two-dimensional paraboloid performance surface. The MSE as a function of two weight components w_0 and w_1. Minimum error, i.e. optimum is found at the bottom of the bowl, i.e. at W^*, the Wiener weight vector*

performance function will be bowl shaped and form a hyperparaboloid in the general case, which only has **one global minimum**. Figure 3.5 shows an example of a two-dimensional (paraboloid) quadratic performance surface. The minimum MSE point, i.e. the optimum combination of weights, can be found at the bottom of the bowl. This set of weights is sometimes called the Wiener weight vector $\mathbf{W}^*$. The optimum solution, i.e. the Wiener weight vector can be found by finding the point where the gradient of the performance function is zero.

Differentiating equation (3.9), we obtain the gradient

$$\nabla(\xi) = \frac{\partial \xi}{\partial \mathbf{W}} = \left[\frac{\partial \xi}{\partial w_0} \quad \frac{\partial \xi}{\partial w_1} \quad \cdots \quad \frac{\partial \xi}{\partial w_L} \right]^{\mathrm{T}} = 2\mathbf{R}\mathbf{W} - 2\mathbf{P} \qquad (3.12)$$

The optimum weight vector is found where the gradient is zero, hence

$$\nabla(\xi) = \mathbf{0} = 2\mathbf{R}\mathbf{W}^* - 2\mathbf{P} \qquad (3.13)$$

Assuming that $\mathbf{R}$ is non-singular the Wiener–Hopf equation in matrix form is

$$\mathbf{W}^* = \mathbf{R}^{-1}\mathbf{P} \qquad (3.14)$$

From this, we realize that adaptation (optimization), i.e. finding the optimum weight vector, is an iterative method of finding the inverse (or pseudoinverse) of the input correlation matrix.

The minimum mean square error (at optimum) is now obtained by substituting $\mathbf{W}^*$ from equation (3.14) for $\mathbf{W}$ in equation (3.9) and using the symmetry of the input correlation matrix, i.e. $\mathbf{R}^T = \mathbf{R}$ and $(\mathbf{R}^{-1})^T = \mathbf{R}^{-1}$

$$\begin{aligned}\xi_{min} &= \mathrm{E}[d_k^2] + \mathbf{W}^{*T}\mathbf{R}\mathbf{W}^* - 2\mathbf{P}^T\mathbf{W}^* \\ &= \mathrm{E}[d_k^2] + (\mathbf{R}^{-1}\mathbf{P})^T \mathbf{R}\mathbf{R}^{-1}\mathbf{P} - 2\mathbf{P}^T\mathbf{R}^{-1}\mathbf{P} \\ &= \mathrm{E}[d_k^2] - \mathbf{P}^T\mathbf{R}^{-1}\mathbf{P} = \mathrm{E}[d_k^2] - \mathbf{P}^T\mathbf{W}^*\end{aligned} \qquad (3.15)$$

Finally, a useful and important statistical condition exists between the error signal ε_k and the components of the input signal vector $\mathbf{X}_k$ when $\mathbf{W} = \mathbf{W}^*$. Multiplying equation (3.7) by $\mathbf{X}_k$ from the left, we obtain

$$\varepsilon_k \mathbf{X}_k = d_k \mathbf{X}_k - y_k \mathbf{X}_k = d_k \mathbf{X}_k - \mathbf{X}_k \mathbf{X}_k^T \mathbf{W} \qquad (3.16)$$

Taking the expected value of equation (3.16) yields

$$\mathrm{E}[\varepsilon_k \mathbf{X}_k] = \mathrm{E}[d_k \mathbf{X}_k] - \mathrm{E}[\mathbf{X}_k \mathbf{X}_k^T]\mathbf{W} = \mathbf{P} - \mathbf{R}\mathbf{W} \qquad (3.17)$$

Inserting equation (3.14) into equation (3.17) yields

$$\mathrm{E}[\varepsilon_k \mathbf{X}_k]_{\mathbf{W}=\mathbf{W}^*} = \mathbf{P} - \mathbf{R}\mathbf{R}^{-1}\mathbf{P} = \mathbf{P} - \mathbf{P} = \mathbf{0} \qquad (3.18)$$

This result corresponds to the result of Wiener-filter theory. When the impulse response (weights) of a filter is optimized, the error signal is uncorrelated (orthogonal) to the input signals.

3.3 Adaptation algorithms

From the above, we have realized that the mean square error (MSE) performance surface for the linear combiner is a quadratic function of the weights (input signal and desired response statistically stationary). The task of the adaptation algorithm is to locate the optimum setting $\mathbf{W}^*$ of the weights, hence obtaining the best performance from the adaptive system. Since the parameters of the performance surface may be unknown and analytical descriptions are not available, the adaptation process is an iterative process, searching for the optimum point.

Many different adaptation schemes have been proposed. In this text a few common generic types will be discussed. The ideal adaptation algorithm converges quickly, but without oscillatory behavior, to the optimum solution. Once settled at the optimum, the ideal algorithm should track the solution, if the shape of the performance surface changes over time. The adaptation algorithm should preferably also be easy to implement, and not be excessively demanding with regards to the computations.

The shape of the **learning curve**, i.e. a plot of the MSE ξ_i as a function of the iteration number i, is in many cases a good source of information about the properties of an adaptation algorithm. Note! For every iteration period i there is commonly a large number of sample instants k, since ξ is the average over k (time).

3.3.1 The method of steepest descent

The idea of the steepest descent method is as follows: starting in an arbitrary point $\mathbf{W}_0$ of the performance surface, estimate the gradient and change the weights in the negative direction of the gradient ∇_0. In this way, the weight vector point will proceed "downhill" until reaching the bottom of the performance surface "bowl" and the optimum point. The steepest descent algorithm can be expressed as the following

$$\mathbf{W}_{i+1} = \mathbf{W}_i + \mu(-\nabla_i) \tag{3.19}$$

where μ is the step size at each iteration. The larger the step size, the faster the convergence. An excessively large step size would, on the other hand, cause instability and oscillatory problems in the solution. This problem is common to all closed-loop control systems. Finding a good compromise on μ may, therefore, be a problem.

Another problem is obtaining the gradient ∇_i. Since the gradient is normally not known in advance, it has to be estimated by adjusting the weights one by one by a small increment $\pm\delta$ and measuring the MSE ξ. Hence, the gradient is expressed as

$$\nabla(\xi) = \frac{\partial \xi}{\partial \mathbf{W}} = \left[\frac{\partial \xi}{\partial w_0} \quad \frac{\partial \xi}{\partial w_1} \quad \cdots \quad \frac{\partial \xi}{\partial w_L} \right]^{\mathrm{T}} \tag{3.20}$$

where the respective derivatives can be estimated by

$$\frac{\partial \xi}{\partial w_n} \approx \frac{\xi(w_n + \delta) - \xi(w_n - \delta)}{2\delta} \tag{3.21}$$

One has to remember that the performance surface is a "noisy" surface in the general case. To obtain an MSE with a small variance, many samples are required. There will be two perturbation points for each dimension of the weight vector, $2L$ averaged measurements, each consisting of a number of samples are needed. There is, of course, a trade-off. The larger the number of averaged points, i.e. the longer the estimation time, the less noise there will be in the gradient estimate. A noisy gradient estimate will result in an erratic adaptation process.

Disregarding the noise, it should also be noted that the steepest descent "path" is not necessarily the straightest route to the optimum point in the general case. An example is shown in Figure 3.6(a). Looking down into the bowl-shaped performance function of Figure 3.5, the movement of the weight vector during the iterations is shown. Starting at a random point, the change of the weight vector is always in the direction of the negative gradient (steepest descent).

3.3.2 Newton's method

Newton's method may be seen as an improved version of the steepest descent method discussed above. In this method, the steepest descent route is not used, but the weight vector moves "directly" towards the optimum point. This is achieved by adding information about the shape of the surface to the iterative adaptation algorithm

$$\mathbf{W}_{i+1} = \mathbf{W}_i + \mu \mathbf{R}^{-1}(-\nabla_i) \tag{3.22}$$

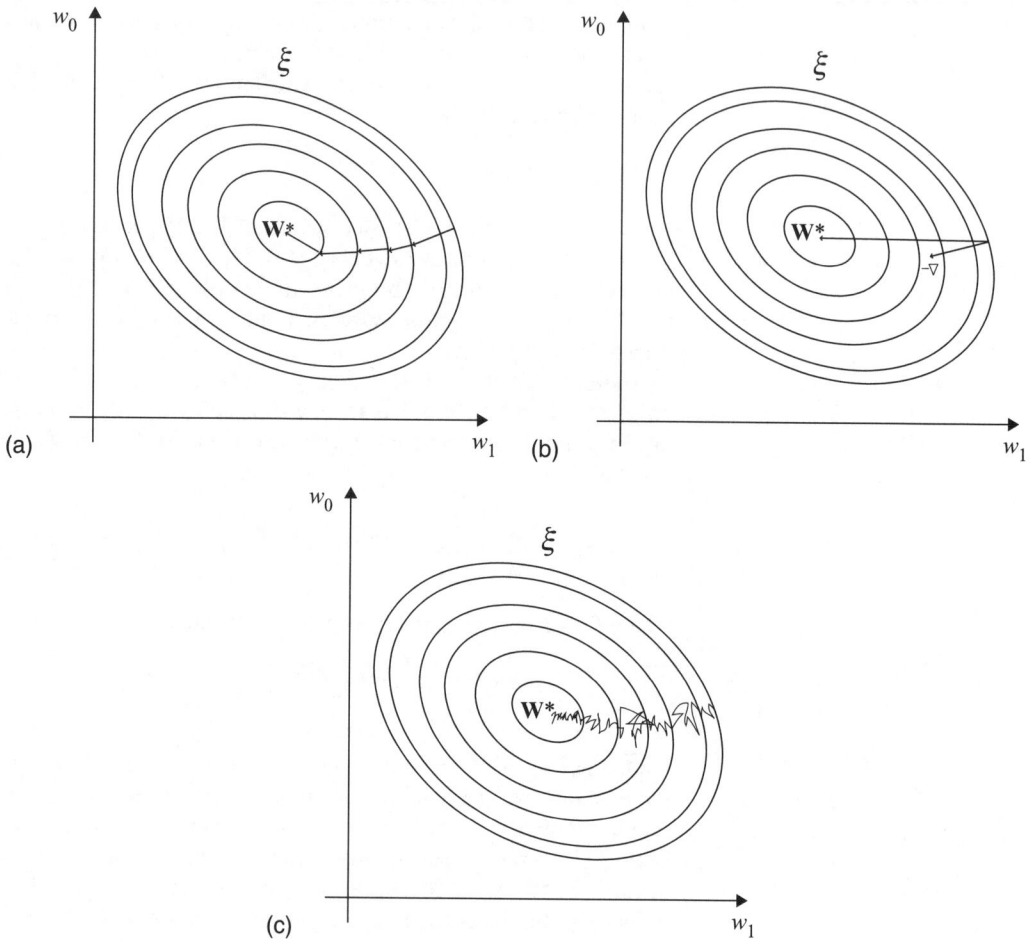

Figure 3.6 *Looking down into the performance "bowl" (see Figure 3.5), iteration of the weight vector towards optimum at **W*** for (a) steepest descent, (b) Newton's method and (c) LMS*

The extra information is introduced into equation (3.22) by including $\mathbf{R}^{-1}$, the inverse of the input correlation matrix. When the gradient is multiplied by this matrix, the resulting move of the weight vector will not be in the negative direction of the gradient, but rather in the direction towards the optimum point.

Due to the additional matrix multiplication, Newton's method requires more computational power than the method of steepest descent. In addition, knowledge of the inverse of the input correlation matrix is required, which in turn requires averaging of a number of samples.

The advantage of this method is that it often finds the optimum point in very few iterations. It is easy to show that if we have perfect knowledge of the gradient and the inverse of the input correlation matrix, Newton's method will find the optimum solution in only one iteration. Setting $\mu = 0.5$ in equation

(3.22) and inserting equation (3.13) we obtain

$$
\begin{aligned}
\mathbf{W}_{i+1} &= \mathbf{W}_i - 0.5\mathbf{R}^{-1}\nabla_i \\
&= \mathbf{W}_i - 0.5\mathbf{R}^{-1}(2\mathbf{R}\mathbf{W}_i - 2\mathbf{P}) \\
&= \mathbf{R}^{-1}\mathbf{P} = \mathbf{W}^*
\end{aligned}
\tag{3.23}
$$

where the last equality is given by equation (3.14). Hence, it does not matter where we start on the performance surface; if the gradient and input signal inverse correlation matrix are known, the optimum point will be reached in only **one** iteration (see Figure 3.6(b)). In practice, this is not possible in the general case, since the gradient estimate and the elements of the input correlation matrix will be noisy.

3.3.3 The least mean square algorithm

The methods presented above, the method of steepest descent and Newton's method, both require an estimation of the gradient at each iteration. In this section, the **least mean square** (LMS) algorithm will be discussed. The LMS algorithm uses a special estimate of the gradient that is valid for the adaptive linear combiner. Thus, the LMS algorithm is more restricted in its use than the other methods.

On the other hand, the LMS algorithm is important because of its ease of computation, and because it does not require the repetitive gradient estimation. The LMS algorithm is quite often the best choice for many different adaptive signal processing applications. Starting from equation (3.7) we recall that

$$
\varepsilon_k = d_k - y_k = d_k - \mathbf{X}_k^{\mathrm{T}}\mathbf{W}
\tag{3.24}
$$

In the previous methods, we would estimate the gradient using $\xi = \mathrm{E}\left[\varepsilon_k^2\right]$, but the LMS algorithm uses ε_k^2 itself as an estimate of ξ. This is the main point of the LMS algorithm. At each iteration in the adaptive process, we have a gradient estimate of the form

$$
\hat{\nabla}_k =
\begin{bmatrix}
\dfrac{\partial \varepsilon_k^2}{\partial w_0} \\[2ex]
\dfrac{\partial \varepsilon_k^2}{\partial w_1} \\[1ex]
\vdots \\[1ex]
\dfrac{\partial \varepsilon_k^2}{\partial w_L}
\end{bmatrix}
= 2\varepsilon_k
\begin{bmatrix}
\dfrac{\partial \varepsilon_k}{\partial w_0} \\[2ex]
\dfrac{\partial \varepsilon_k}{\partial w_1} \\[1ex]
\vdots \\[1ex]
\dfrac{\partial \varepsilon_k}{\partial w_L}
\end{bmatrix}
= -2\varepsilon_k \mathbf{X}_k
\tag{3.25}
$$

where the derivatives of ε_k follow from equation (3.24). Using this simplified gradient estimate, we can now formulate a steepest descent type algorithm

$$
\mathbf{W}_{k+1} = \mathbf{W}_k - \mu\hat{\nabla}_k = \mathbf{W}_k + 2\mu\varepsilon_k\mathbf{X}_k
\tag{3.26}
$$

This is the LMS algorithm. The gain constant μ governs the speed and stability of the adaptation process. The weight change at each iteration is based on imperfect, noisy gradient estimates, which implies that the adaptation process does not follow the true line of steepest descent on the performance surface (see Figure 3.6(c)). The noise in the gradient estimate is, however, attenuated with time by the adaptation process, which acts as a low-pass filter.

The LMS algorithm is elegant in its simplicity and efficiency. It can be implemented without squaring, averaging or differentiation. Each component of the gradient estimate is obtained from a single data sample input; no perturbations are needed.

The LMS algorithm takes advantage of prior information regarding the quadratic shape of the performance surface. This gives the LMS algorithm considerable advantage over the previous adaptation algorithms when it comes to adaptation time. The LMS algorithm converges to a solution much faster than the steepest descent method, particularly when the number of weights is large.

3.4 Applications

In this section some example applications of adaptive system will be briefly presented. The presentation is somewhat simplified, and some practical details have been omitted thus simplifying the understanding of the system's main feature.

3.4.1 Adaptive interference canceling

Adaptive interference canceling devices can be used in many situations where a desired signal is disturbed by additive background noise. This could be, for instance, the signal from the headset microphone used by the pilot in an aircraft. In this case, the speech signal is disturbed by the background noise created by motors, propellers and air flow. It could also be the weak signal coming from electrocardiographic (ECG) electrodes, being disturbed by 50 or 60 Hz interference caused by nearby power lines and appliances.

The idea behind the interference canceling technique is quite simple. Assume that our desired signal s is disturbed by additive background noise n_0. The available signal is

$$x = s + n_0 \tag{3.27}$$

If we could only obtain the noise signal n_0, it could easily be subtracted from the available signal and the noise would be canceled completely. In most cases, the noise signal n_0 is not available, but if a correlated noise signal n_1 can be obtained, n_0 may be created by filtering n_1. Since the filter function needed is not known in most cases, this is a perfect job for an adaptive filter. An adaptive noise-canceling circuit is shown in Figure 3.7. As can be seen from the figure, the output signal is fed back as an error signal ε to the adaptation algorithm of the adaptive filter. Assume that s, n_0, n_1 and y are statistically stationary and have zero means. Further, assume that s is uncorrelated with n_0 and n_1, and that n_1 is correlated with n_0, then

$$\varepsilon = s + n_0 - y \tag{3.28}$$

Squaring equation (3.28) gives

$$\varepsilon^2 = s^2 + (n_0 - y)^2 + 2s(n_0 - y) \tag{3.29}$$

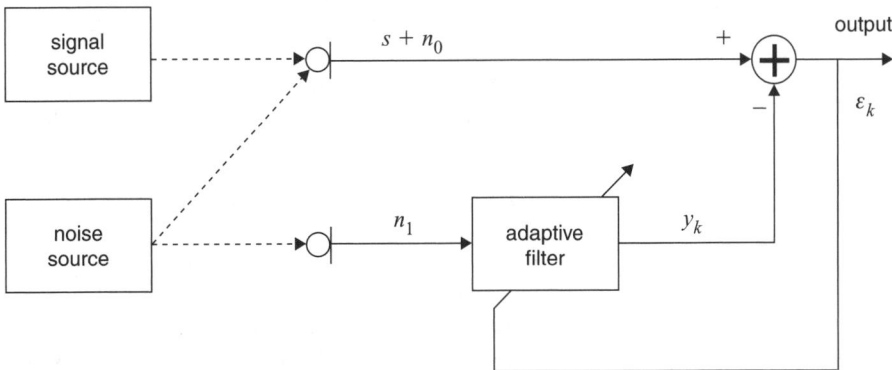

Figure 3.7 *An adaptive noise-canceling circuit, using an adaptive filter*

Taking expectations of both sides and making use of the fact that s is uncorrelated with n_0 and y, we get

$$E[\varepsilon^2] = E[s^2] + E[(n_0 - y)^2] + 2E[s(n_0 - y)]$$
$$= E[s^2] + E[(n_0 - y)^2] \tag{3.30}$$

From this equation it can be seen that adapting the filter by minimizing the mean square error $E[\varepsilon^2]$, i.e. the output signal power, is done by minimizing the term $E[(n_0 - y)^2]$, since the input signal power $E[s^2]$ cannot be affected by the filter. Minimizing the term $E[(n_0 - y)^2]$ implies that the output y of the filter is a best least squares estimate of unknown noise term n_0, which is needed to cancel the noise in the input signal. Hence, from equation (3.28) we realize that minimizing $E[(n_0 - y)^2]$ implies that $E[(\varepsilon - s)^2]$ is also minimized and the noise power of the output signal is minimized.

Let us pursue the previous example of canceling the background noise n_0 picked up by the pilot's headset microphone. In a practical situation, we would set up a second microphone on the flight deck in such a way that it only picks up the background noise n_1. Using a device like in Figure 3.7, the background noise n_0 disturbing the speech signal s could be suppressed. The same technique has been adopted for cellular telephones as well, using an extra microphone to pick up background noise.

Other areas where adaptive interference canceling systems are used is for echo cancellation (Widrow and Stearns, 1985) in telephone lines where echo may occur as a result of unbalanced hybrids and for interference suppression in radio systems.

The performance of the adaptive noise-canceling system depends on three factors. First, there must be a fair correlation between the noise n_1 picked up by the "reference microphone" and the disturbing background noise n_0. If they are completely uncorrelated, the adaptive filter will set all its weights to zero (i.e. "switch-off") and will have no effect (no benefit or damage).

Second, the filter must be able to adapt to such a filter function such that n_0 can be well created from n_1. This implies that the processor model used in the filter is relevant and that the adaptation algorithm is satisfactory. Note that we have not assumed that the adaptive filter necessarily converges to a linear filter.

Third, it is preferable that there may not be any correlation between the signal n_1 picked up by the reference microphone, and the desired signal s. If there is a correlation, the system will also try to cancel the desired signal, i.e. reducing $E[s^2]$. Hence, we have to choose the reference noise signal carefully.

3.4.2 Equalizers

Equalizers or adaptive inverse filters are frequently used in many telecommunication applications, where a transmitted signal is being distorted by some filtering function inherent in the transmission process. One example is the limited bandwidth of a telephone line (filter effect) that will tend to distort high-speed data transmissions. Now, if the transfer function $H(\omega)$ from transmitter (TX) to receiver (RX) of the telephone line is known, we can build a filter having the inverse transfer function, $G(\omega) = H^{-1}(\omega)$ a so-called **inverse filter** or **zero-forcing filter**. If the received signal is fed to the inverse filter, the filtering effect of the telephone line can be neutralized, such that

$$H(\omega)\,G(\omega) = H(\omega)\,H^{-1}(\omega) = 1 \tag{3.31}$$

The transfer function of a communication channel (e.g. telephone line) can be modeled as a **channel filter** $H(\omega)$. Usually, the transfer function is not known in advance and/or varies over time. This implies that a good fixed inverse filter cannot be designed, but an adaptive inverse filter would be usable. This is what an equalizer is, an adaptive inverse filter that tries to neutralize the effect of the channel filter at all times in some optimum way. Today's high-speed telephone modems would not be possible without the use of equalizers.

Figure 3.8 shows a block diagram of a transmission system with a channel filter having transfer function $H(z)$ and an equalizer made up of an adaptive filter used as inverse filter having transfer function $G(z)$. The signal s_k entering the transmission channel is distorted by the transfer function of the channel filter, but on top of this, additive noise n_k is disturbing the transmission, hence the received signal is

$$r_k = s_k * h(k) + n_k \tag{3.32}$$

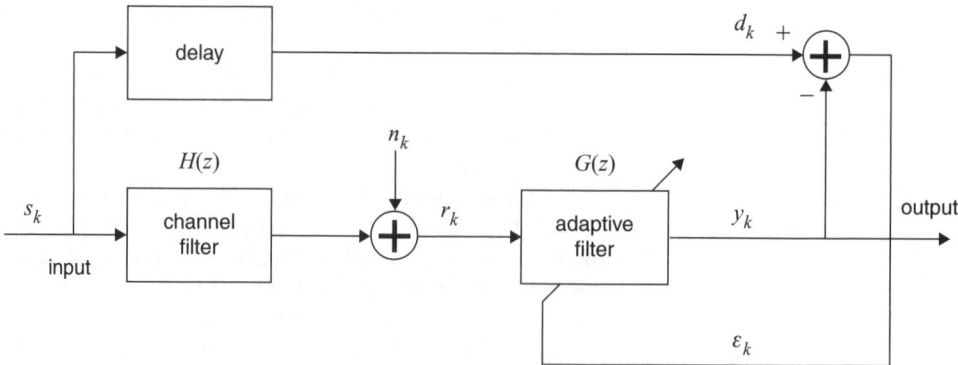

Figure 3.8 *A transmission situation with channel filter, additive noise and an equalizer implemented by means of an adaptive filter*

In the **channel model** above, $*$ denotes convolution and $h(k)$ is the impulse response corresponding to the channel filter transfer function $H(z)$. In order to make the adaptive filter converge resulting in a good inverse transfer function, a desired signal d_k is needed. If the transmitted signal s_k is used as the desired signal, the adaptive filter will converge in a way to minimize the mean square error

$$\mathrm{E}\big[\varepsilon_k^2\big] = \mathrm{E}\big[(s_k - y_k)^2\big] \tag{3.33}$$

implying that the output y_k will resemble the input s_k as closely as possible in the least square sense, and the adaptive filter has converged to a good inverse $G(z)$ of the channel filter transfer function $H(z)$. The inverse filter will eventually also track variations in the channel filter.

Finding a desired signal d_k is not entirely easy at all times. In the present example, we have used the transmitted signal s_k, which is not known in the general case at the receiving site. A common method is to transmit known "training signals" now and then, to be used for adaptation of the inverse filter. There are also other ways of defining the error signal. For instance, in a digital transmission system (where s_k is "digital"), the error signal can be defined as

$$\varepsilon_k = \hat{s}_k - y_k \tag{3.34}$$

where $\hat{s}_k$ is the digital **estimate** of the transmitted (digital) signal s_k and $\hat{s}_k$ is the output of the non-linear detector using the analog signal y_k as input. In this case, the detector simply consists of an analog comparator. Hence, the estimate of the transmitted signal is used as the desired signal instead of the transmitted signal itself (see Figure 3.9). As the equalizer converges, the analog signal y_k will be forced stronger and stronger when $\hat{s}_k = 1$, and weaker and weaker for $\hat{s}_k = 0$.

There are two problems associated with equalizers. First, since the zeros in the channel filter transfer function will be reflected as poles in the inverse filter, we may not be able to find a stable inverse filter for all channel filters. Second, assume that there is a deep notch in the channel filter function at a specific frequency. This notch will be counteracted by a sharp peak, having a high gain at the same frequency in the transfer function of the inverse filter. The additive noise, possibly being stronger than the desired signal at this specific frequency, will then be heavily amplified. This may cause a very poor signal-to-noise ratio (SNR), rendering the equalized output signal unreadable.

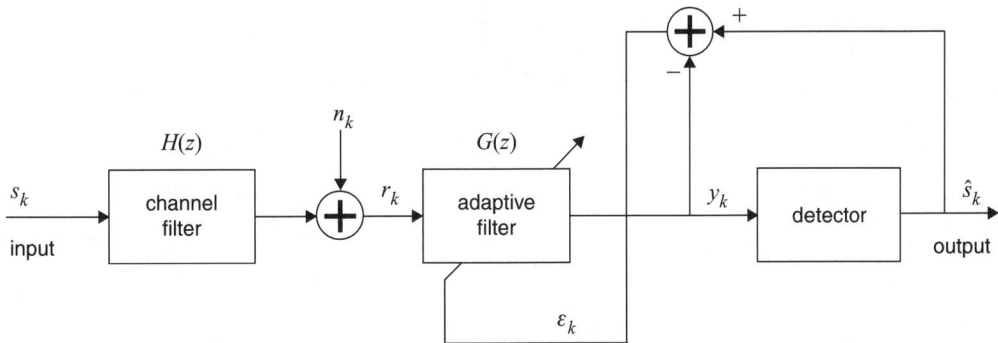

Figure 3.9 *An equalizer using the estimated transmitted signal as desired signal*

Equalizers are sometimes also used in high performance audio systems to counteract the transfer function caused by room acoustics and to avoid feedback problems. In this situation, however, the transmitted signal is "precompensated" before being sent to the loudspeakers. Hence, the order of the inverse filter and the channel filter is reversed when compared with Figure 3.8.

For equalizing telephone lines and room acoustics, equalizers using a simple adaptive linear inverse filter will often do well. When it comes to equalizers for data transmission over radio links, more complex equalizers are often required to achieve acceptable performance. Different types of **decision-feedback equalizers** are common in digital radio transmission systems (Proakis, 1989; Ahlin and Zander, 1998).

In a radio link (Ahlin and Zander, 1998), the signal traveling from the transmitter to the receiver antenna will be propagated over many different paths simultaneously, with all the paths having different lengths. The multitude of received signal components compounded in the receiver antenna will hence be delayed by different amounts of time, which will give the components different phase shifts. This in turn results in a filtering effect of the transmitted signal (the channel filter). The phenomenon is denoted **multipath propagation** or **Rayleigh fading**, causing **frequency selective fading**. Due to the transfer function of the channel filter, the short data bits will be "smeared" out, causing interference in the following data bit time slots. This is called **intersymbol interference (ISI)**, and can be counteracted by including an equalizer in the receiver.

Figure 3.10 shows an example of a decision-feedback equalizer. This is mainly the same equalizer as in Figure 3.9, but extended with a feedback loop consisting of another adaptive filter. The latter filter "remembers" the past detected digital symbols, and counteracts the residuals of the intersymbol interference. Both filters are adapted synchronously using a common adaptation algorithm.

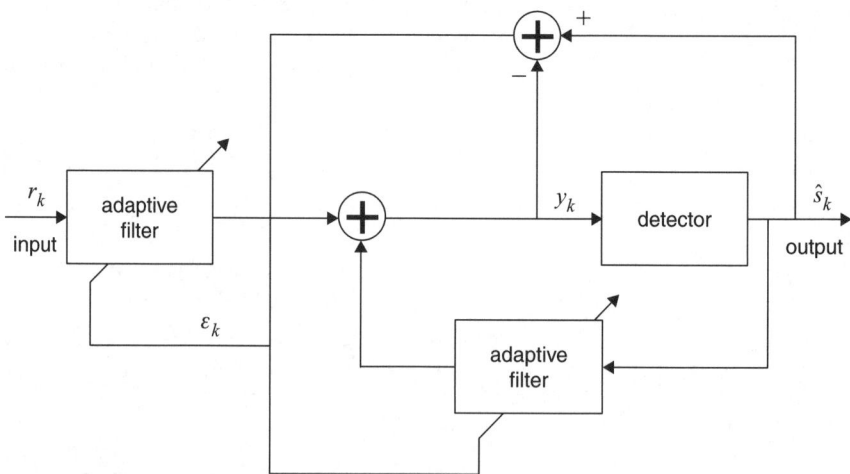

Figure 3.10 *A decision-feedback equalizer (channel filter and additive noise not shown)*

This type of non-linear equalizer works well for combating intersymbol interference present in highly time-dispersive radio channels. If we assume that both filters are adaptive FIR filters, the output of the equalizer will be

$$y_k = \sum_{l=0}^{L} w_{lk} r_{k-l} + \sum_{j=0}^{J} v_{jk} \hat{s}_{j-k} \tag{3.35}$$

where w_{lk} are the weights of the first filter, and v_{jk} the weight of the feedback filter at time instant k.

3.4.3 Adaptive beamforming

So far, we have only discussed the use of adaptive systems in the time and frequency domains, in the single input processor model, according to equations (3.5) and (3.6). In this section, applications using the more general **multiple-input** linear combiner model will be addressed and signal conditioning in the **spatial domain** will be treated.

The underlying signal processing problem is common to the reception of many different types of signals, such as electromagnetic, acoustic or seismic. In this signal processing oriented discussion, the difference between these situations is simply the choice of sensors: radio antennas, microphones, hydrophones, seismometers, geophones, etc. Without sacrificing generality, we will use the reception of electromagnetic signals by radio antennas as an example in the following discussion.

Assume that we are interested in receiving a fairly weak signal in the presence of a strong interfering signal. The desired signal and the interfering signal originate from different locations and at the receiving site, the signals have different angles of arrival. A good way of dealing with this problem is to design a directive antenna, an antenna having a maximum sensitivity in one direction (the "beam" or "lobe") and a minimum sensitivity in other directions. The antenna is directed in such a way that the maximum sensitivity direction coincides with the direction of the location of the desired signal, and minimum sensitivity ("notch") angle in the direction of the interferer (see Figure 3.11).

There are, however, some problems. The physical size of a directive antenna depends on the wavelength of the signal. Hence, for low-frequency signals, the wavelength may be so long that the resulting directive antenna would be huge and not possible to build in practice. Another problem occurs if the desired signal source or the interferer or both are mobile, i.e. (fast) moving. In such a case, the directive antenna may need to be redirected continuously (for instance, radar antennas). If the directive antenna structure needs to be moved (quickly), this may create challenging mechanical problems.

An alternative way of building a directive antenna is to use a number of omnidirectional, fixed antennas mounted in an **antenna array**. The output of the antennas is fed to a signal processing device, and the resulting directivity pattern of all the antennas in unison is created "electronically" by the signal processing device. Hence, the shape and direction of the lobes and notches can be changed quickly, without having to move any part of the physical antenna system. The antenna system may, however, still be large for low frequencies, but for many cases easier to build in practice than in the previous case.

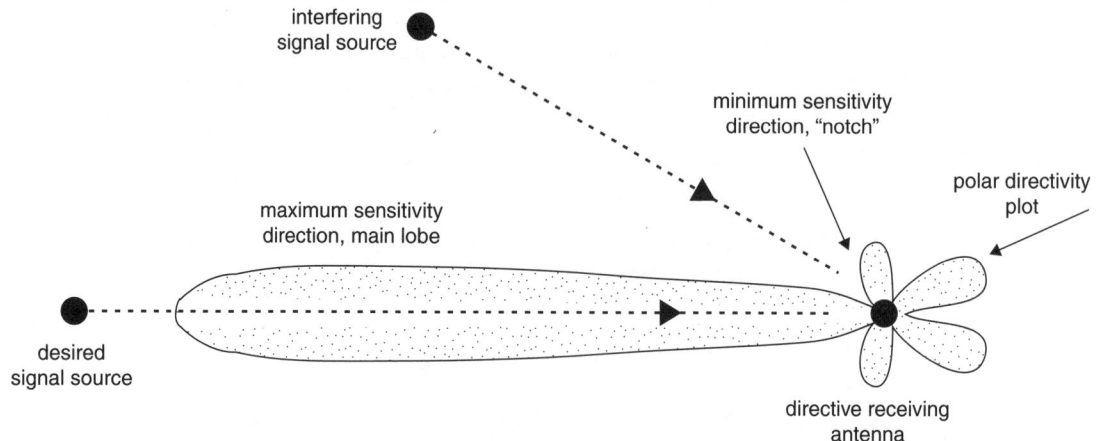

Figure 3.11 *Enhancing the desired signal and attenuating the interference using a directive antenna (the directivity plot is a polar diagram showing the relative sensitivity of the antenna as a function of the angle)*

The drawback of the "electronic" directive antenna is the cost and complexity. The advent of fast and inexpensive digital signal processing components has, however, made the approach known as **beamforming** (Widrow and Stearns, 1985) more attractive during the past few years. The steady increase in processing speed of digital signal processing devices has also made it possible to use this technology for high-frequency signal applications, which was otherwise an impossible task a few years back.

The following simplified example (see Figure 3.12) illustrates the use of a multiple-input, adaptive linear combiner in a beam-forming "electronic" directive antenna system. Assume that we are using two omnidirectional antennas A_1 and A_2. The distance between the two antennas is l. We are trying to receive a weak desired narrow-band signal $s \cos(\omega t)$ but unfortunately we are disturbed by a strong narrow-band, jamming signal of the same frequency $u \cos(\omega t)$. The desired signal is coming from a direction perpendicular to the normal of the antenna array plane, which implies that the phase front of the signal is parallel to the antenna array plane. This means that the received desired signal in the two antennas will have the same phase.

The undesired jamming signal is coming from another direction, with the angle α relative to the normal of the antenna array plane. In this case, the jamming signal will reach the antennas at different times, and there will be a phase shift between the jamming signal received by antenna A_1 compared to the signal received by A_2. It is easily shown that the difference in phase between the two will be

$$\phi = \frac{2\pi l \sin(\alpha)}{\lambda} = \frac{\omega l \sin(\alpha)}{c} \tag{3.36}$$

where c is the propagation speed of the signal, in this case equal to the speed of light, and ω is the angular frequency. Hence, the compound signals received by antennas A_1 and A_2, respectively, can be expressed as

$$x_1 = s \cos(\omega t) + u \cos(\omega t) \tag{3.37}$$

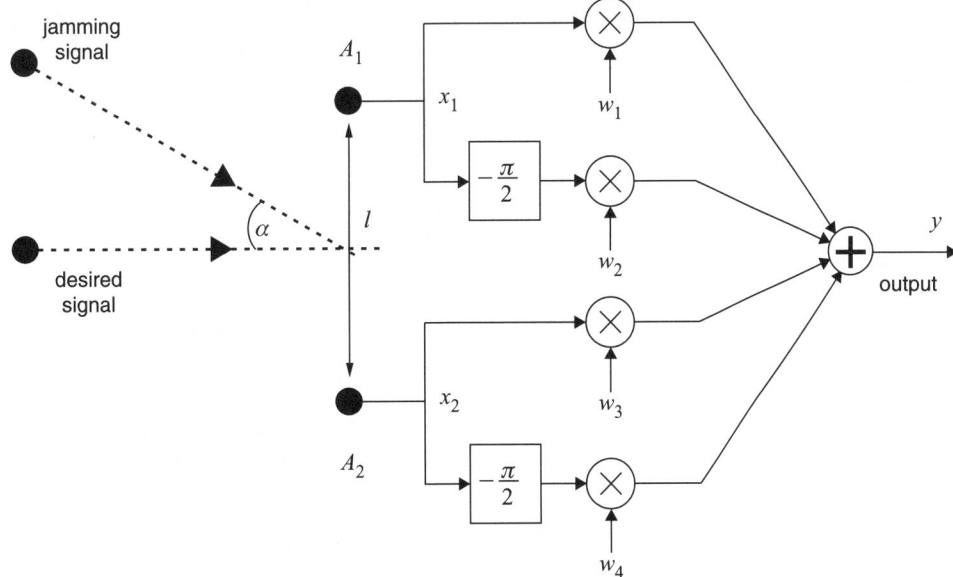

Figure 3.12 *An example simplified narrow-band directive antenna array, using two antennas and a multiple-input adaptive linear combiner and quadrature filters*

$$x_2 = s\cos(\omega t) + u\cos(\omega t - \phi) \tag{3.38}$$

The signal processing system consists of a four input adaptive linear combiner, having the weights w_1 through w_4. The signal coming from each antenna is branched, and passed directly to the combiner in one branch and via a **quadrature filter** (a Hilbert transform) in the other. The quadrature filter is an all-pass filter, introducing a $-\frac{\pi}{2}$ radian phase shift of the signal, but not changing the amplitude. The quadrature filter will be further discussed in Chapter 5. The output of the combiner can now be written as

$$
\begin{aligned}
y = {} & w_1(s\cos(\omega t) + u\cos(\omega t)) \\
& + w_2\left(s\cos\left(\omega t - \frac{\pi}{2}\right) + u\cos\left(\omega t - \frac{\pi}{2}\right)\right) \\
& + w_3(s\cos(\omega t) + u\cos(\omega t - \phi)) \\
& + w_4\left(s\cos\left(\omega t - \frac{\pi}{2}\right) + u\cos\left(\omega t - \phi - \frac{\pi}{2}\right)\right)
\end{aligned} \tag{3.39}
$$

Using the fact that $\cos(\beta - (\pi/2)) = \sin(\beta)$ and rearranging the terms we obtain

$$
\begin{aligned}
y = {} & s((w_1 + w_3)\cos(\omega t) + (w_2 + w_4)\sin(\omega t)) + u(w_1\cos(\omega t) \\
& + w_3\cos(\omega t - \phi) + w_2\sin(\omega t) + w_4\sin(\omega t - \phi))
\end{aligned} \tag{3.40}
$$

Now, the objective is to cancel the jamming signal u, i.e. creating a notch in the directivity pattern of the antenna array in the direction of the jammer. Hence, we want the jamming signal term of equation (3.40) to be zero, i.e.

$$u(w_1 \cos(\omega t) + w_3 \cos(\omega t - \phi) + w_2 \sin(\omega t) + w_4 \sin(\omega t - \phi)) = 0$$

(3.41)

Using the well-known trigonometric relations: $\cos(\beta - \gamma) = \cos(\beta)\cos(\gamma) + \sin(\beta)\sin(\gamma)$ and $\sin(\beta - \gamma) = \sin(\beta)\cos(\gamma) - \cos(\beta)\sin(\gamma)$, equation (3.41) can be rewritten as

$$u(w_1 + w_3 \cos(\phi) - w_4 \sin(\phi)) \cos(\omega t)$$
$$+ u(w_2 + w_3 \sin(\phi) + w_4 \cos(\phi)) \sin(\omega t) = 0 \qquad (3.42)$$

In this case, solutions, i.e. the optimum weights $\mathbf{W}^*$, can be found by inspection. There are many possible solutions

$$w_1 = -w_3 \cos(\phi) + w_4 \sin(\phi) \tag{3.43a}$$

$$w_2 = -w_3 \sin(\phi) - w_4 \cos(\phi) \tag{3.43b}$$

We choose a solution that is simple from a practical point of view, by setting

$$w_1^* = 1 \quad \text{and} \quad w_2^* = 0$$

Hence, two weights and one quadrature filter can be omitted from the structure shown in Figure 3.12. Antenna A_1 is simply connected directly to the summing point and the hardware (software) that constitutes the weights w_1 and w_2 and the upper quadrature filter can be removed.

Using equations (3.43a) and (3.43b) the components of one optimum weight vector $\mathbf{W}^*$ can be calculated

$$w_1^* = 1$$
$$w_2^* = 0$$
$$w_3^* = -\cos(\phi)$$
$$w_4^* = \sin(\phi)$$

Inserting $\mathbf{W}^*$ in equation (3.40) the output will be

$$y = s((1 - \cos(\phi)) \cos(\omega t) + \sin(\phi) \sin(\omega t))$$
$$= s(\cos(\omega t) - \cos(\omega t + \phi)) \tag{3.44}$$

The result is not surprising. The jamming signal is canceled and the output is a sum of two components, the original signal received by antenna A_1 and a phase-shifted version of the signal received at antenna A_2. As expected, the phase shift is exactly the shift needed to cancel the jamming signal.

Now, in this case it was quite easy to find a solution analytically. In a more complex system, having, for instance, 30 or even more antennas and a number of interfering, jamming signals coming from different directions, there is no guarantee that an analytical solution can be found. In such a case, an adaptive system can iteratively find an optimum solution in the mean square sense, using, for instance, some kind of LMS-based adaptation algorithm. Further, if the signal sources are moving, an adaptive system would be required.

Working with wide-band signals, the input circuits of the linear combiner may consist of not only quadrature filters, but more elaborate filters, e.g. adaptive FIR filters.

In this chapter only a few applications of adaptive digital processing systems have been discussed. There are numerous areas (Widrow and Stearns, 1985), for instance in process identification, modeling and control theory, where adaptive processing is successfully utilized.

Summary

In this chapter the following main topics have been addressed:

- The structure of a generic closed-loop adaptive system
- Basic processor structures, the linear combiner and the FIR filter
- The theory behind performance functions
- Different iterative methods of finding the optimum weights: steepest descent, Newton and LMS
- Some applications: adaptive interference canceling, the equalizer and adaptive beamforming.

Review questions

R3-1 Draw a block diagram of a generic closed-loop adaptive system showing signals and function blocks. Explain the function briefly.

R3-2 Draw the block diagram of a linear combiner and explain the associated equations.

R3-3 What is the performance function and what is it used for?

R3-4 If you find the cross-correlation vector $\mathbf{P}$ in a proposed adaptive filter to be zero, what is the problem?

R3-5 Three adaptation algorithms are discussed in this chapter: steepest descent, Newton and LMS. What path does the weight vector travel towards optimum when using the different algorithms?

R3-6 What are the pros and cons of LMS?

R3-7 What is an equalizer good for?

R3-8 Why is the error signal uncorrelated to the input signals when the adaptation is completed and the Wiener vector is found?

Solved problems

P3-1 Assume we are designing an adaptive interference canceling system as in Figure 3.7. The interference component is $n_{0k} = \sin(\pi(k/7))$ and the reference noise $n_{1k} = \cos(\pi(k/7))$. The transfer function of the adaptive filter is $H(z) = w_0 + w_1 z^{-1}$. Find an expression for the MSE performance surface ξ as a function of the filter weights.

P3-2 Formulate the LMS algorithm for the system in P3-1 as two separate functions showing how to iterate the weights.

P3-3 Using the LMS algorithm and the adaptive interference canceling system in P3-1 above, write a MATLAB™ program to simulate the adaptive filter and to plot y_k, the output of the interference canceling system ε_k, the weights w_{0k} and w_{1k} and the MSE learning curve ξ as a function of the time index k. Use $\mu = 0.05$ and 0 initial conditions for the weights.

P3-4 For the system in P3-1, calculate the Wiener-vector solution for the filter weights. Compare to the results obtained in P3-3 above.

4 Non-linear applications

Background There are an infinite number of non-linear signal processing applications. In this chapter, a few examples, such as **the median filter**, **artificial neural networks (ANN)** and **fuzzy logic** will be discussed. Some of these examples are devices or algorithms that are quite easy to implement using digital signal processing techniques, but could be very difficult or almost impossible to build in practice using classical "analog" methods.

Objectives In this chapter we will discuss:

- The median filter and threshold decomposition
- Feedforward neural networks
- Training algorithms for neural networks
- Feedback neural networks and simulated annealing
- Fuzzy control systems.

4.1 The median filter

4.1.1 Basics

A median filter is a non-linear filter used for signal smoothing. It is particularly good for removing **impulsive type noise** from a signal. There are a number of variations of this filter, and a two-dimensional variant is often used in digital image processing systems to remove noise and speckles from images. The non-linear function of the median filter can be expressed as

$$y(n) = \mathrm{med}[x(n-k),\ x(n-k+1),\ldots,x(n),\ldots,$$
$$x(n+k-1),\ x(n+k)] \tag{4.1}$$

where $y(n)$ is the output and $x(n)$ is the input signals. The filter "collects" a window containing $N = 2k + 1$ samples of the input signal and then performs the **median** operator on this set of samples. Taking the median means, sort the samples by magnitude and then select the mid-value sample (the median). For this reason, N is commonly an odd number. If for some reason an even number of samples must be used in the window, the median is defined as shown below. Assuming the samples are sorted in such a way that x_1 is the smallest and x_{2k+1} the largest value

$$\mathrm{med}[x_1, x_2, \ldots, x_N] = \begin{cases} x_{k+1} & \text{for } N = 2k+1 \\ \frac{1}{2}(x_k + x_{k+1}) & \text{for } N = 2k \end{cases} \tag{4.2}$$

Different methods have been proposed to analyze and characterize median filters. The technique of **root signal** analysis (Mitra and Kaiser, 1993) deals with signals that are invariant to median filtering and defines the "passband" for median filters. A root signal of a median filter of length $N = 2k + 1$ is a signal that passes through the filter unaltered; this signal satisfies the following equation

$$x(n) = \text{med}[x(n - k), \; x(n - k + 1), \ldots,$$
$$x(n), \ldots, \; x(n + k - 1), \; x(n + k)] \tag{4.3}$$

The median filter itself is simple, and in the standard form there is only one design parameter, namely the filter length $N = 2k + 1$. There are some terms used which pertain to root signals (Gallagher Jr. and Wise, 1981).

A **constant neighborhood** is a region of at least $k + 1$ consecutive, identically valued points.

An **edge** is a monotonically rising or falling set of points surrounded on both sides by constant neighborhoods.

An **impulse** is a set of at least one but less than $k + 1$ points whose values are different from the surrounding regions and whose surrounding regions are identically valued constant neighborhoods.

So, in essence, a **root signal** is a signal consisting of only constant neighborhoods and edges. This definition implies that a signal which is a root signal to a median filter of length N is also a root signal of any median filter whose length is less than N.

It is also interesting to note that a median filter preserves edges, both positive and negative, provided they are separated by a constant neighborhood. The longer the filter length, the farther apart the edges have to be, but the actual magnitude of the slopes is irrelevant. This means that a median filter filters out impulses and oscillations, but preserves edges. This will be demonstrated below.

4.1.2 Threshold decomposition

Analyzing combinations of linear filters by using the principle of superposition is in many cases easier than analyzing combinations of non-linear devices like the median filter. However, by using a method called **threshold decomposition**, we divide the analysis problem of the median filter into smaller parts.

Threshold decomposition of an integer M-valued signal $x(n)$, where $0 \leq x(n) < M$ means decomposing it into $M - 1$ **binary** signals $x^1(n)$, $x^2(n), \ldots, x^{M-1}(n)$

$$x^m(n) = \begin{cases} 1 & \text{if } x(n) \geq m \\ 0 & \text{else} \end{cases} \tag{4.4}$$

Note! The upper index is only an index, and does not imply "raised to". Our original M-valued signal can easily be reconstructed from the binary signal by adding them together

$$x(n) = \sum_{m=1}^{M-1} x^m(n) \tag{4.5}$$

Now, a very interesting property of a median filter (Fitch *et al.*, 1984) is that instead of filtering the original M-valued signal, we can decompose it into $M - 1$ "channels" (equation (4.4)), each containing a **binary** median filter. Then we can add the outputs of all the filters (equation (4.5)) to obtain an M-valued output signal (see Figure 4.1).

The threshold decomposition method is not only good for analyzing purposes, but it is also of great interest for implementing median filters. A binary median filter is easy to implement, since the median operation can be replaced by a simple vote of majority. If there are more "ones" than "zeros", the filter output should be "one". This can be implemented in many different ways. For example, a binary median filter of length $N = 3$ can be implemented using simple Boolean functions

$$y(n) = x(n-1) \cap x(n) \cup x(n-1) \cap x(n+1) \cup x(n) \cap x(n+1) \qquad (4.6)$$

where $x(n)$ and $y(n)$ are binary variables and the Boolean operations $\cap$ is AND and $\cup$ is OR. For large filter lengths, this method may result in complex Boolean calculations, and a simple counter can be used as an alternative. The pseudo-code below shows one implementation example:

```
binmed:        counter:=0
               for i:=1 to N do
                           if x[i] then counter++
                           else counter--
               if counter>0 then y[n]:=1
               else y[n]:=0
```

Now, one may argue that if there are a large number of thresholds M, there are going to be a large number of binary filters as well. This is of course true, but if the filter is implemented as software for a digital signal processor (DSP), the

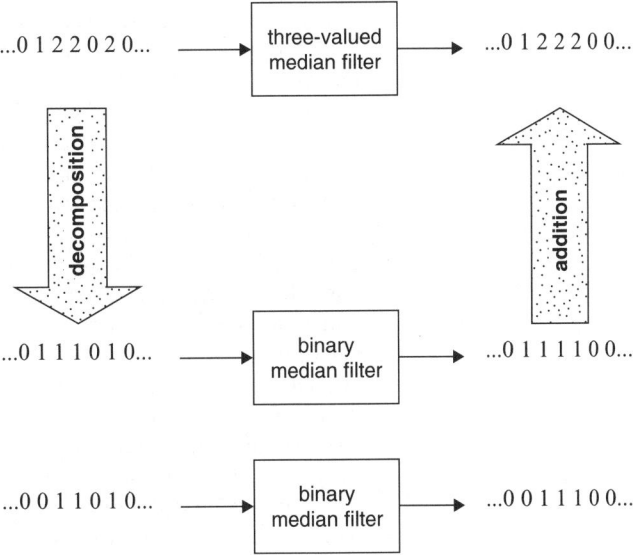

Figure 4.1 *Median filtering of a three-valued signal by threshold decomposition,* N = 3

same piece of computer code can be reused in all binary filters (as a function or subroutine). Hence, the problem deals with the processing speed. Depending both on the kind of signals being filtered and the features of the output signal of primary interest, we may not need to process all thresholds or/and non-equidistant thresholds may be used.

The technique of threshold decomposition and using binary, Boolean filters has led to a new class of filters, the so-called **stacked filters**.

4.1.3 Performance

If a median filter is compared to a conventional linear mean filter, implemented as a finite impulse response (FIR) filter having constant tap weights (moving average, MA), the virtues of the median filter (running median) become apparent. Consider the following example, where we compare the output of a moving average filter of length $N = 9$, and a median filter of the same length. The mean filter with constant weights (a linear smoothing filter) can be expressed as

$$y_1(n) = \sum_{i=1}^{9} \frac{1}{9} x(n + i - 5) \tag{4.7a}$$

and the median filter (a non-linear smoothing filter)

$$y_2(n) = \text{med}[x(n - 4), \ldots, x(n), \ldots, x(n + 4)] \tag{4.7b}$$

Assume that the input signal is a square wave signal that is to be used for synchronization purposes, i.e. the edges of the signal are of primary interest, rather than the absolute level. Unfortunately, the signal is distorted by additive Gaussian noise, and impulsive type noise transients, typically generated by arcs and other abrupt discharge phenomena (see Figure 4.2).

As can be seen from Figure 4.2, when comparing the outputs of the two filters, the median filter performs quite well preserving the edges of the signal but suppressing the noise transients. The standard linear mean filter, however, suffers from two basic problems. First, when a quite large transient occurs in the filtering window, it will affect the output as long as it is in the averaging window. This means that a narrow but strong peak will be "smeared" out and basically create a new pulse that can be mistaken for a valid square wave signal. Due to the long time constant of the linear FIR filter, the edges of the desired signal will also be degraded. This may be a significant problem if timing is crucial, e.g. when the signal is used for synchronization purposes. Using a hard limiter-type device to "sharpen up" the edges again is of course possible. However, the slope of the edges does not only depend on the response time of the mean filter, but also on the actual (varying) amplitude of the input signal. This will again result in uncertainty of the true timing.

The uncertainty of the edges in the median filter case is mainly due to the additive noise, which will be suppressed by the filter. The signal form is quite similar to the input signal with the addition of noise. For a square wave without noise, the median filter will present a "perfect" edge.

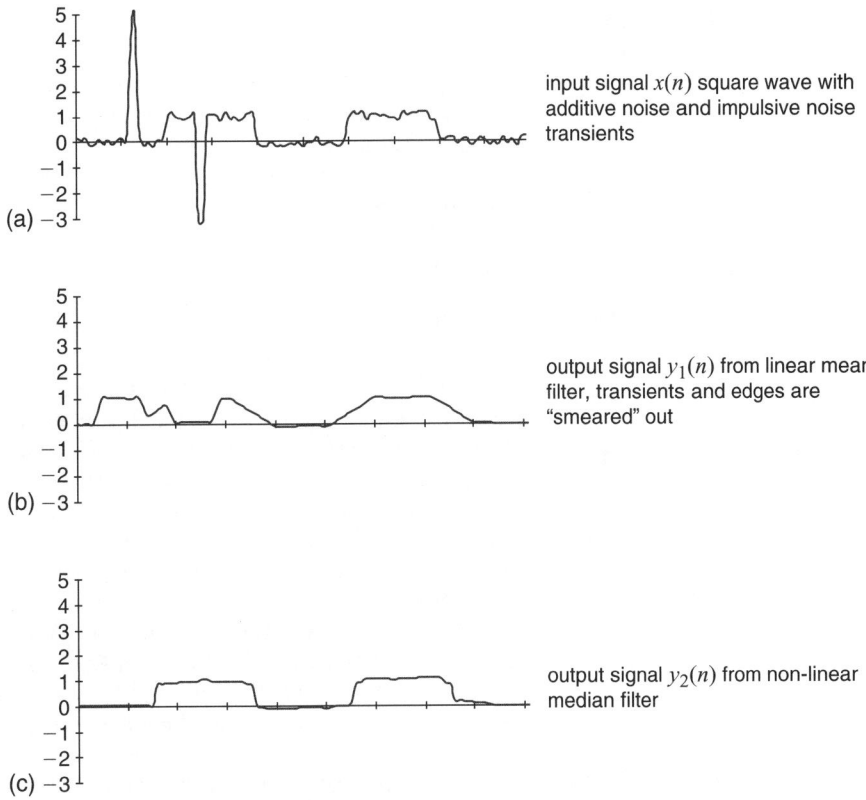

Figure 4.2 *(a) Input signal filtered by (b) linear running average filter and (c) non-linear running median filter, filter length* N = 9 *in both cases*

4.1.4 Applications

If the median filter is implemented as software on a digital computer, any standard sorting algorithm like "bubblesort" or "quick sort", etc. (Wirth, 1976) can be used to sort the values (if a stacked filter approach is not taken). When calculating the expected sorting time, it should be noted that we do not need to sort all the values in the filter window of the median filter. We only need to continue sorting until we have found the mid-valued sample.

The use of median filters was first suggested for smoothing statistical data. This filter type has, however, found most of its applications in the area of digital image processing. An edge preserving filter like the median filter can remove noise and speckles without blurring the picture. Removing artifacts from imperfect data acquisition, for instance horizontal stripes sometimes produced by optical scanners, is done successfully using median filters. Median filters are also used in radiographic systems in many commercial tomographic scan systems and for processing electroencephalogram (EEG) signals and blood pressure recordings. This type of filter is also likely to be found in commercial digital television sets in the future because of the very good cost-to-performance ratio.

4.2 Artificial neural networks

4.2.1 Background

"**Neural networks**" is a somewhat ambiguous term for a large class of massively parallel computing models. The terminology in this area is quite confused in that scientific well-defined terms are sometimes mixed with trademarks and sales bull. A few examples are: "connectionist's net", "artificial neural systems (ANS)", "parallel distributed systems (PDS)", "dynamical functional systems", "neuromorphic systems", "adaptive associative networks", "neuron computers", etc.

4.2.2 The models

In general, the models consist of a large number of typically non-linear computing **nodes**, interconnected with each other via an even larger number of adaptive **links** (weights). Using more or less crude models, the underlying idea is to mimic the function of neurons and nerves in a biological brain.

Studying neural networks may have two major purposes: either we are interested in modeling the behavior of biological nerve systems, or we want to find smart algorithms to build computing devices for technical use. There are many interesting texts dealing with neural networks from the point of perception, cognition and psychology (Hinton and Anderson, 1981; McClelland and Rumelhart, 1986; Grossberg, 1987a, b; Rumelhart and McClelland, 1987). In this book, however, only the "technical" use of neural network models will be treated, hence the term "**artificial neural networks (ANN)**" will be used from here on.

"Computers" that are built using artificial neural network models have many nice features:

- The "programming" can set the weights to appropriate values. This can be done adaptively by "training" (in a similar way to teaching humans). No procedural programming language is needed. The system will learn by examples.
- The system will be able to generalize. If an earlier unknown condition occurs, the system will respond in the most sensible way based on earlier knowledge. A "conventional" computer would simply "hang" or exhibit some irrelevant action in the same situation.
- The system can handle incomplete input data and "soft" data. A conventional computer is mainly a number cruncher, requiring well-defined input figures.
- The system is massively parallel and can easily be implemented on parallel hardware, thus obtaining high processing capacity.
- The system will contain a certain amount of redundancy. If parts of the system are damaged, the system may still work, but with degraded performance ("graceful descent"). A 1-bit error in a conventional computer will in many cases "crash".

Systems of the type outlined above have been built for many different purposes and in many different sizes during the past 50 years. Some examples are systems for classification, pattern-recognition, content addressable memories (CAM), adaptive control, forecasting, optimization and signal processing. Since appropriate hardware is still not available, most of the systems have been implemented in software on conventional sequential computers. This unfortunately implies that the true potential of the inherently, parallel artificial neural network algorithms has not been very well exploited. The systems built so far have been rather slow and small, typically of the size of 10^4–10^6 nodes having 10^1–10^2 links each. Taking into account that the human brain has some 10^{10}–10^{11} nodes with 10^3–10^6 links each, it is easy to realize that the artificial systems of today are indeed small. In terms of complexity, they are comparable to the brain of a fly.

4.2.3 Some historical notes

The history of artificial neural networks can be traced early in twentieth century history. The first formal model was published by W.S. McCulloch and W. Pitts in 1943. A simple node type was used and the network structures were very small. However, it was possible to show that the network could actually perform meaningful computations. The problem was finding a good way of "training" (i.e. "adapting" or "programming") the network. In 1949, D. Hebb published *The Organization of Behavior*, where he presented an early version of a correlation-based algorithm to train networks. This algorithm later came to be known as **Hebb's rule**. In the original paper, the algorithm was not well analyzed or proved, so it came to be regarded as an unproved hypothesis until 1951. In the same year, M. Minsky and D. Edmonds built a "learning machine", consisting of 300 electron tubes and miscellaneous surplus equipment from old bombers, which illustrated a practical example.

It is worth noting that the artificial neural network ideas are as old as the digital computer (e.g. electronic numerical integrator and computer (ENIAC) in 1944). The digital computer was however developed faster, mainly because it was easier to implement and analyze.

In the late 1950s and early 1960s, B. Widrow and M.E. Hoff introduced a new training algorithm called the **delta rule** or the **Widrow–Hoff algorithm**. This algorithm is related to the least mean square (LMS) algorithm used in adaptive digital filters. Widrow also contributed a new network node type called the **adaptive linear neuron (ADALINE)** (Widrow and Lehr, 1990). At about the same time, F. Rosenblatt worked extensively with a family of artificial neural networks called **perceptrons**. Rosenblatt was one of the first researchers to simulate artificial neural networks on conventional digital computers instead of building them using analog hardware. He formulated "the perceptron convergence theorem" and published his *Principles of Neurodynamics* in 1962.

M. Minsky and S.A. Papert also investigated the perceptron, but they were not as enthusiastic as Rosenblatt. In 1969 they published the book *Perceptrons* (Minsky and Papert, 1969), which was a pessimistic and may be somewhat unfair presentation, focusing on the shortcomings of the perceptrons. Unfortunately, this book had a depressing impact on the entire artificial neural network research society. Funding drained, and for the next decade only a few researchers continued working in this area. One of them was S. Grossberg, working with "competitive learning". In general, however, not very much happened in the area of artificial neural networks during the 1970s. Conventional digital von Neumann-based computers and symbol-oriented artificial intelligence (AI) research dominated.

In the beginning of the 1980s, a renaissance took place. The artificial neural network was "reinvented" and a number of new and old research groups started working on the problems again; this time armed with powerful digital computers. In 1980, "**feature maps**" was presented by T. Kohonen (Finland) and in 1982, J. Hopfield published a couple of papers on his **Hopfield Net**. Perhaps too well marketed, this network was able to find solutions to some famous non-polynomial (NP)-complete optimization problems. In 1983, S. Kirkpatrick *et al*. introduced "**simulated annealing**", a way of increasing the chances to find a **global optimum** when using artificial neural networks in optimization applications.

In 1984, P. Smolensky presented "harmony theory", a network using probabilistic node functions. In the same year, K. Fukushima demonstrated his "neocognitron", a network able to identify complex handwritten characters, e.g. Chinese symbols.

The network training method "**back-propagation**" had been around in different versions since the late 1960s, but was further pursued by D. Rumelhart *et al*. in 1985. This year, the "**Boltzmann machine**", a probabilistic type network, was also presented by G.E. Hinton and T.J. Sejnowski.

In the following years, many new variants of artificial neural network systems and new areas of application occurred. Today, the area of artificial neural networks, an extremely interdisciplinary technology, is used for instance in (Lippmann, 1987; Chichocki and Unbehauen, 1993) signal processing, optimization, identification, estimation, prediction, control, robotics, databases, medical diagnostics, biological classification (e.g. blood and genes), chemistry and economy. However, there are not very many "new" ideas presented today, rather extensions and enhancements of old theories. Unfortunately, we still lack the ideal hardware. Extensive work in the area of application specific integrated circuit (ASIC) is in progress, but there are some basic problems hard to overcome when trying to build larger networks. Many of these problems have to do with the large number of weights (links). For example, how should all the weights be stored and loaded into a chip? How to find the settings, in other words, how to perform the training? How should these examples be chosen since a large number of weights will require an even larger number of training examples?

4.2.4 Feedforward networks

The class of feedforward networks is characterized by having separate inputs and outputs and no internal feedback signal paths. Hence, there are no stability problems. The nodes of the network are often arranged in one or more discrete **layers** and are commonly easy to implement. If the nodes do not possess any dynamics like integration or differentiation, the entire network mainly performs a non-linear mapping from one vector space to another. After the training of the net is completed, meaning that the weights are determined, no iteration process is needed. A given input vector results in an output vector in a one-step process.

4.2.4.1 Nodes

The node j in a feedforward artificial neural network has a basic node function of the type

$$x_j = g\left(f\left(\sum_{i=1}^{N} w_{ij}x_i + \varphi_j\right)\right) \tag{4.8}$$

where x_j is the output of node j and φ_j is the bias of node j. The N inputs to node j are denoted x_i where $i = 1, 2, \ldots, N$ and the weights (links) are w_{ij} from input i to node j. For pure feedforward networks $w_{ii} \equiv 0$. The function $f(\ )$ is the **activation function** sometimes also called the "squashing function". This function can be chosen in a variety of ways, often depending on the ease of implementation. The function should, however, be **monotonically non-decreasing** to avoid unambiguous behavior. Some training methods will also require the activation function to be **differentiable**. The activation function is commonly a non-linear function, since a linear function will result in a trivial network. Some examples of common activation functions are as follows:

- **Hard limiter**

$$f(x) = \begin{cases} a & x \geq 0 \\ b & x < 0 \end{cases} \qquad (4.9)$$

where a and b are constants. If $a = 1$ and $b = -1$, equation (4.9) turns into the sign function.

- **Soft limiter**

$$f(x) = \begin{cases} a & x \geq a \\ x & b \leq x < a \\ b & x < b \end{cases} \qquad (4.10)$$

The function is linear in the range $a \cdots b$, where a and b are constants. The function saturates upwards at a and downwards at b.

- **Sigmoid ("logistic function")**

$$f(x) = \frac{1}{1 + e^{-x/T}} \qquad (4.11)$$

In this expression, the parameter T is referred to as the "computational temperature". For small temperatures, the function "freezes" and the shape of equation (4.11) approaches the shape of a hard limiter as equation (4.9). Sometimes the function $f(x) = \tanh(x/T)$ or similar is also used.

Finally, the **output function** $g(\ )$ may be an integration or summation or alike if the node is supposed to have some kind of memory. In most feedforward networks, however, the nodes are without memory, hence a common output function is

$$g(x) = x \qquad (4.12)$$

4.2.4.2 Network topology

When dealing with networks consisting of a number of nodes N, matrix algebra seems to be handy in general. All weights w_{ij} in a network can, for instance, be collected into a weight matrix $\mathbf{W}$ $(N \times N)$, and all signals into a vector $\mathbf{X}$ $(N \times 1)$. The jth column vector $\mathbf{W}_j$ of matrix $\mathbf{W}$ hence represents the weights of node j. Now, it is straightforward to see that a node basically performs a non-linear scalar product between the corresponding weight vector and the signal vector (assuming output function as in equation (4.12))

$$x_j = f(\mathbf{X}^\mathrm{T}\mathbf{W}_j + \varphi_j) \tag{4.13}$$

Using a weight matrix in this way allows arbitrary couplings between **all** nodes in the network. Unfortunately, the size of the matrix grows quickly as the number of elements increases as N^2. When dealing with feedforward-type networks, this problem is counteracted by building the network in **layers**, thereby reducing the number of allowed couplings. A layered network is commonly divided into three (or more) layers denoted as the input layer, the hidden layer(s) and the output layer. Figure 4.3 shows a simple three-layer feedforward network. In this structure, the input signals (input vector) enter the network at the input layer. The signals are then only allowed to proceed in one direction to the hidden layer and then finally to the output layer, where the signals (output vector) exit the network. In this way, the weight matrix can be divided into three considerably smaller ones, since the nodes in each layer only need to have access to a limited number of the signal components in the total signal vector $\mathbf{X}$. Questions arise with regards to the number of layers and the number of nodes in each of these layers that should be used. While the number of nodes per layer is hard to determine, the number of layers can be calculated with the following method.

Using a layered structure, impose some restrictions on the possible mappings from the input vector space to the output vector space. Consider the following example. Assume that we have a two-dimensional input vector and require a scalar output. We use two signal levels "1" and "0". The output should be "1" if **one** of the two input vector components is "1". The output should be "0"

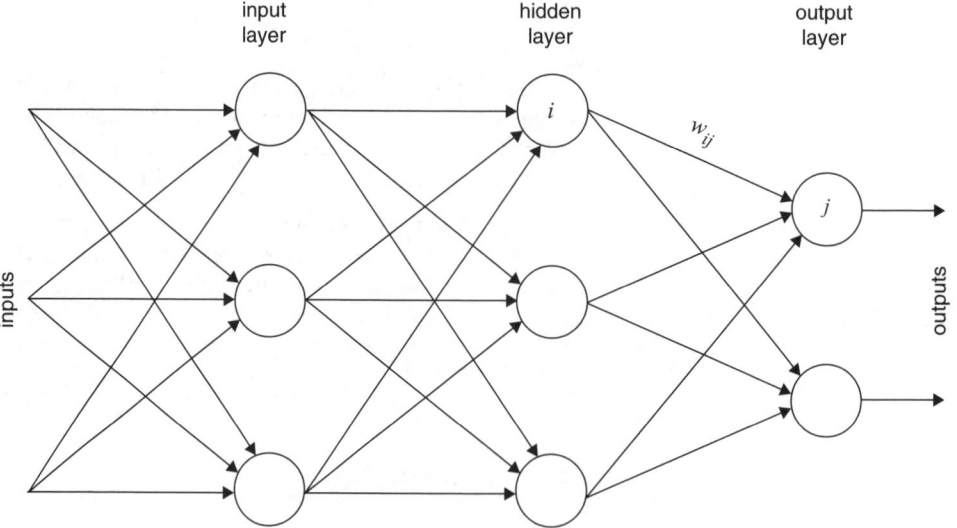

Figure 4.3 *A simple three-layer feedforward network*

otherwise. This is a basic exclusive OR function (XOR or modulo 2 addition or "parity function"). The task of the artificial neural network is to divide the two-dimensional input vector space into two **decision regions**. Depending on in which decision region the input vector is located, a "1" or a "0" should be presented at the output.

Starting out with the simplest possible one-layer network, consisting of one node with a hard limiter-type activation function, we can express the input–output function as

$$x_3 = f\left(\sum_{i=1}^{3} w_{i3}x_1 + \varphi_3\right) = \begin{cases} 1 & w_{13}x_1 + w_{23}x_2 + \varphi_3 \geq 0 \\ 0 & w_{13}x_1 + w_{23}x_2 + \varphi_3 < 0 \end{cases} \tag{4.14}$$

where we assume that $w_{33} = 0$ and that the hard limiter has the function

$$f(x) = \begin{cases} 1 & x \geq 0 \\ 0 & x < 0 \end{cases} \tag{4.15}$$

Table 4.1 *Truth table for XOR function*

x_1	x_2	x_3
0	0	0
0	1	1
1	0	1
1	1	0

The desired response is shown in the truth table (see Table 4.1).

Now, the task is to find the weights and the bias needed to implement the desired function. It is possible to implement an AND or OR function, but an XOR function cannot be achieved. The reason is that a node of this type is only able to divide the input vector space with a (hyper)plane. In our example (from equation (4.14)) the borderline is

$$x_2 = -\frac{\varphi_3}{w_{23}} - \frac{w_{13}}{w_{23}}x_1 \tag{4.16}$$

In other words, we cannot achieve the required shape of the decision regions, since the XOR function requires two disjunct areas like A and A (see Figure 4.4). Since many more complex decision tasks can be traced back to the XOR function, this function has a fundamental importance in computation. The basic perceptron presented by Rosenblatt was mainly a one-layer network of the type presented in the example above. The main criticism presented in the book *Perceptrons* by Minsky and Papert (Minsky and Papert, 1969) was concerned with the inability of the perceptron to solve the basic XOR problem. This was a bit unfair, since the perceptron can easily be modified to handle the XOR problem too. This can be done by adding a second layer of nodes to our system. With respect to the example above it means adding another two nodes, and obtaining a **two-layer** artificial neural network. The network contains two nodes in the input layer, no hidden layers and one node in the output layer. The nodes are of the same simple type as in the example above. Now, each of the two nodes in the input layer can divide the input vector space with hyperplanes. In our two-dimensional example, this is two straight lines as in equation (4.16). If one node has its decision line to the left of the two desired areas A (see Figure 4.4), the output of the node will be "1" when an input vector corresponding to A is present. The other node has its line to the right of the desired decision region and gives a "1" output signal for A. The desired response can now be obtained by "ANDing" the output of the two nodes. AND and OR functions can easily be achieved using a single node, so the node in the output layer of our example network will do the "ANDing" of the outputs from the nodes in the input layer. Hence, the XOR problem is solved.

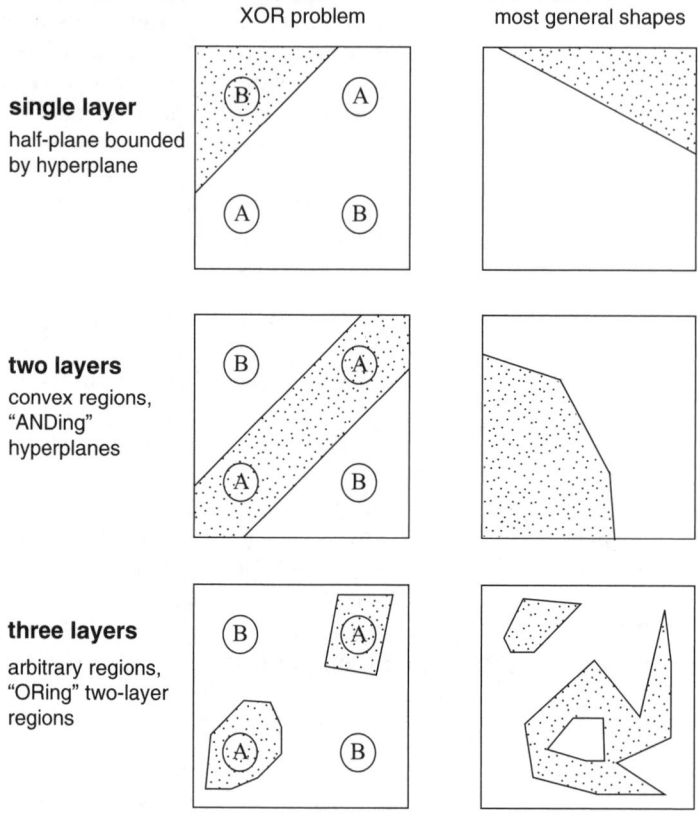

Figure 4.4 *Possible decision region shapes for different numbers of layers*

Figure 4.5 shows a proposed solution to the XOR problem. Since the input signals have the indices 1 and 2, respectively, the nodes are numbered as 3 and 4 in the input layer and 5 in the output layer. The weights and biases of the nodes may be chosen as

$$
\begin{array}{llll}
\text{node 3:} & w_{13} = 1 & w_{23} = 1 & \varphi_3 = -0.5 \\
\text{node 4:} & w_{14} = -1 & w_{24} = -1 & \varphi_4 = 1.5 \\
\text{node 5:} & w_{35} = 1 & w_{45} = 1 & \varphi_5 = -1.5
\end{array}
$$

Note! There are many possible solutions, try it yourself.

Going back to Figure 4.4 again, we can now draw some general conclusions about the possible shapes of decision regions as a function of the number of layers. A one-layer network can only split the input vector space with hyperplanes. Using a second layer, having the outputs of the first layer as inputs, a number of hyperplanes can be "ANDed" together and convex decision regions can now be formed. The number of nodes in each layer determines the number and complexity of the regions. Finally, incorporating a third layer, a number of convex regions, created by the preceding second layer, can now be "ORed" together and arbitrary complex decision regions can be obtained. So, from theoretical point of view, there is no need for more than three layers in a layered feedforward artificial neural network. When implementing these nets, however, more

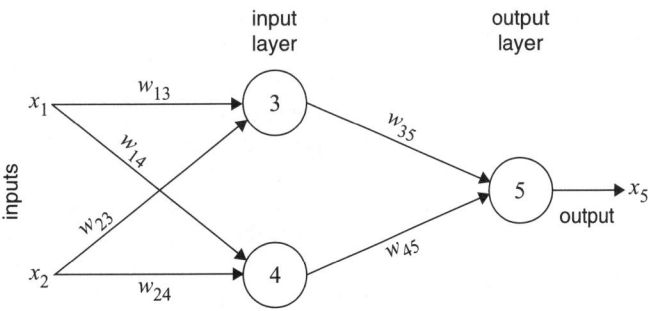

Figure 4.5 *Example of a two-layer feedforward ANN of perceptron type, capable of solving the XOR problem*

layers may sometimes be used. This may facilitate reuse of some intermediate computing results and reduce the total number of nodes in the network and/or simplify the training, i.e. the process of determining the weights and biases in the nodes.

Returning to the XOR example, there is another way of solving the problem using only **one** perceptron-type node. The underlying principle is the way in which the input signals or "features" are chosen. This is a very important, often forgotten topic in many cases. Choosing "bad" input signals may make it impossible even for an advanced artificial neural network to perform a simple task, while choosing "smart" input signals may solve complex problems using fairly simple networks. If we put a non-linear "preprocessor" in front of the inputs of our original single-layer network, creating for instance the two input signals

$$y_1 = x_1 + x_2$$
$$y_2 = (x_1 + x_2)^2 \tag{4.17}$$

the XOR problem can now be solved using the original single-perceptron node (index 3) with the weights and bias

node 3: $w_{13} = 2$ $w_{23} = -1$ $\varphi_3 = -0.5$

Note that y_2 represents a circular area, i.e. a convex region.

The importance of a wise selection of input signals or features for a given problem and network cannot be overemphasized.

4.2.4.3 Training and adaptation

Training or adaptation is used to determine the weights and biases of an artificial neural network, i.e. to determine the function of the network. By means of a number of algorithms, this can be done in a number of ways. Before going into a more detailed discussion, some general problems should be addressed.

Training by examples is of course a nice way of "programming" the network, however, the more weights and biases that need to be determined, the more input data examples will be needed (and the more training time will be required). Adaptive systems like this "eat" input data. Once used as an example, the data is normally not good for any further training.

Secondly, finding a significant subset of all possible input data suitable for training is not easy in the general case. Using a "bad" training set, the network may "learn" the wrong mappings and poor performance may result (like teenagers?). Normally, training is performed using a **training set** of input data, while performance testing uses another set of input data. If the same input data set is used for both testing and training, we can only be sure that the network has learnt how to handle the **known** data set properly. If this is the primary task of the system, it would probably be better to use a look-up table (LUT) algorithm than an artificial neural network.

One of the main advantages of an artificial neural network is that it is able to generalize, i.e. if earlier **unknown** input data is presented, the network can still give a "reasonable" answer. A "conventional" computing system would in many cases present a useless error message in the same situation. The network can in this respect be viewed as a huge lookup table, where only a fraction of all the possible entries have been initialized (by the training data set). The network itself will then "interpolate" all the non-initialized table entries. Hence, a proper choice of training data is crucial to achieve the desired function.

Training methods can be divided into two groups, **unsupervised** and **supervised**. In both cases, the network being trained is presented with training data. In the unsupervised (Rumelhart and McClelland, 1987) case, the network itself is allowed to determine a proper output. The rule for this is built into the network. In many cases, this makes the network behavior hard to predict. This type of network is known as "clustering" devices. The problem with unsupervised learning is that the result of the training is somewhat uncertain. Sometimes the network may come up with training results that are almost incomprehensible. Another problem is that the order in which the training data is presented may affect the final training result.

In the case of supervised training, not only is the training data presented to the network, but also the corresponding desired output. There are many different supervised training algorithms around, but most of them stem from a few classical ones, which will be presented below. The bias term φ of a node can be treated as a weight connected to an input signal with a constant value of 1.

The oldest training algorithm is probably **Hebb's rule**. This training algorithm can only be used for single-layer networks. There are many extensions to this rule of how to update the weights, but in its basic form it can be expressed as the following

$$w_{ij}(n+1) = w_{ij}(n) + \mu x_i(n) x_j(n) \tag{4.18}$$

where μ is the "learning rate", x_i is the output of node i, which is also one of the inputs to node j and x_j is the output of node j. The underlying idea of the Hebbian rule is that if both nodes i and j are "active" (positive output) or "inactive" (negative output) the weight in between them, w_{ij}, should be increased, otherwise it should be decreased. In a one-layer network, x_i is a component in the training input data vector and x_j corresponds to the desired output.

This simple rule has two advantages, it is **local**, i.e. it only needs local information x_i and x_j, and it does not require the non-linearity $f(\ )$ of the node to be differentiable. A drawback is that to obtain good training results, the input vectors in the training set need to be **orthogonal**. This is because all the training input vectors are added to the weight vectors of the active nodes, hence two or more nodes may obtain almost the same weight vectors if they become "active" often. Further, the magnitude of the weight vectors may grow very large. This will result in "cross-talk" and erroneous outputs. If on the other hand, the input data vectors are orthogonal, it is very unlikely that two nodes will be active about equally often and hence obtain similar weight vectors.

The **Widrow–Hoff rule** or **delta rule** is closely related to the LMS algorithm discussed in Chapter 3. This training algorithm can only be used for single-layer networks.

In this algorithm, an error signal ε_j is calculated as (for node j)

$$\varepsilon_j = d_j - u_j = d_j - \sum_i w_{ij} x_i - \varphi_j \tag{4.19}$$

Note that u_j is what comes out of the summation stage in the node and goes into the activation function $f(\ )$. The term d_j is the desired output signal when the training data vector consisting of $x_1, x_2, \ldots, x_N$ is applied. The weights are then updated using

$$w_{ij}(n + 1) = w_{ij}(n) + \mu x_i(n)\, \varepsilon_j(n) \tag{4.20}$$

In this case, the input vectors in the training set do not need to be orthogonal. The **difference** between the actual output and the desired output, i.e. the error signal ε_j, is used to control the addition of the input data vector to the weight vector. This rule is also local and it does not require the non-linearity $f(\ )$ of the node to be differentiable.

The **perceptron learning rule** is a variation of the delta rule described above. This training algorithm can, in its standard form, only be used for single-layer networks. The idea is to adjust the weights in the network in such a way that minimizes the total sum of square errors; the total error for the network is

$$E = \sum_j \varepsilon_j^2 = \sum_j (d_j - x_j)^2 = \sum_j \left(d_j - f\left(\sum_i w_{ij} x_i - \varphi_j \right) \right)^2 \tag{4.21}$$

A simple steepest descent approach is now to take the derivative of the error with respect to the weight w_{ij}. From this, we can tell in which direction to change this particular weight to reduce the total square error for the network. This procedure is repeated for all weights, hence we calculate the gradient of the weight vector on the quadratic error hypersurface. For w_{ij} we obtain

$$\frac{\partial E}{\partial w_{ij}} = -2(d_j - x_j)\frac{\partial x_j}{\partial w_{ij}} = -2\varepsilon_j \frac{\partial x_j}{\partial u_j} \frac{\partial u_j}{\partial w_{ij}} = -2\varepsilon_j f'(u_j) x_i \tag{4.22}$$

As can be seen, this algorithm requires the activation function $f(\ )$ of the node to be differentiable. This is the delta rule, but we have now also included the activation function. The weights are now updated as

$$w_{ij}(n + 1) = w_{ij}(n) + \mu x_i(n)\, \varepsilon_j(n) f'(u_j(n)) \tag{4.23}$$

where (as before)

$$u_j = \sum_i w_{ij} x_i + \varphi_j \tag{4.24}$$

So far, we have only discussed training algorithms for single-layer feedforward artificial neural networks. Training multi-layer networks is harder, since in normal cases, we lack desired output signals for all nodes that are not in the output layer.

For training multi-layer feedforward networks, there is an algorithm called **the generalized perceptron learning rule**, sometimes denoted **the back-propagation algorithm**. This algorithm is a generalization of the perceptron learning rule discussed above.

The idea is to start training the output layer. Since we do have access to desired output values d_k, this can be done by using the standard perceptron learning rule as outlined above. Now, we go backwards in the layer and start training the first layer preceding the output layer. Here we have to calculate the errors on the outputs of the hidden nodes by propagating the errors on the output through the output layer. This is where "back-propagation" becomes useful.

Assume that there are M nodes in the output layer, and that we are to adjust the weights of node j in the layer just below the output layer. Firstly, the error on the output of node j will contribute to the errors of all output nodes. Let us calculate this coupling first by taking the derivative of the total output squared error E with respect to the output of node j

$$\frac{\partial E}{\partial x_j} = \sum_{k \in M} \frac{\partial E}{\partial u_k} \frac{\partial u_k}{\partial x_j} = -2 \sum_{k \in M} (d_k - x_k) \frac{\partial x_k}{\partial u_k} \frac{\partial u_k}{\partial x_j}$$

$$= -2 \sum_{k \in M} (d_k - x_k) f'(u_k) w_{jk} \tag{4.25}$$

When this link between the error on the output and the error of the output of node j is known, we can find the derivative of the total error with respect to the weights we are to update in node j (see also equation (4.22) above)

$$\frac{\partial E}{\partial w_{ij}} = \frac{\partial E}{\partial x_j} \frac{\partial x_j}{\partial w_{ij}} = \frac{\partial E}{\partial x_j} \frac{\partial x_j}{\partial u_j} \frac{\partial u_j}{\partial w_{ij}} = \frac{\partial E}{\partial x_j} f'(u_j) x_i$$

$$= -2 f'(u_j) x_i \sum_{k \in M} \varepsilon_k f'(u_k) w_{jk} \tag{4.26}$$

The weights can now be updated using the standard form (see also equation (4.23) above)

$$w_{ij}(n + 1) = w_{ij}(n) + \mu x_i(n) f'(u_j(n)) \sum_{k \in M} \varepsilon_k(n) f'(u_k(n)) w_{jk} \tag{4.27}$$

This procedure is repeated for the next layer and so on, until the weights of the nodes in the input layer have been updated.

4.2.4.4 Applications

Feedforward artificial neural networks are typically used in applications like **pattern recognition** (see Chapter 7), **pattern restoration** and **classification**. The "patterns" are the input vectors, where the components are "features" relevant in the application. If we are, for instance, dealing with a speech recognition system, e.g. the features can represent signal power in different frequency bands of a speech signal. The features can also be pixel values in an artificial neural network-based image processing system for optical character reading (OCR). Other areas are processing of radar and sonar echo signals, matching fingerprints, correcting transmission errors in digital communication systems, classifying blood samples, genes and electrocardiograph (ECG) signals in biological and medical applications, troubleshooting of electronic systems, etc. In most cases, the applications belong to some basic system types as detailed below.

Pattern associator: In this system type, the network mainly performs a mapping from the input vector space to the output vector space. The mechanism here is to map (or "translate") an input vector to an output vector in a way that the network has been trained. If the desired mapping is simple and can be expressed algebraically, no artificial neural network is normally needed. If, on the other hand, the mapping cannot be easily formulated, and/or only samples of input and output vectors are available, training an artificial neural network is possible.

Networks of this type have, for instance, been tried for weather forecasting (Hu, 1963). Since the coupling between temperature, humidity, air pressure, wind speed and direction in different places affects the weather in a very complicated way, it is easier to train a network than to formulate mathematical relations.

Auto-associator or content addressable memory (CAM): This system type can be viewed as a special case of the pattern associator above, in the sense that the input vector is mapped back on itself. The system acts as a content addressable memory, where the previously trained patterns are stored in the weights. If an incomplete or distorted version of an earlier known pattern is presented to the network, it will respond with a restored, error-free pattern.

The human memory works as a content addressable memory. Once we get some "clues", or parts of the requested set of information, the rest of this information will be recalled by associations. Networks of this type have been tried for database search and for error correction of distorted digital signals in telecommunications systems.

Classifier, identifiers: This system type can also be regarded as a special case of the pattern associator above. In this situation, our aim is to categorize the input pattern and to identify to which class of patterns it belongs. This type of network typically has few outputs and in some cases also a special type of output layer having "lateral feedback" (treated below) that assures that only one output (class) at a time can be active.

Examples of this type of networks are systems for classifying ECG signals (Specht, 1964), and sonar and radar echo signals. The outputs may be "healthy/unhealthy", "submarine/no submarine" and so on.

Regularity detectors: This system type can be viewed as a variant of the classifier above, but in this case there is no *a priori* set of categories into which the input patterns are to be classified. The network is trained using unsupervised training, i.e. the system must develop its own featured representation of the input patterns. In this way, the system is used to explore statistically salient features of the population of input patterns. "Competitive learning" (Rumelhart and McClelland, 1987) is an algorithm well suited for regularity detectors. Examples of applications in this area are finding significant parameters in a large set of data or finding good data compression and error-correction coding schemes.

4.2.5 Feedback networks

Feedback networks, also known as **recurrent networks**, have all outputs internally fed back to the inputs in the general case. Hence, a network of this type commonly does not have dedicated inputs and outputs. Besides the non-linear activation function, nodes in such a network also have some kind of dynamics built in, for instance, as in an integrator or accumulator.

The main idea of this class of networks is **iteration**. The input vector to this system is applied as initial states of the node outputs and/or as bias values. After these initial conditions are set, the network is iterated until convergence, when the components of the output vector can be found on the node outputs. To achieve convergence and to avoid stability problems, the weights, i.e. the feedback parameters, have to be chosen carefully. It is very common that the weights are constant, set *a priori* in feedback networks, hence this type of networks is not "trained" in the same way as feedforward-type networks.

Common networks of this type are **the Hopfield net**, **The Boltzmann machine** and **Kohonen's feature maps**.

4.2.5.1 Nodes

The node functions in a feedback network have many similarities to the node functions used in feedforward networks. An important difference, however, is that in the feedback network case, the node functions include dynamics or memory, e.g. an integrator. This is necessary, as the network will be iterated to obtain the final output. (In the feedforward case, obtaining the output is a one-step process.) A common node model used by, e.g. Hopfield, has the differential equation

$$\frac{du_j}{dt} = \sum_i w_{ij} x_i + \varphi_j - u_j = \sum_i w_{ij} f(u_i) + \varphi_j - u_j \tag{4.28}$$

where $f(\ )$ is the activation function. As can be seen, we have taken the node function (4.8) and inserted an integrator in between the summation point and the non-linear activation function. Hence, the feedback network is but a system of N non-linear differential equations, which can be expressed in a compact matrix form

$$\frac{d\mathbf{U}}{dt} = \mathbf{W}^T \mathbf{X} + \mathbf{\Theta} - \mathbf{U} = \mathbf{W}^T \mathbf{F}(\mathbf{U}) + \mathbf{\Theta} - \mathbf{U} \tag{4.29}$$

where

$$\mathbf{U} = [u_1 \quad u_2 \quad \cdots \quad u_N]^T$$
$$\mathbf{F}(\mathbf{U}) = [f(u_1) \quad f(u_2) \quad \cdots \quad f(u_N)]^T = \mathbf{X}$$
$$\mathbf{\Theta} = [\varphi_1 \quad \varphi_2 \quad \cdots \quad \varphi_N]^T$$

Now, it is straightforward to realize that when iterating the network it will converge to a stable state when equation (4.29) is equal to the zero vector

$$\frac{d\mathbf{U}}{dt} = 0$$

If we assume that the weight matrix $\mathbf{W}$ is symmetric, i.e. $w_{ij} = w_{ji}$ and that we integrate equation (4.29), we can define the computational "energy" function for the network

$$H(\mathbf{X}) = -\frac{1}{2}\mathbf{X}^{\mathrm{T}}\mathbf{W}^{\mathrm{T}}\mathbf{X} - \mathbf{X}^{\mathrm{T}}\mathbf{\Theta} + \sum_{j=1}^{N}\int_{0}^{x_j}f^{-1}(x)\,\mathrm{d}x \qquad (4.30)$$

Now, if the activation function is a sigmoid equation (4.11), the inverse function will be

$$u = f^{-1}(x) = -T\ln\left(\frac{1}{x} - 1\right) \qquad (4.31)$$

If we, like Hopfield, use the hard limiter activation function, this corresponds to a sigmoid with a low computational temperature T, implying that the third term of equation (4.30) will be small. Hence, it can be neglected and equation (4.30) can be simplified to

$$H(\mathbf{X}) = -\frac{1}{2}\mathbf{X}^{\mathrm{T}}\mathbf{W}^{\mathrm{T}}\mathbf{X} - \mathbf{X}^{\mathrm{T}}\mathbf{\Theta} \qquad (4.32)$$

Hence, when the network settles to a stable state (equation (4.30)) all derivatives, i.e. the gradient of the energy function (4.32), is zero. This means that the stable states of the network and the output patterns correspond to the local minimum of the energy function. To "program" the network means formulating the problem to be solved by the network in terms of an energy function of the form (4.32). When this energy function has been defined, it is straightforward to identify the weights and biases of the nodes in the feedback artificial neural network.

A more general way to design a feedback network is of course to start by defining an energy function and obtaining the node functions by taking the derivatives of the energy function. This method is, in the general case, much harder since node functions may turn out to be complicated and convergence problems may occur. The energy function must be a Lyapunov (Åström and Wittenmark, 1984) type function, i.e. it should be monotonically decreasing in time. Further, Hopfield has shown that the weight matrix $\mathbf{W}$ must be symmetric, in other words $w_{ij} = w_{ji}$ and have vanishing diagonal elements $w_{ii} = 0$ to guarantee stability.

4.2.5.2 Network topology

In the general case, all nodes are connected with all other nodes in a feedback network (see Figure 4.6). There are however some special cases; one such case is **lateral feedback** (Lippmann, 1987; Stranneby, 1990). This type of feedback is commonly used within a layer of a feedforward network. An example is the **MAXNET**. This lateral feedback-type network assures that one and only one output of a layer is active, and that the active node is the one having the largest magnitude of the input signal of all nodes in the layer. The MAXNET is hence a device to find the maximum value within all elements in a vector. Quite often,

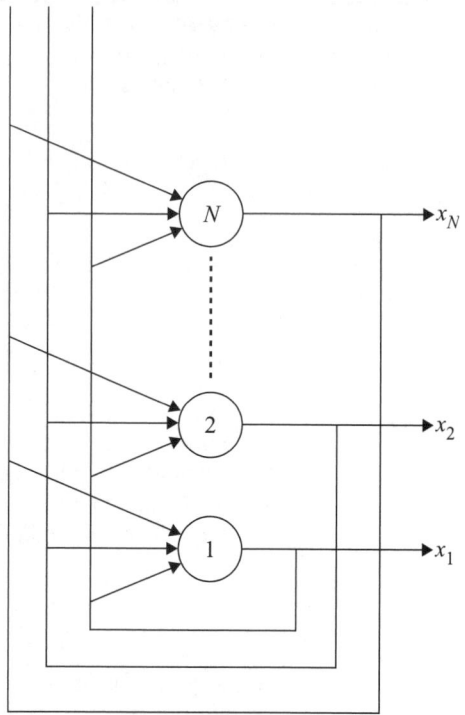

Figure 4.6 *Example of a small, general feedback ANN*

a structure like this is used on the output of a classifying feedforward-type artificial neural network, to guarantee that only one class is selected.

4.2.5.3 Local and global minimum

As was seen in the previous section, when the feedback artificial neural network is iterated, it will finally settle in one of the minimum points of the energy function of the type as equation (4.32); which minimum point depends on the initial state of the network, i.e. the input vector. The network will simply "fall down" in the "closest" minimum after iteration is started. In some applications this is a desired property, in others it is not.

If we like the network to find the **global** minimum, in other words the "best" solution in some sense, we need to include some extra mechanism to assure that the iteration is not trapped in the closest **local** minimum. One network having such a mechanism is the **Boltzmann machine**.

In this type of feedback artificial neural network, **stochastic node functions** are used. Every node is fed additive zero-mean-independent (uncorrelated) noise. This noise has the effect of making the node outputs noisy, thus "boiling" the hypersurface of the energy function. In this way, if the state of the network happens to "fall down" in a local minimum, there is a certain probability that it will "jump up" again and fall down in another minimum. The trick is then gradually to decrease the effect of the noise, making the network finally end in the "deepest" minimum, i.e. the global one. This can be performed by slowly decreasing the "temperature" T in the sigmoid activation function in the network nodes.

The trick of decreasing the "temperature" or cooling the network until it "freezes" to a solution is called **simulated annealing**. The challenge here is to find the smartest annealing scheme that will bring the network to the global minimum with a high probability in as short a time as possible. Quite often networks using simulated annealing converge very slowly.

The term "annealing" is borrowed from crystallography. At high temperatures, the atoms of a metal lose the solid-state phase, and the particles position themselves randomly according to statistical mechanics. The particles of the molten metal tend toward the minimum energy state, but the high thermal energy prevents this. The minimum energy state means a highly ordered state such as a defect-free crystal lattice. To achieve defect-free crystals, the metal is annealed, i.e. it is first heated to a temperature above the melting point and then slowly cooled. The slow cooling is necessary to prevent dislocations and other crystal lattice disruptions.

4.2.5.4 Applications

Artificial neural networks have traditionally been implemented in two ways, either by using analog electronic circuits, or as software on traditional digital computers. In the latter case, which is relevant for DSP applications, we of course have to use the discrete time equivalents of the continuous-time expressions present in this chapter. In the case of feedforward artificial neural networks, the conversion to discrete time is trivial. For feedback networks having node dynamics, standard numerical methods like Runge–Kutta may be used.

Feedback artificial neural networks are used in content addressable memories (CAM) and for solving miscellaneous optimization problems.

In **CAM**, each minimum in the energy function corresponds to a memorized pattern. Hence, in this case, "falling" in the closest local minimum is a desirable property. Given an incomplete version of a known pattern, the network can accomplish **pattern completion**. Unfortunately, the packing density of patterns in a given network is not very impressive. Research is in progress in an attempt to find better energy functions that are able to harbor more and "narrow" minimums, but still result in stable networks.

Solving **optimization problems** is probably the main application of feedback artificial neural networks. In this case, the energy function is derived from the objective function of the underlying optimization problem. The global minimum is of primary interest, hence simulated annealing and similar procedures are often used. For some hard optimization problems and/or problems with real-time requirements (e.g. in control systems), a local minimum, i.e. sub-optimum solution found in a reasonable time, may be satisfactory.

Many classical NP-complete optimization problems, e.g. "The traveling salesman" (Hopfield and Tanks, 1986) and "The 8-queens problem" (Holmes, 1989), have been solved using feedback artificial neural networks.

It is also possible to solve optimization problems using a feedforward artificial neural network with an external feedback path. This method has been proposed for transmitter (TX) power and frequency assignment in radio networks (Stranneby, 1996).

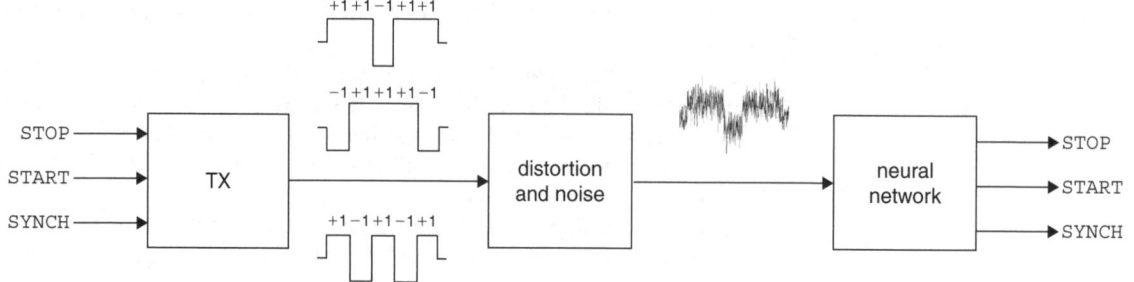

Figure 4.7 *Example of a telecommunications system using PAM pulse trains and a neural nework-based RX*

4.2.6 An example application

4.2.6.1 The problem

Suppose we have an application as in Figure 4.7. The transmitter (TX) is sending random sequences of the three commands: STOP, START and SYNCH. These commands are coded as pulse amplitude modulation (PAM) pulse trains representing five bipolar (±1) symbols as in the figure. Unfortunately, the transmission conditions are poor, so the pulse trains reaching the receiver (RX) are weak, distorted and noisy. The task of the neural network-based RX is to determine which command was most probably transmitted. Since the distortion mechanism changes from time to time, the RX needs to be adaptive, i.e. it needs to be "trained" at regular intervals to learn how to interpret the noisy received pulses. The training is accomplished by the TX sending a predetermined, known sequence of commands. This system is of course simplified, since the main purpose of this example is to demonstrate a neural network application. Digital transmission systems are covered in Chapter 8.

4.2.6.2 The Hamming net

There are many possible solutions of how to design the receiver in Figure 4.7. In this example, a hybrid type of neural network called a **Hamming net** (Lippmann, 1987; Stranneby, 1990) will be used. The input layer is a one-layer, adaptive feedforward network, while the second layer is a lateral feedback network having fixed weights (see Figure 4.8). The latter part is sometimes referred to as the MAXNET.

The input pulse train is sampled at five instants of time, corresponding to the five bipolar symbols. At the transmitting site, these symbols are of course clean and benign, having only the levels ±1 (see Figure 4.7). At the receiver, however, almost any pulse amplitudes can be found due to the distortion and noise introduced in the transmission process. The five sampled input values are denoted (from left to right) x_1, x_2, x_3, x_4, x_5 and referred to as the **input vector**.

The first network layer (left) in Figure 4.8, nodes 6–8, is a feedforward network performing a cross-correlation operation between the incoming input vector and the "learnt" prototype vectors (expected pulse trains) corresponding to the three commands. Hence, the output of node 6, i.e. x_6, is the degree of

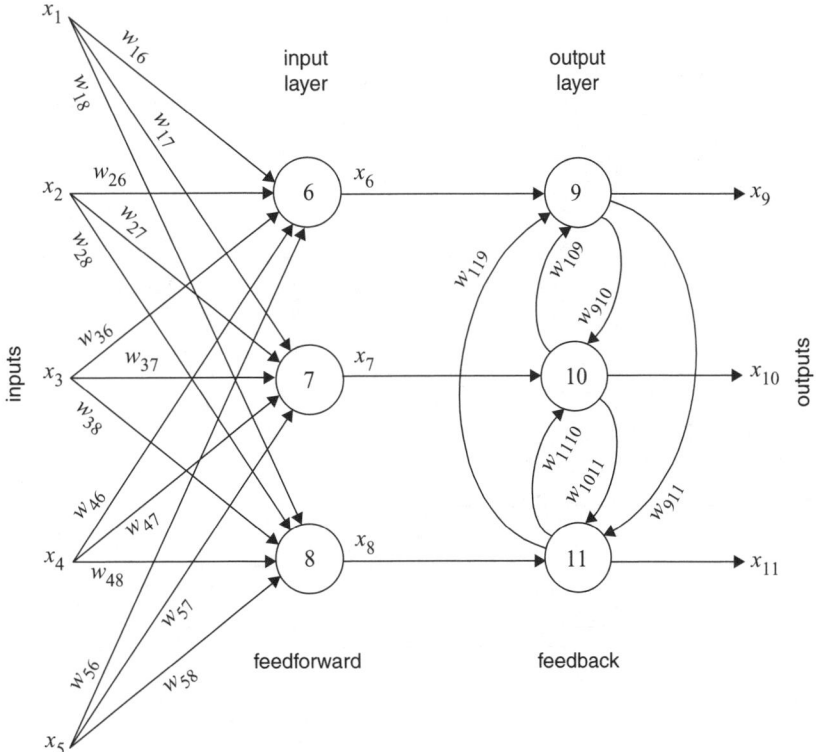

Figure 4.8 *Architecture of the Hamming net used in the example*

correlation between the input vector and the expected vector for the command STOP. Node 7, with output x_7 corresponds to START, and node 8, output x_8 to the SYNCH command. To be able to determine which symbol is being received, we finally need to find the input node having maximum output. This task is accomplished in the second layer (right) of the network in Figure 4.8. This layer is a feedback network, consisting of nodes 9–11. The feedback network nodes are initialized to the output values of the first layer, and are thereafter iterated. As the iteration is completed, only one of the nodes 9–11 will have a **non-zero** output signal, which will indicate which command was received. Obviously, node 9, output x_9 corresponds to STOP, node 10, x_{10} to START and node 11, x_{11} to the SYNCH command. There are, of course, easier ways to find the maximum output of three outputs, but in this example our purpose is to demonstrate a feedback network.

4.2.6.3 *The feedforward input layer*

The input layer consists of nodes as depicted in Figure 4.9, having the node function (compare to equation (4.8)) as

$$x_j = f(u_j) = f\left(\sum_{i=1}^{5} w_{ij}x_i + \phi_j\right) \quad \text{for } j = 6, 7, 8 \tag{4.33}$$

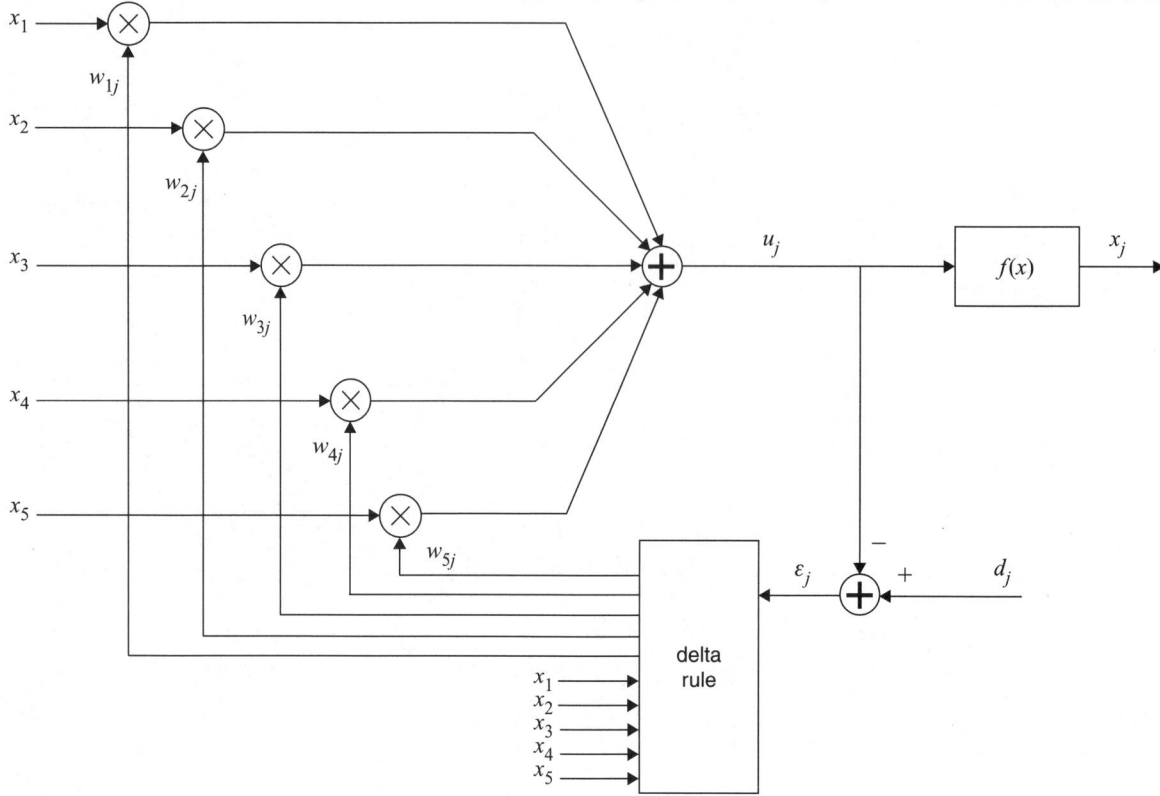

Figure 4.9 *An input-layer feedforward node with delta rule adaptation*

where the bias is $\phi_j = 0$ and the activation function is

$$f(x) = \begin{cases} x & x \geq 0 \\ 0 & x < 0 \end{cases} \tag{4.34}$$

which is basically the soft limiter (equation (4.10)) with $b = 0$ and $a \to \infty$.

The node is trained using the Widrow–Hoff rule (delta rule) as in equations (4.19) and (4.20). Inserting equation (4.33) we obtain

$$w_{ij}(n+1) = w_{ij}(n) + \mu x_i(n)\varepsilon_j(n) = w_{ij}(n) + \mu x_i(n)(d_j(n) - u_j(n))$$

$$= w_{ij}(n) + \mu x_i(n)\left(d_j(n) - \sum_{i=1}^{5} w_{ij}(n)x_i(n)\right) \tag{4.35}$$

where $d_j(n)$ is the desired output $u_j(n)$ (input to the non-linear activation function) when the input vector $x_1(n), x_2(n), x_3(n), x_4(n), x_5(n)$ is presented. In most cases, the value of the learning rate μ must be determined heuristically. It is easy to see that this type of training is the same as that taking place in an adaptive filter using the LMS algorithm (see Chapter 3).

Figure 4.10 shows a typical training sequence. The figure shows how the weight values of node 6 converge. In this example, all the weights were initialized to 0.5 ("cold start") and μ was set conservatively to 0.1. As can be seen, after approximately 20 training runs, the weights are settled. It is possible to increase μ and achieve a faster training, but since the input vector is subject to noise, a too fast training may result in poor weights. Often, a longer training time means less impact of the input signal noise on the final weight values.

When the network is run under "normal" conditions, i.e. not training, the weights are fixed to the values obtained during the training phase. The weights are not changed until a new training sequence starts. In most cases, the following training sequences will be faster since the starting values of the weights are already quite good.

4.2.6.4 *The feedback layer, MAXNET*

The output layer consists of nodes 9–11 as shown in Figure 4.11. The nodes have partly the same node function as in the input layer, but also have a delay (memory) to achieve a dynamic process when iterating

$$x_j(m+1) = f\left(x_j(m) + \sum_{i=9}^{11} w_{ij}x_i(m)\right) \quad \text{for } j = 9,\ 10,\ 11 \tag{4.36}$$

where the fixed weights are set to

$$w_{ij} = \begin{cases} 0 & i = j \\ -\sigma & i \neq j \end{cases} \tag{4.37}$$

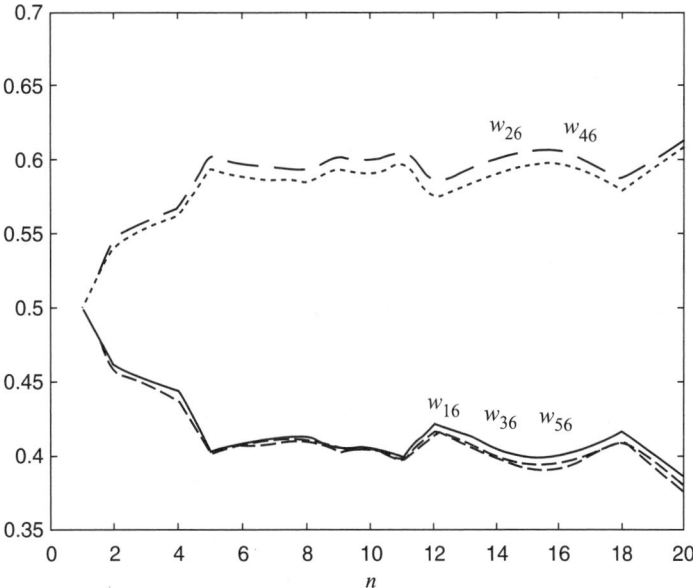

Figure 4.10 *Weights of node 6 during a training sequence*

and σ is an iteration constant, for stability set to $\sigma < 1/M$ where M is the number of nodes in the feedback network, in our case equal to 3. Before iteration is started, the network nodes are initialized to $x_9(0) = x_6, x_{10}(0) = x_7, x_{11}(0) = x_8$. Figure 4.12 shows the iteration of the feedback network. Note that the process continues until all outputs but one are "dead" (zero). The "surviving" output belongs to the node initialized by the largest starting level, i.e. the largest output of x_6, x_7, x_8. In this simulation $\sigma = 0.1$ is used for demonstration purposes.

Simulating the entire network, it can be shown to recognize the correct pulse trains in most cases, even when considerable random background noise is present. This example, a non-linear correlation receiver, is probably not

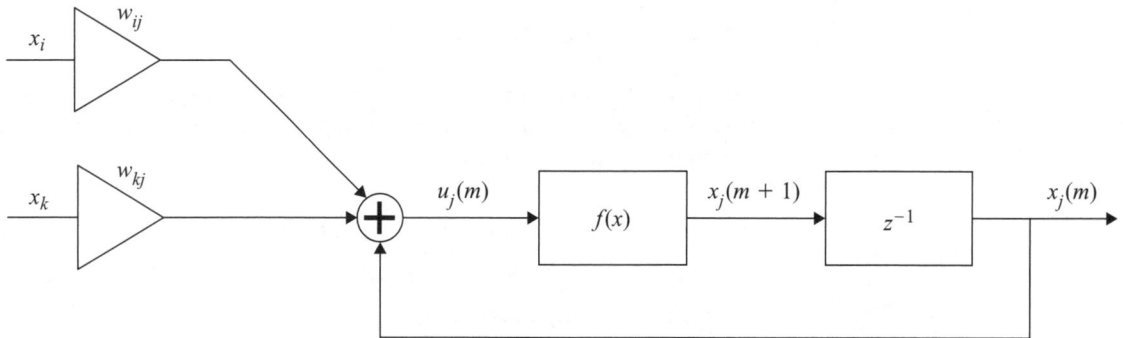

Figure 4.11 *An output-layer feedback node with one sample delay*

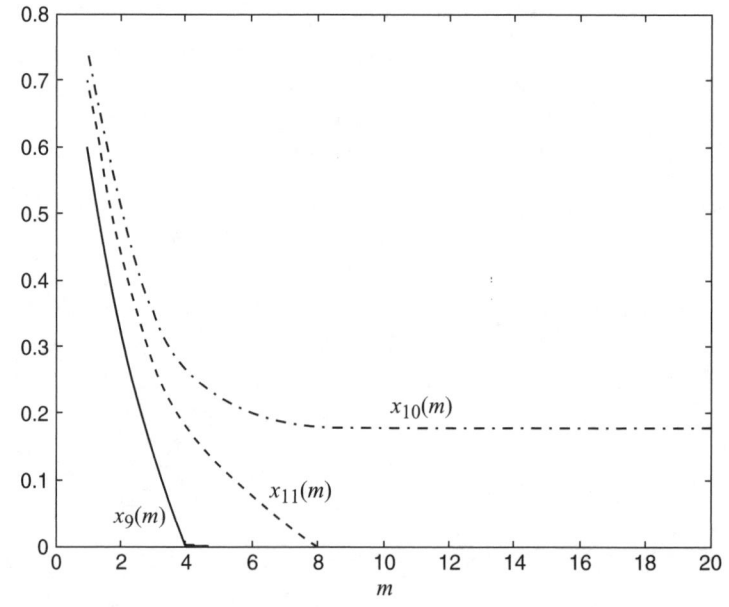

Figure 4.12 *Iteration of the MAXNET feedback network, all outputs "die" except for one*

the optimum implementation for the stated problem, but merely meant as an example. Further ideas on digital transmission systems can be found in Chapter 8.

4.3 Fuzzy logic

4.3.1 General

A **fuzzy logic** or **fuzzy control** (Palm *et al.*, 1996; Passino and Yurkovich, 1998) system performs a static, non-linear mapping between input and output signals. It can in some respects be viewed as a special class of feedforward artificial neural networks. The idea was originally proposed in 1965 by Professor Lofti Zadeh at the University of California, but has not been used very much until the past decade. In ordinary logic, only false or true is considered, but in a fuzzy system we can also deal with intermediate levels, e.g. one statement can be 43% true and another one 89% false.

Another interesting property is that the behavior of a fuzzy system is not described using algorithms and formulas, but rather as a set of **rules** that may be expressed in natural language. Hence, this kind of system is well suited in situations where no mathematical models can be formulated or when only heuristics are available. Using this approach, practical experience can be converted into a systematic, mathematical form, in, for instance, a control system.

A simple fuzzy logic system is shown in Figure 4.13. The **fuzzifier** uses **membership functions** to convert the input signals to a form that the **inference engine** can handle. The inference engine works as an "expert", interpreting the input data and making decisions based on the rules stored in the **rule database**. The rule database can be viewed as a set of "If–Then" rules. These rules can be linguistic descriptions, formulated in a way similar to the knowledge of a human expert. Finally, the **defuzzifier** converts the conclusions made by the inference engine into output signals.

As an example, assume we are to build a smart radar-assisted cruise control for a car. The input signals to our simplified system are the *speed* of our car and the *distance* to the next car in front of us. (The distance is measured using radar equipment mounted on the front bumper.) The continuous output control signals are *accelerate* and *brake*. The task of the fuzzy controller is to keep the car at a

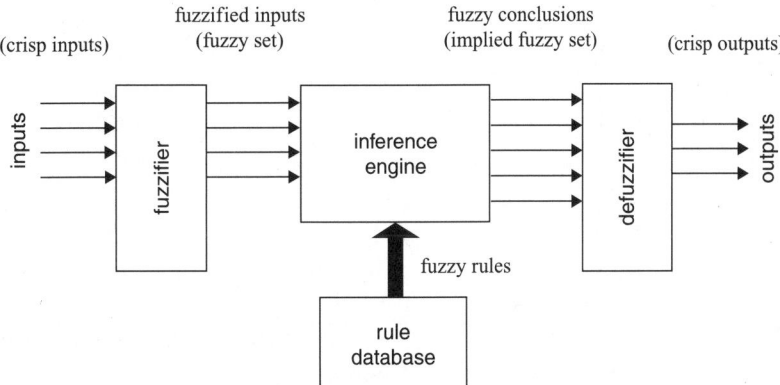

Figure 4.13 *Example of a simple fuzzy logic system*

constant speed, but to avoid crashing into the next car in a traffic jam. Further, to get a smooth ride, the accelerator and brake should be operated gently.

4.3.2 Membership functions

The outputs of the fuzzifier are called **linguistic variables**. In the above example, we could introduce two such variables SPEED and DISTANCE. The **linguistic value** of such a variable is described using adjectives like *slow, fast, very fast,* etc. The task of a membership function is to interpret the value of the continuous input signal to **degree of membership** with respect to a given linguistic value. The degree of membership can be any number between zero and one, and it can be regarded as a measure of to what extent the input signal has the property represented by the linguistic value. It is common to use 3, 5 or 7 linguistic values for every linguistic variable, hence the same number of membership functions are required. Figure 4.14 shows the membership functions for our example given above. Here we have used three linguistic values for SPEED – *slow, medium* and *fast* and three values for DISTANCE – *close, ok* and *far*.

From the figure it can be seen that the set of membership functions converts the value of the continuous input signal to degree of membership for every linguistic value. In this example, the membership functions overlap, so the SPEED can, for instance, be both *medium* and *fast* but with different degrees of membership. Further, we have chosen to "saturate" the outermost membership functions; for instance, if the distance goes to infinity or zero it is regarded as *far* and *close*, respectively. There are many possible shapes for a membership function; the simplest and most common one is a symmetric triangle, which we have used in our example. Other common shapes are trapezoid, Gaussian or similar "bell-shaped" functions, peak shapes and skewed triangles. If a purely rectangular shape is used, the degree of membership can only be one or zero and nothing in between. In such a case, we are said to have a **crisp** set representation rather than a fuzzy set representation.

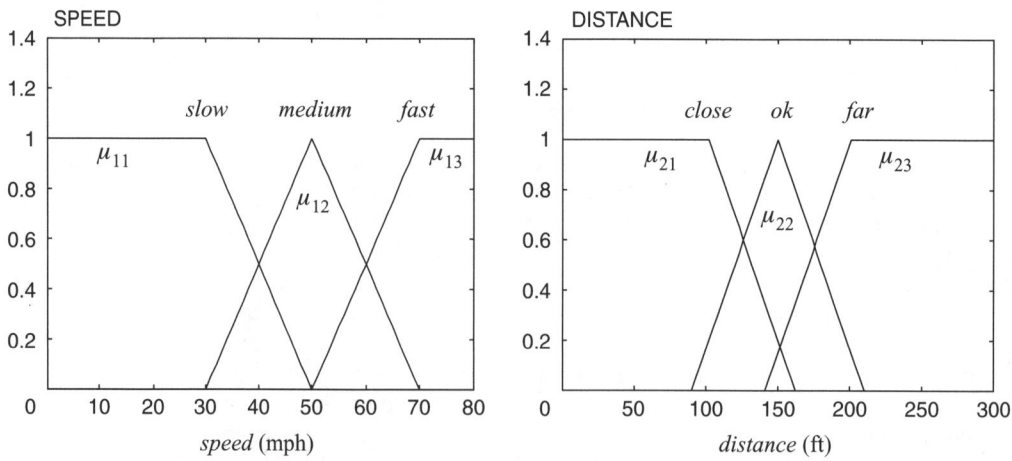

Figure 4.14 *Input signal membership functions used in the example*

4.3.3 Fuzzy rules and inference

The input to the inference engine is the linguistic variables produced by the membership functions in the fuzzifier (the fuzzy set). The output linguistic variables of the inference engine are called the **conclusions** (the implied fuzzy set). The mapping between the input and output variables is specified in natural language by rules having the form

$$\textbf{If } \text{premise } \textbf{Then } \text{consequent} \tag{4.38}$$

The rules can be formulated in many forms. Of these forms, two are commonly standard, multi-input single-output (MISO) and multi-input multi-output (MIMO). In this text, we will deal with MISO rules only. (An MIMO rule is equivalent to a number of MISO rules.) The premise of a rule is in the multi-input case a "logic" combination of conditions. Two common "logic" operators are **AND** and **OR**. Two simple examples of rules containing three conditions are

$$\textbf{If } \text{condition}_1 \textbf{ AND } \text{condition}_2 \textbf{ AND } \text{condition}_3 \textbf{ Then } \text{consequent} \tag{4.39a}$$

$$\textbf{If } \text{condition}_1 \textbf{ OR } \text{condition}_2 \textbf{ OR } \text{condition}_3 \textbf{ Then } \text{consequent} \tag{4.39b}$$

Now, if we had been using Boolean variables (one or zero), the definition of the operators AND and OR would be obvious. In this case, however, the conditions are represented by degree of membership, i.e. any real number between zero and one. Hence, we need an extended definition of the operators. There are different ways to define AND and OR in fuzzy logic systems. The most common way to define the AND operation between two membership values, μ_{ij} and μ_{kl}, is by using the minimum

$$\mu_{ij} \text{ AND } \mu_{kl} = \min\{\mu_{ij}, \mu_{kl}\} \tag{4.40}$$

where μ_{ij} means the membership value of function j of the linguistic variable i. For example, 0.5 AND 0.7 = 0.5. The most common way to define the OR operation is by using the maximum

$$\mu_{ij} \text{ OR } \mu_{kl} = \max\{\mu_{ij}, \mu_{kl}\} \tag{4.41}$$

For example, 0.5 OR 0.7 = 0.7. An alternative way of defining the operators is to use algebraic methods

$$\mu_{ij} \text{ AND } \mu_{kl} = \mu_{ij}\mu_{kl} \tag{4.42}$$

$$\mu_{ij} \text{ OR } \mu_{kl} = \mu_{ij} + \mu_{kl} - \mu_{ij}\mu_{kl} \tag{4.43}$$

An interesting question is how many rules are needed in the rule database? Assuming we have n linguistic variables (inputs to the inference engine) and that the number of membership functions for variable i is N_i, then the total number of rules will be

$$N_R = \prod_{i=1}^{n} N_i = N_1 \cdot N_2 \cdot \cdots \cdot N_n \tag{4.44}$$

Assuming this, we need to consider **all** possible combinations of input signals. Further, if MISO rules are used, N_R rules may be needed for every output linguistic variable in the worst case. From equation (4.44) it is easy to see that the rule database and the computational burden grow quickly if too many variables and membership functions are used. Hence, the selection of good input signals and an appropriate number of linguistic values is crucial to system performance.

To pursue our example, we need to define output linguistic variables, values and membership functions for the conclusions produced by the inference engine. Let us introduce two output variables. Firstly, ACCELERATE having three values, *release*, *maintain* and *press* (for the accelerator pedal). Secondly, BRAKE with values, *release*, *press* and *press hard* (for the brake pedal). We are assuming automatic transmission and are ignoring "kick-down" features and so on. The associated membership functions used are triangular (see Figure 4.15). In this case, "saturating" functions cannot be used since infinite motor power and accelerating brakes are not physically possible. From Figure 4.15 we can also see that the center of a membership function corresponds to a given value of thrust or braking force.

ACCELERATE:	*release*	−25% thrust	(braking using motor)
	maintain	+25% thrust	(to counteract air drag)
	press	+75% thrust	(to accelerate)
BRAKE:	*release*	0% braking force	
	press	25% braking force	(gentle stopping)
	press hard	75% braking force	(panic!)

The next step is to formulate the rules. According to equation (4.44) we will need $3 \times 3 = 9$ rules for every output (using MISO rules) if all combinations are considered. The following rules are suggested.

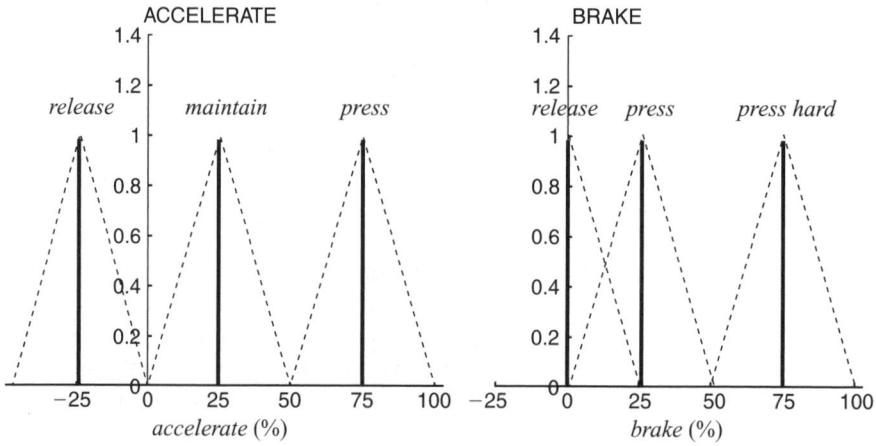

Figure 4.15 *Output signals and associated membership functions used in the example*

If SPEED is *slow* **AND** DISTANCE is *close* **Then** ACCELERATE
is *release* (4.45a)

If SPEED is *slow* **AND** DISTANCE is *ok* **Then** ACCELERATE
is *press* (4.45b)

If SPEED is *slow* **AND** DISTANCE is *far* **Then** ACCELERATE
is *press* (4.45c)

If SPEED is *medium* **AND** DISTANCE is *close* **Then** ACCELERATE
is *release* (4.45d)

If SPEED is *medium* **AND** DISTANCE is *ok* **Then** ACCELERATE
is *maintain* (4.45e)

If SPEED is *medium* **AND** DISTANCE is *far* **Then** ACCELERATE
is *maintain* (4.45f)

If SPEED is *fast* **AND** DISTANCE is *close* **Then** ACCELERATE
is *release* (4.45g)

If SPEED is *fast* **AND** DISTANCE is *ok* **Then** ACCELERATE
is *release* (4.45h)

If SPEED is *fast* **AND** DISTANCE is *far* **Then** ACCELERATE
is *release* (4.45i)

If SPEED is *slow* **AND** DISTANCE is *close* **Then** BRAKE
is *press* (4.46a)

If SPEED is *slow* **AND** DISTANCE is *ok* **Then** BRAKE
is *release* (4.46b)

If SPEED is *slow* **AND** DISTANCE is *far* **Then** BRAKE
is *release* (4.46c)

If SPEED is *medium* **AND** DISTANCE is *close* **Then** BRAKE
is *press* (4.46d)

If SPEED is *medium* **AND** DISTANCE is *ok* **Then** BRAKE
is *release* (4.46e)

If SPEED is *medium* **AND** DISTANCE is *far* **Then** BRAKE
is *release* (4.46f)

If SPEED is *fast* **AND** DISTANCE is *close* **Then** BRAKE
is *press hard* (4.46g)

If SPEED is *fast* **AND** DISTANCE is *ok* **Then** BRAKE is *press* (4.46h)

If SPEED is *fast* **AND** DISTANCE is *far* **Then** BRAKE
is *release* (4.46i)

Having defined all the rules needed, we may now be able to find simplifications
to reduce the number of rules. For instance, it is possible to reduce the rules

(4.45g)–(4.45i) to just one rule, since the value of DISTANCE does not matter, and in all cases ACCELERATE should be *release* if SPEED is *fast*. The reduced rule is

If SPEED is *fast* **Then** ACCELERATE is *release* $\qquad$ (4.47a)

Reasoning in the same way, rules (4.45a) and (4.45d) can be reduced

If DISTANCE is *close* **Then** ACCELERATE is *release* $\qquad$ (4.47b)

Combining rules (4.47a) and (4.47b) we obtain

If SPEED is *fast* **OR** DISTANCE is *close* **Then** ACCELERATE is *release* $\qquad$ (4.47c)

Keeping rules (4.45b), (4.45c), (4.45e) and (4.45f) as is, we are now left with five rules for ACCELERATE. Trying some reduction for the rules generating BRAKE, one solution is to join rules (4.46c), (4.46f) and (4.46i) to

If DISTANCE is *far* **Then** BRAKE is *release* $\qquad$ (4.48)

Leaving the other rules for BRAKE as is, we now have seven rules, so the total database will contain 12 rules (MISO). Further reductions are possible for both ACCELERATE and BRAKE, but they will result in complex rules and will probably not reduce the computational demands. Finally, the rules are rewritten in "mathematical" form and stored in the rule database. If we adopt the following notations

μ_{11} membership value for SPEED, *slow*
μ_{12} membership value for SPEED, *medium*
μ_{13} membership value for SPEED, *fast*
μ_{21} membership value for DISTANCE, *close*
μ_{22} membership value for DISTANCE, *ok*
μ_{23} membership value for DISTANCE, *far*

For conclusion (implied fuzzy set) about thrust

η_{3i} implied membership value for ACCELERATE, by rule i
b_{3i} position of peak of recommended membership function in ACCELERATE, by rule i

For conclusion (implied fuzzy set) about braking force

η_{4j} implied membership value for BRAKE, by rule j
b_{4j} position of peak of recommended membership function in BRAKE, by rule j

We can now formulate our reduced set of rules above in a mathematical form. For ACCELERATE we get

$$(4.47c): \quad \eta_{31} = \max\{\mu_{13}, \mu_{21}\} \quad \text{for } b_{31} = -0.25 \qquad (4.49a)$$

$$(4.45b): \quad \eta_{32} = \min\{\mu_{11}, \mu_{22}\} \quad \text{for } b_{32} = 0.75 \qquad (4.49b)$$

$$(4.45c): \quad \eta_{33} = \min\{\mu_{11}, \mu_{23}\} \quad \text{for } b_{33} = 0.75 \qquad (4.49c)$$

(4.45e): $\eta_{34} = \min\{\mu_{12}, \mu_{22}\}$ for $b_{34} = 0.25$ (4.49d)

(4.45f): $\eta_{35} = \min\{\mu_{12}, \mu_{23}\}$ for $b_{35} = 0.25$ (4.49e)

and for BRAKE we get

(4.48): $\eta_{41} = \mu_{23}$ for $b_{41} = 0$ (4.50a)

(4.46a): $\eta_{42} = \min\{\mu_{11}, \mu_{21}\}$ for $b_{42} = 0.25$ (4.50b)

(4.46b): $\eta_{43} = \min\{\mu_{11}, \mu_{22}\}$ for $b_{43} = 0$ (4.50c)

(4.46d): $\eta_{44} = \min\{\mu_{12}, \mu_{21}\}$ for $b_{44} = 0.25$ (4.50d)

(4.46e): $\eta_{45} = \min\{\mu_{12}, \mu_{22}\}$ for $b_{45} = 0$ (4.50e)

(4.46g): $\eta_{46} = \min\{\mu_{13}, \mu_{21}\}$ for $b_{46} = 0.75$ (4.50f)

(4.46h): $\eta_{47} = \min\{\mu_{13}, \mu_{22}\}$ for $b_{47} = 0.25$ (4.50g)

4.3.4 Defuzzification

The output conclusions from the inference engine must now be combined in a proper way to obtain a useful, continuous output signal. This is what takes place in the defuzzification interface (going from a fuzzy set to a crisp set). There are different ways of combining the outputs from the different rules. The implied membership value η_{ij} could be interpreted as the degree of certainty that the output should be b_{ij} as stated in the rule database.

The most common method is the **center of maximum (CoM)**. This method mainly takes the weighted mean of the output membership function peak position with respect to the implied membership value produced by M rules, i.e.

$$y_i = \frac{\sum_{j=1}^{M} b_{ij}\eta_{ij}}{\sum_{j=1}^{M} \eta_{ij}} \tag{4.51}$$

where y_i is the output signal, a continuous signal that can take any value between the smallest and largest output membership function peak positions. Another defuzzification method is **center of area (CoA)** also known as the **center of gravity (CoG)**. In this case, the respective output membership function is "chopped-off" at the level equal to the implied membership value η_{ij}. In this case, the area of the "chopped-off" (Figure 4.16) membership function A_{ij} is used as a weighting coefficient

$$y_i = \frac{\sum_{j=1}^{M} b_{ij}A_{ij}}{\sum_{j=1}^{M} A_{ij}} \tag{4.52}$$

Yet another method of defuzzification is **mean of maximum (MoM)**. In this method, the output is chosen to be the one corresponding to the highest membership value, i.e. the b_{ij} corresponding to the largest η_{ij}. MoM is often used in managerial decision-making systems and not very often in signal processing and control systems. CoG requires more computational power than CoM, hence CoG is rarely used while CoM is common.

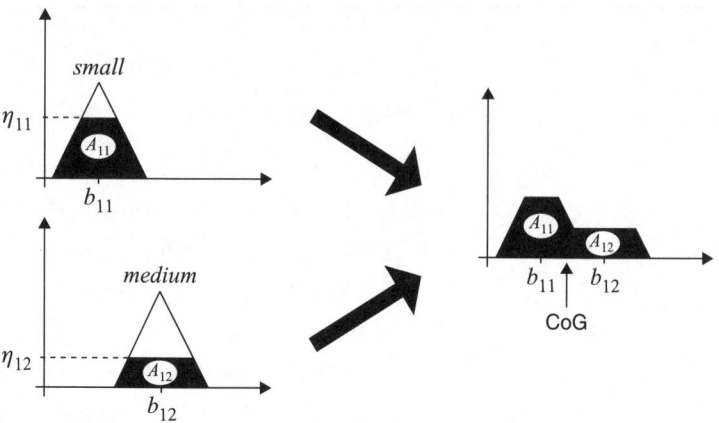

Figure 4.16 *Defuzzification using CoG*

Let us use CoM to complete our example. There are two continuous output signals from our cruise control, *accelerate* and *brake*. Assume *accelerate* is denoted as y_3 and *brake* as y_4. Using equation (4.51) it is straightforward to derive the defuzzification process

$$
\begin{aligned}
y_3 &= \frac{\sum_{j=1}^{5} b_{3j}\eta_{3j}}{\sum_{j=1}^{5} \eta_{3j}} \\
&= \frac{-0.25\eta_{31} + 0.75\eta_{32} + 0.75\eta_{33} + 0.25\eta_{34} + 0.25\eta_{35}}{\eta_{31} + \eta_{32} + \eta_{33} + \eta_{34} + \eta_{35}} \\
&= \frac{0.75(\eta_{32} + \eta_{33}) + 0.25(\eta_{34} + \eta_{35} - \eta_{31})}{\eta_{31} + \eta_{32} + \eta_{33} + \eta_{34} + \eta_{35}}
\end{aligned}
\tag{4.53}
$$

$$
\begin{aligned}
y_4 &= \frac{\sum_{j=1}^{7} b_{4j}\eta_{4j}}{\sum_{j=1}^{7} \eta_{4j}} \\
&= \frac{0\eta_{41} + 0.25\eta_{42} + 0\eta_{43} + 0.25\eta_{44} + 0\eta_{45} + 0.75\eta_{46} + 0.25\eta_{47}}{\eta_{41} + \eta_{42} + \eta_{43} + \eta_{44} + \eta_{45} + \eta_{46} + \eta_{47}} \\
&= \frac{0.75\eta_{46} + 0.25(\eta_{42} + \eta_{44} + \eta_{47})}{\eta_{41} + \eta_{42} + \eta_{43} + \eta_{44} + \eta_{45} + \eta_{46} + \eta_{47}}
\end{aligned}
\tag{4.54}
$$

In Figure 4.17 plots of y_3 and y_4 are shown. It should be remembered, however, that this example is simplified and that there are many alternative ways of designing fuzzy systems.

4.3.5 Applications

Fuzzy logic is used in some signal and image processing systems and in image identification and classifying systems. Most fuzzy systems today are, however, control systems. Fuzzy regulators are suitable for applications where

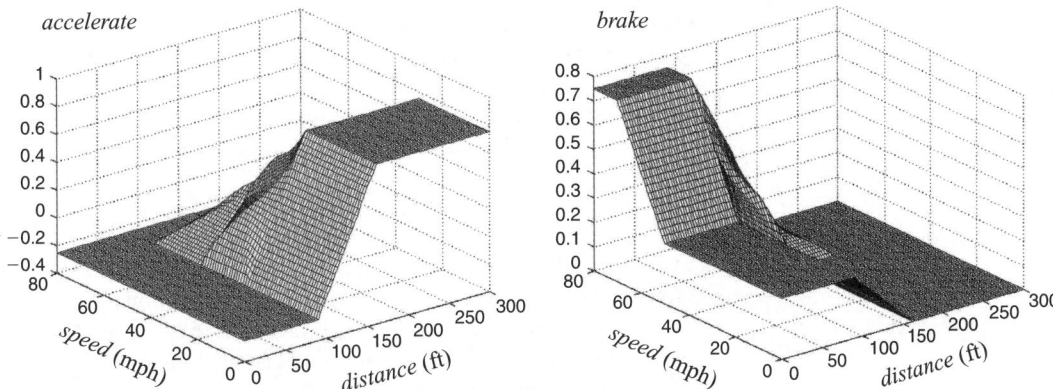

Figure 4.17 *Output signals* accelerate *and* brake *as functions of input signals* speed *and* distance

mathematical models are hard or impossible to formulate. The process subject to control may be time varying or strongly non-linear, requiring elaborate theoretical work to be understood. Another suitable situation arises when there is an abundant amount of practical knowledge from manual control such that experience can be formulated in natural language rather than mathematical algorithms. It is common that people having limited knowledge in control theory find fuzzy control systems easier to understand than traditional control systems.

One drawback that exists is that there are no mathematical models available, and no computer simulations can be done. Numerous tests have to be performed in practice to prove performance and stability of a fuzzy control system under all conditions. A second drawback is that the rule database has a tendency to grow large, requiring fast processing hardware to be able to perform in real time.

Fuzzy controllers can be found not only in space ships but also in air conditioners, refrigerators, microwave ovens, automatic gearboxes, cameras (auto-focus), washing machines, copying machines, distilling equipment, industrial baking processes and many other everyday applications.

Summary

In this chapter the following main topics have been addressed:

- The median filter and threshold decomposition
- Neural network architectures: feedforward, layered networks, feedback and lateral feedback
- Neural network node models, hard limiter, soft limiter and sigmoid
- Unsupervised and supervised learning: Hebb's rule, the Widrof–Hoff rule, the perceptron learning rule and back-propagation
- Applications: pattern recognition, classification, pattern associator, content addressable memory regularity detectors and optimization
- The Boltzmann machine computational temperature and simulated annealing
- Fuzzy control systems.

Review questions

R4-1 What are the main advantages of a median filter?
R4-2 What are the main features of feedforward and feedback neural networks?

R4-3 Draw a block diagram of a feedforward neural network node and explain the terms: weight, bias, activation function, hard limiter, soft limiter and sigmoid.

R4-4 Why are many feedforward neural networks partitioned into layers? What limitations does the layer structure have on the shape of the decision regions?

R4-5 Give a brief description of the training algorithms: Hebb's rule, the Widrow–Hoff rule, the perceptron learning rule and the back-propagation algorithm. In what type of networks are the respective algorithms usable?

R4-6 For feedback neural networks used in optimization, local minimum in the energy function may present a problem. Explain why. How can simulated annealing ease this problem? What is the drawback?

R4-7 Draw a block diagram of a simple fuzzy logic system. What are the tasks of the four blocks? Explain the terms: linguistic variable, linguistic value, membership function, degree of membership, crisp set, center of maximum, center of area and mean of maximum.

R4-8 In which applications can fuzzy systems be used? Pros and cons?

Solved problems

P4-1 Write a program in MATLAB^{TM} implementing a median filter and plot the input X and output Y vectors for filter lengths $N = 3, 5, 7$. Use the input signal

```
X = [0 0 0 0 1 2 3 4 4 4 3 -8 1 0 0 1 1 0 0 0
      1 1 1 0 0 0 0].
```

Explain the obtained results.

P4-2 In the text in Section 4.2.4.2, it is claimed that the XOR problem can be solved with only **one** feedforward neural network node using proper preprocessing of the input signals (see equation (4.17)). Draw a diagram of such a node, and make a truth table to prove that it really works.

P4-3 For Hopfield networks the energy function (4.30) can be approximated by equation (4.32), since the network uses a hard limiter activation function. Show that the sigmoid function actually can be approximated by a hard limiter for small computational temperatures. What are the parameters a and b in such a case?

5 Spectral analysis and modulation

Background In many cases when analyzing signals or trying to find features of signals, transformation to the frequency plane is advantageous, i.e. dealing with, for instance, kilohertz and megahertz rather than milliseconds and microseconds. In this chapter, some common methods for spectral analysis of temporal signals will be presented.

Another topic addressed in this chapter is modulation. Quite often in analog and digital telecommunication systems, the information signals cannot be transmitted as is. They have to be encoded (modulated) onto a carrier signal, suited for the transmission media in question. Some common modulation methods are demonstrated.

Objectives In this chapter we will discuss:

- Discrete Fourier transform (DFT) and fast Fourier transform (FFT)
- Windowing techniques, spectrum estimation and periodogram averaging
- Auto-correlation, cross-correlation, auto-covariance and cross-covariance
- Parametric spectrum analysis, auto-regressive (AR), moving average (MA) and auto-regressive moving average (ARMA) models
- Wavelet analysis
- Amplitude shift keying (ASK), frequency shift keying (FSK) and phase shift keying (PSK)
- Phasors, complex modulation and in phase/quadrature phase (I/Q)-modulators
- The Hilbert transform.

5.1 Discrete Fourier transform and fast Fourier transform

The **discrete Fourier transform (DFT)** (Burrus and Parks, 1985) from the time domain to the frequency domain representation, is derived from the time DFT

$$X(\omega) = \sum_{n=-\infty}^{+\infty} x(n) \, e^{-j2\pi(\omega/\omega_s)n} \tag{5.1}$$

The spectrum produced using this transform is periodic with the sampling frequency ω_s and for real input signals $x(n)$, the spectrum always has "even" symmetry along the real axis and "odd" symmetry on the imaginary axis. In practice, we cannot calculate this sum, since it contains an infinite number of samples. This problem is only solved by taking a section of N samples of the sequence $x(n)$. To achieve this, $x(n)$ is multiplied by a **windowing sequence** $\Psi(n)$ obtaining the windowed input signal $x_N(n)$. Since multiplication in the time domain corresponds to convolution in the frequency domain, the time DFT of the windowed sequence will be

$$X_N(\omega) = X(\omega) * \Psi(\omega) \tag{5.2}$$

where $\Psi(\omega)$ is the spectrum of the windowing sequence $\Psi(n)$ and $*$ denotes convolution. The ideal windowing sequence would have a rectangular spectrum, distorting the desired spectrum as little as possible and avoiding spectral "leakage". Unfortunately, a rectangular frequency response is practically impossible to achieve, therefore we must settle for some compromise. For example, commonly used windowing sequences are Rectangular, Bartlett, Hanning, Hamming and Kaiser–Bessel windows (Oppenheimer and Schafer, 1975).

Now, let us assume we have chosen an appropriate windowing sequence. Next we must determine how many frequency points should be used for calculation of the transform in order to maintain a reasonable accuracy. There is no simple answer, but in most cases good results are obtained using as many equally spaced frequency points as the number of samples in the windowed input signal, i.e. N. Hence the spacing between the frequency points will be ω_s/N. Now, inserting this into equation (5.1), the DFT in its most common form can be derived

$$X_N\left(k\frac{\omega_s}{N}\right) = \sum_{n=0}^{N-1} x_N(n)\,\mathrm{e}^{-j2\pi(kn/N)} = \sum_{n=0}^{N-1} x_N(n)\,W_N^{kn} \tag{5.3}$$

where the **twiddle factor** is

$$W_N = \mathrm{e}^{-j(2\pi/N)} \tag{5.4}$$

Unfortunately, the number of complex computations needed to perform the DFT is proportional to N^2. The acronym FFT (**fast Fourier transform**), refers to a group of algorithms, all very similar, which uses fewer computational steps to efficiently compute the DFT. The number of steps are typically proportional to $N\,\mathrm{lb}(N)$, where $\mathrm{lb}(x) = \log_2(x)$ is the logarithm base 2. Reducing the number of computational steps is of course important if the transform has to be computed in a real-time system. Fewer steps implies faster processing time, hence higher sampling rates are possible. Now, there are essentially two tricks employed to obtain this "sped-up version" of DFT:

(1) When calculating the sum (equation (5.3)) for $k = 0, 1, 2, \ldots, N$, many complex multiplications are repeated. By doing the calculations in a "smarter" order, many calculated complex products can be stored and reused.
(2) Further, the twiddle factor (5.4) is periodic, and only N factors need to be computed. This can, of course, be done in advance and the values can be stored in a table.

Let us illustrate the ideas behind the FFT algorithm by using a simple example having $N = 4$. If we use the original DFT transform as in equation (5.3), the following computations are needed

$$k = 0: \; X_4(0) = x_4(0)W_4^0 + x_4(1)W_4^0 + x_4(2)W_4^0 + x_4(3)W_4^0$$
$$k = 1: \; X_4(1) = x_4(0)W_4^0 + x_4(1)W_4^1 + x_4(2)W_4^2 + x_4(3)W_4^3$$
$$k = 2: \; X_4(2) = x_4(0)W_4^0 + x_4(1)W_4^2 + x_4(2)W_4^4 + x_4(3)W_4^6 \tag{5.5}$$
$$k = 3: \; X_4(3) = x_4(0)W_4^0 + x_4(1)W_4^3 + x_4(2)W_4^6 + x_4(3)W_4^9$$

As can be seen from equation (5.5), 16 complex multiplications and 12 complex additions are needed. Now, let us see how these numbers may be reduced. Firstly, if we put the odd and even numbered terms in two groups and divide the odd terms by W_4^k, equation (5.5) can be rewritten as

$$X_4(0) = \left(x_4(0)W_4^0 + x_4(2)W_4^0\right) + W_4^0\left(x_4(1) + x_4(3)\right)$$
$$X_4(1) = \left(x_4(0)W_4^0 + x_4(2)W_4^2\right) + W_4^1\left(x_4(1) + x_4(3)W_4^2\right)$$
$$X_4(2) = \left(x_4(0)W_4^0 + x_4(2)W_4^4\right) + W_4^2\left(x_4(1) + x_4(3)W_4^4\right) \tag{5.6}$$
$$X_4(3) = \left(x_4(0)W_4^0 + x_4(2)W_4^6\right) + W_4^3\left(x_4(1) + x_4(3)W_4^6\right)$$

Secondly, we use the periodicity of the twiddle factor (5.4)

$$W_4^4 = W_4^0, \quad W_4^6 = W_4^2$$

further, we know that $W_4^0 = 1$. Inserting this into equation (5.6), we obtain

$$X_4(0) = \left(x_4(0) + x_4(2)\right) + \left(x_4(1) + x_4(3)\right) = A + C$$
$$X_4(1) = \left(x_4(0) + x_4(2)W_4^2\right) + W_4^1\left(x_4(1) + x_4(3)W_4^2\right) = B + W_4^1 D$$
$$X_4(2) = \left(x_4(0) + x_4(2)\right) + W_4^2\left(x_4(1) + x_4(3)\right) = A + W_4^2 C \tag{5.7}$$
$$X_4(3) = \left(x_4(0) + x_4(2)W_4^2\right) + W_4^3\left(x_4(1) + x_4(3)W_4^2\right) = B + W_4^3 D$$

From equation (5.7) we can now see that our computations can be executed in two steps. In step one we calculate the terms A–D, and in the second step, we calculate the transform values $X_4(k)$, hence step 1

$$A = x_4(0) + x_4(2)$$
$$B = x_4(0) + x_4(2)W_4^2 = x_4(0) - x_4(2)$$
$$C = x_4(1) + x_4(3) \tag{5.8}$$
$$D = x_4(1) + x_4(3)W_4^2 = x_4(1) - x_4(3)$$

This step requires two complex additions and two complex subtractions. If all input signals are real, we only need real additions and subtractions

$$X_4(0) = A + C$$
$$X_4(1) = B + W_4^1 D$$
$$X_4(2) = A + W_4^2 C = A - C \tag{5.9}$$
$$X_4(3) = B + W_4^3 D = B - W_4^1 D$$

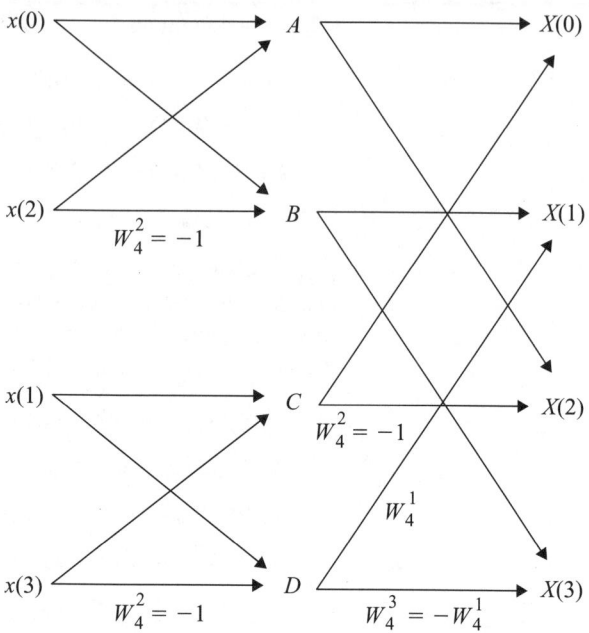

Figure 5.1 *FFT Butterfly signal flow diagram, showing the example having* N = 4

Two complex multiplications, two complex additions and two subtractions are needed. In total, we now have two complex multiplications and eight complex additions (subtractions). This is a considerable savings compared to the 16 multiplications and 12 additions required for computing the DFT in its original form.

The steps of the FFT, in our example (equations (5.8) and (5.9)), are often described in signal flow chart form, denoted "FFT butterflies" (see Figure 5.1). For the general case, the FFT strategy can be expressed as

$$X_N \left(k \frac{\omega_s}{N} \right) = \sum_{n=0}^{N-1} x_N(n) W_N^{kn} = \sum_{n=0}^{(N/2)-1} x_N(2n) W_{N/2}^{kn}$$

$$+ W_N^k \sum_{n=0}^{(N/2)-1} x_N(2n+1) W_{N/2}^{kn} \tag{5.10}$$

and

$$W_N^{kn} = W_N^{(kn+jN)} \quad \text{where } j = 1, 2, \ldots \tag{5.11}$$

5.2 Spectral analysis

Spectral analysis, by estimation of the **power spectrum** or **spectral power density** of a deterministic or random signal, involves performing a squaring function. Obtaining a good estimate of the spectrum, i.e. the signal power contents as a function of the frequency, is not entirely easy in practice. The main problem is that in most cases we only have access to a limited set of samples of the signal; in another situation, we are forced to limit the number of samples in order to be able to perform the calculations in a reasonable time. These limitations introduce errors in the estimate. If the signal is a random-type signal, we may also obtain large fluctuations in power estimates based on samples from different populations. Hence, in the general case there is no way to obtain the **true** power spectrum of a signal unless we are allowed to observe it infinitely. This is why the term **estimation** is frequently used in this context.

5.2.1 Discrete Fourier transform and fast Fourier transform approaches

Spectral analysis using the Fourier transform, a non-parametric method, was originally proposed in 1807 by the French mathematician and Baron J.B.J. Fourier. The discrete version of this transform, commonly used today in digital signal processing (DSP) applications, is called **discrete Fourier transform (DFT)**. A smart algorithm for calculating the DFT, causing less computational load for a digital computer, is the **fast Fourier transform (FFT)**. From a purely mathematical point of view, DFT and FFT do the same job as shown above. Considering a sampled signal $x(n)$ for $-\infty < n < \infty$ further, assume that the signal has finite energy, i.e.

$$\sum_{n=-\infty}^{\infty} x^2(n) < \infty \tag{5.12}$$

As seen above, the DFT of the signal can then be expressed as

$$X(\omega) = \sum_{n=-\infty}^{\infty} x(n)\,e^{-j2\pi(\omega/\omega_s)n} \tag{5.13}$$

where ω_s is the sampling frequency. By Parseval's relation, the energy of the signal can then be expressed as

$$\sum_{n=-\infty}^{\infty} |x(n)|^2 = \frac{1}{2\pi}\int_{-\pi}^{\pi} |X(\omega)|^2\,d\omega = \frac{1}{2\pi}\int_{-\pi}^{\pi} S_{xx}(\omega)\,d\omega \tag{5.14}$$

where $S_{xx}(\omega)$ is referred to as the **spectral density** of the signal $x(n)$. The spectral density is hence the squared magnitude of the Fourier transform of the signal. As mentioned above, in most cases the signal is not defined for $-\infty < n < \infty$, or we may not be able to handle an infinite length sequence in a practical application. In such a case, the signal is assumed to be non-zero only for $n = 0, 1, \ldots, N-1$ and assumed to be zero otherwise. The spectral density is then written as

$$S_{xx}(\omega) = |X(\omega)|^2 = \left| \sum_{n=0}^{N-1} x(n)\,e^{-j2\pi(\omega/\omega_s)n} \right|^2 \tag{5.15}$$

The power of the signal as a function of the frequency is called a **periodogram**, which A. Schuster defined in 1898 as

$$P(\omega) = \frac{1}{N} S_{xx}(\omega) \tag{5.16}$$

In the case that the signal $x(n)$ exists over the entire interval $(-\infty, \infty)$ and its energy is infinite (e.g. sinusoidal or other periodic signals), it is convenient to define the spectral density as (Mitra and Kaiser, 1993)

$$\Phi_{xx}(\omega) = \lim_{N \to \infty} \frac{1}{2N+1} \left| \sum_{n=-N}^{N} x(n) \, e^{-j2\pi(\omega/\omega_s)n} \right|^2 \tag{5.17}$$

Now, in many practical cases, to use an FFT algorithm, N is chosen to be a power of 2 (e.g. 64, 128, 256, ...). This means that the spectral density can only be calculated at N discrete points, which in many cases turns out to be too sparse. Quite often, we need a finer frequency spacing, i.e. we need to know the spectrum at L points where $L > N$. This can be accomplished by **zero padding**, so that the data sequence consisting of N samples is extended by adding $L - N$ zero value samples to the end of the sequence. The "new" L point sequence is then transformed using DFT or FFT. The effective frequency spacing is now

$$\frac{2\pi}{L\omega_s} < \frac{2\pi}{N\omega_s} \tag{5.18}$$

It is important to remember that zero padding **does not** increase the true resolution in frequency, it merely **interpolates** the spectrum at more points. However, for many applications, this is sufficient. So far, we have assumed that the signal $x(n)$ has been identical to zero outside the interval $n = 0, 1, \ldots, N - 1$. This is equivalent to multiplying the sequence $x(n)$ with a rectangular window sequence

$$s(n) = x(n) \, w(n) \tag{5.19}$$

where the **rectangular window** can be expressed as

$$w(n) = \begin{cases} 1 & \text{for } 0 \leq n < N \\ 0 & \text{else} \end{cases} \tag{5.20}$$

We recall from basic signal theory that a multiplication in the time domain like equation (5.19) corresponds to a convolution in the frequency domain as in equation (5.2). Therefore, the result of the windowing process has the "true" spectrum of the signal convoluted with the Fourier transform of the window sequence. What we obtain is not the spectrum of the signal, but rather the spectrum of the signal $x(n)$ **distorted** by the transform of the window sequence. This distortion results in "smearing" of the spectrum, which implies that narrow spectral peaks can neither be detected nor distinguished from each other (Mitra and Kasier, 1993; Lynn and Fuerst, 1998). This "leakage" is an inherent limitation when using conventional Fourier techniques for spectral analysis. The only way to reduce this effect is to observe the signal for a longer duration, i.e. gather more samples. The quality of the spectral estimate can, however, be somewhat improved by using a more "smooth" window sequence than the quite "crude" rectangular window. Some examples of windows with different distortion properties (Oppenheimer and Schafer, 1975; Mitra and Kaiser, 1993) are given below:

- **The Bartlett window (triangular window)**

$$w(n) = \begin{cases} 2n/(N-1) & 0 \leq n < (N-1)/2 \\ 2 - (2n/(N-1)) & (N-1)/2 \leq n < N \end{cases} \tag{5.21}$$

- **The Hann window**

$$w(n) = \frac{1}{2}\left(1 - \cos\left(\frac{2n\pi}{N-1}\right)\right) \quad 0 \le n < N \tag{5.22}$$

- **The Hamming window**

$$w(n) = 0.54 - 0.46\cos\left(\frac{2n\pi}{N-1}\right) \quad 0 \le n < N \tag{5.23}$$

- **The Blackman window**

$$w(n) = 0.42 - 0.5\cos\left(\frac{2n\pi}{N-1}\right) + 0.08\cos\left(\frac{4n\pi}{N-1}\right) \quad 0 \le n < N \tag{5.24}$$

5.2.2 Using the auto-correlation function

In the previous section, we concluded that a spectral power estimate could be obtained by taking the square of the magnitude of the Fourier transform (5.15) of the signal $x(n)$. There is an alternative way of achieving the same result by using the **auto-correlation function** of the signal $x(n)$. Let us start this section by a brief discussion on correlation functions (Schwartz and Shaw 1975; Denbigh 1998; Papoulis and Pillai 2001) of time discrete signals. Assume that the mean of the two complex signal sequences $x(n)$ and $y(n)$ are zero

$$E[x(n)] = \eta_x = 0 \quad \text{and} \quad E[y(n)] = \eta_y = 0$$

The **cross-correlation** between these signals are then defined as

$$R_{xy}(n, m) = E[x(n)\, y^*(m)] \tag{5.25}$$

where $*$ denotes complex conjugate and E[ ] is the expected mean operator (ensemble average). If the mean of the signals is not zero, the **cross-covariance** is

$$C_{xy}(n, m) = R_{xy}(n, m) - \eta_x(n)\, \eta_y^*(m) \tag{5.26}$$

Note! If the mean is zero, the cross-correlation and cross-covariance are equal. If the signal is correlated by itself, **auto-correlation** results

$$R_{xx}(n, m) = E[x(n)\, x^*(m)] \tag{5.27}$$

As a consequence of above, the **auto-covariance** is defined as

$$C_{xx}(n, m) = R_{xx}(n, m) - \eta_x(n)\, \eta_x^*(m) \tag{5.28}$$

If the signal $x(n)$ has zero mean and the auto-correlation only depends on the **difference** $k = m - n$, not on the actual values of n or m themselves, the signal $x(n)$ is said to be **wide-sense stationary** from a statistical point of view. This basically means that regardless of which point in the sequence the statistical properties of the signal is studied, they all turn out to be similar. For a stationary signal (which is a common assumption with signals), the **auto-correlation function** is then defined as

$$R_{xx}(k) = E[x(n)x^*(n+k)] \tag{5.29}$$

Note in particular that $R_{xx}(0) = \sigma^2$ is the **variance** of the signal and represents the power of the signal. Now, starting from equation (5.17), we study the square magnitude of the Fourier-transformed signal sequence $x(n)$ as $N \to \infty$

$$\left| \sum_{n=-\infty}^{\infty} x(n) e^{-j2\pi(\omega/\omega_s)n} \right|^2$$

$$= \left(\sum_{n=-\infty}^{\infty} x(n) e^{-j2\pi(\omega/\omega_s)n} \right) \left(\sum_{n=-\infty}^{\infty} x(n) e^{-j2\pi(\omega/\omega_s)n} \right)^*$$

$$= \sum_{n=-\infty}^{\infty} \sum_{m=-\infty}^{\infty} x(n) x^*(m) e^{-j2\pi(\omega/\omega_s)(m-n)}$$

$$= \sum_{n=-\infty}^{\infty} \sum_{k=-\infty}^{\infty} x(n) x^*(n+k) e^{-j2\pi(\omega/\omega_s)k}$$

$$= \sum_{k=-\infty}^{\infty} R_{xx}(k) e^{-j2\pi(\omega/\omega_s)k} \tag{5.30}$$

Hence, we now realize that the spectral density can be obtained in an **alternative way**, namely by taking the DFT of the auto-correlation function

$$\Phi_{xx}(\omega) = \sum_{k=-\infty}^{\infty} R_{xx}(k) e^{-j2\pi(\omega/\omega_s)k} \tag{5.31}$$

This relationship is known as the Wiener–Khintchine theorem.

5.2.3 Periodogram averaging

When trying to obtain spectral power estimates of stochastic signals using a limited number of samples, quite poor estimates often result. This is especially true near the ends of the sample window, where the calculations are based on few samples of the signal $x(n)$. These poor estimates cause wild fluctuations in the periodogram. This trend was already observed by A. Schuster in 1898. To get smoother power spectral density estimates, many independent periodograms have to be averaged. This was first studied by M.S. Bartlett and later by P.D. Welch.

Assume that $x(n)$ is a stochastic signal being observed in L points, i.e. for $n = 0, 1, \ldots, L-1$. This is the same as multiplying the signal $x(n)$ by a rectangular window $w(n)$, being non-zero for $n = 0, 1, \ldots, L-1$ (see equation (5.20)).

Using the DFT (5.13) in this product we obtain

$$S(\omega) = \sum_{n=0}^{L-1} x(n) \, w(n) \, \mathrm{e}^{-j2\pi(\omega/\omega_s)n} \tag{5.32}$$

We now consider an estimate of the power spectral density (Mitra and Kaiser, 1993) given by

$$\hat{I}_{xx}(\omega) = \frac{1}{LU} |S(\omega)|^2 \tag{5.33}$$

where U is a normalizing factor to remove any bias caused by the window $w(n)$

$$U = \frac{1}{L} \sum_{n=0}^{L-1} |w(n)|^2 \tag{5.34}$$

The entity $\hat{I}_{xx}(\omega)$ is denoted as a **periodogram** if the window used is the rectangular window, otherwise it is called a **modified periodogram**. Now, **Welch's method** computes periodograms of overlapping segments of data and averages (Mitra and Kaiser, 1993) them. Assume that we have the access to Q contiguous samples of $x(n)$ and that $Q > L$. We divide the data set into P segments of L samples each. Let S be the shift between successive segments, then

$$Q = (P - 1)S + L \tag{5.35}$$

Hence, the "windowed" segment p is then expressed as

$$s^{(p)}(n) = w(n) \, x(n + pS) \tag{5.36}$$

for $0 \leq n < L$ and $0 \leq p < P$. Inserting equation (5.35) into equation (5.32) we obtain the DFT of segment p

$$S^{(p)}(\omega) = \sum_{n=0}^{L-1} s^{(p)}(n) \, \mathrm{e}^{-j2\pi(\omega/\omega_s)n} \tag{5.37}$$

Using equation (5.33), the periodogram of segment p can be calculated

$$\hat{I}_{xx}^{(p)}(\omega) = \frac{1}{LU} \left| S^{(p)}(\omega) \right|^2 \tag{5.38}$$

Finally, the **Welch estimate** is the average of all the periodograms over all segments p as above

$$\hat{I}_{xx}^{W}(\omega) = \frac{1}{P} \sum_{p=0}^{P-1} \hat{I}_{xx}^{(p)}(\omega) \tag{5.39}$$

For the Welch estimate, the bias and variance asymptotically approach zero as Q increases. For a given Q, we should choose L as large as possible to obtain the best resolution, but on the other hand, to obtain a smooth estimate, P should be large, implying L to be small (equation (5.35)). Hence, there is a **trade-off** between high resolution in frequency (large L) and smooth spectral density estimate (small L). Increasing Q is of course always good, but requires longer data acquisition time and more computational power.

The **Bartlett periodogram** is a special case of Welch's method as described above. In Bartlett's method, the segments are **non-overlapping**, i.e. $S = L$ in equation (5.35). The same trade-off described above for Welch's method also applies to Bartlett's method. In terms of variance in the estimate, Welch's method often performs better than Bartlett's method. Implementing Bartlett's method may however in some cases be somewhat easier in practice.

5.2.4 Parametric spectrum analysis

Non-parametric spectral density estimation methods like the Fourier analysis described above are well studied and established. Unfortunately, this class of methods has some drawbacks. When the available set of N samples is small, resolution in frequency is severely limited. Also auto-correlation outside the sample set is considered to be zero, i.e. $R_{xx}(k) = 0$ for $k > N$, which may be an unrealistic assumption in many cases.

Fourier-based methods assume that data outside the observed window is either periodic or zero. The estimate is not only an estimate of the observed N data samples, but also an estimate of the "unknown" data samples outside the windows, under the assumptions above. Alternative estimation methods can be found in the class of **model-based**, **parametric spectrum analysis methods** (Mitra and Kaiser, 1993). Some of the most common methods will be presented briefly below.

The underlying idea of assuming the signal to be analyzed can be generated using a **model filter**. The filter has the causal transfer function $H(z)$ and a white noise input signal $e(n)$, with mean value zero and variance σ_e^2. In this case, the output signal $x(n)$ from the filter is a wide-sense stationary signal with power density

$$\Phi_{xx}(\omega) = \Phi_e(\omega)|H(\omega)|^2 = \sigma_e^2|H(\omega)|^2 \tag{5.40}$$

Hence, the power density can be obtained if the model filter transfer function $H(z)$ is known, i.e. if the model type and its associated filter **parameters** are known. The parametric spectrum analysis can be divided into three steps:

- Selecting an appropriate model and selecting the order of $H(z)$. There are often many different possibilities
- Estimating the filter coefficients, i.e. the parameters from the N data samples $x(n)$ where $n = 0, 1, \ldots, N-1$
- Evaluating the power density, as in equation (5.40) above.

Selecting a model is often easier if some prior knowledge of the signal is available. Different models may give more or less accurate results, but may also be more or less computationally demanding. A common model type is the **auto-regressive moving average (ARMA)** model

$$H(z) = \frac{B(z)}{A(z)} = \frac{\sum_{k=0}^{q} b_k z^{-k}}{1 + \sum_{k=1}^{p} a_k z^{-k}} = \sum_{n=0}^{\infty} h(n) z^{-n} \tag{5.41}$$

where $h(n)$ is the impulse response of the filter and the denominator polynomial $A(z)$ has all its roots inside the unit circle, for the filter to be stable. The ARMA model may be simplified. If $q = 0$, then $B(z) = 1$ and an all-pole filter results, yielding the **auto-regressive (AR)** model

$$H(z) = \frac{1}{1 + \sum_{k=1}^{p} a_k z^{-k}} \tag{5.42}$$

or, if $p = 0$, then $A(z) = 1$ and a filter having only zeros results, a so-called **moving average (MA)** model

$$H(z) = \sum_{k=0}^{q} b_k z^{-k} \tag{5.43}$$

Note, these models are mainly infinite impulse response (IIR) and finite impulse response (FIR) filters (see also Chapter 1). Once a reasonable model is chosen, the next measure is estimating the filter parameters based on the available data samples. This can be done in a number of ways. One way is by using the auto-correlation properties of the data sequence $x(n)$. Assuming an ARMA model as in equation (5.41), the corresponding difference equation (see Chapter 1) can be written as

$$x(n) = -\sum_{k=1}^{p} a_k\, x(n-k) + \sum_{k=0}^{q} b_k\, e(n-k) \tag{5.44}$$

where $e(n)$ is the white noise input sequence. Multiplying equation (5.44) by $x^*(n+m)$ and taking the expected value we obtain

$$R_{xx}(m) = -\sum_{k=1}^{p} a_k\, R_{xx}(m-k) + \sum_{k=0}^{q} b_k\, R_{ex}(m-k) \tag{5.45}$$

where the auto-correlation of $x(n)$ is

$$R_{xx}(m) = \mathrm{E}[x(n)\, x^*(n+m)] \tag{5.46}$$

and the cross-correlation between $x(n)$ and $e(n)$ is

$$R_{ex}(m) = \mathrm{E}[e(n)\, x^*(n+m)] \tag{5.47}$$

Since $x(n)$ is the convolution of $h(n)$ by $e(n)$, $x^*(n+m)$ can be expressed as

$$x^*(n+m) = \sum_{k=0}^{\infty} h^*(k)\, e(n+m-k) \tag{5.48}$$

Inserting into equation (5.47), the cross-correlation can hence be rewritten as

$$R_{ex}(m) = \mathrm{E}\left[e(n) \sum_{k=0}^{\infty} h^*(k)\, e(n+m-k) \right]$$

$$= \sum_{k=0}^{\infty} h^*(k)\, \mathrm{E}[e(n)\, e(n+m-k)] = \sigma_e^2 h^*(-m) \tag{5.49}$$

where we have used the facts that $e(n)$ is a white noise signal, and $h(n)$ is causal, i.e. zero for $n < 0$. We can now express the auto-correlation values for the signal originating from the ARMA model, in terms of the model parameters

$$R_{xx}(m) = \begin{cases} -\sum_{k=1}^{p} a_k R_{xx}(m-k) & \text{for } m \geq q \\ -\sum_{k=1}^{p} a_k R_{xx}(m-k) + \sigma_e^2 \sum_{k=0}^{q-m} h^*(k) b_{k+m} & \text{for } 0 \leq m < q \\ R_{xx}^*(-m) & \text{for } m < 0 \end{cases}$$

(5.50)

The auto-correlation values $R_{xx}(m)$ for $|m| \geq q$ are extrapolated by the filter parameters and the values of $R_{xx}(m)$ for $m = 0, 1, \ldots, q$. Now, for the sake of simplicity, assume that we are using an AR model, i.e. $q = 0$. Then equation (5.50) simplifies to

$$R_{xx}(m) = \begin{cases} -\sum_{k=1}^{p} a_k R_{xx}(m-k) & \text{for } m \geq 0 \\ -\sum_{k=1}^{p} a_k R_{xx}(m-k) + \sigma_e^2 & \text{for } 0 = m \\ R_{xx}^*(-m) & \text{for } m < 0 \end{cases}$$

(5.51)

Thus, if the auto-correlation values $R_{xx}(0)$, $R_{xx}(1)$, $\ldots, R_{xx}(p)$ are known, the filter parameters can be found by solving p linear equations corresponding to $m = 0, 1, \ldots, p$. Note that we only need to know $p + 1$ correlation values to be able to determine all the parameters. From equation (5.51), setting $m = 0$ we can also obtain the variance

$$\sigma_e^2 = R_{xx}(0) + \sum_{k=1}^{p} a_k R_{xx}(-k)$$

(5.52)

Combining equations (5.51) and (5.52), we thus have to solve the following equations to obtain the parameters. These equations are known as the **Yule–Walker equations**

$$\begin{aligned} R_{xx}(0) + a_1 R_{xx}(-1) + \cdots + a_p R_{xx}(-p) &= \sigma_e^2 \\ R_{xx}(1) + a_1 R_{xx}(0) + \cdots + a_p R_{xx}(-p+1) &= 0 \\ \vdots \qquad\qquad \vdots \qquad\qquad \vdots \\ R_{xx}(p) + a_1 R_{xx}(p-1) + \cdots + a_p R_{xx}(0) &= 0 \end{aligned}$$

(5.53)

For the ARMA model, the modified Yule–Walker equations can be used to determine the filter parameters. The Yule–Walker equations can be solved using standard techniques, for instance Gauss elimination, which requires a number of operations proportional to p^3. There are however recursive, smarter algorithms that make use of the regular structure of the Yule–Walker equations, and hence require fewer operations. One such algorithm is the **Levinson–Durbin algorithm** (Mitra and Kaiser, 1993), requiring a number of operations proportional to p^2 only. There are also other ways to obtain the parameters. One such way is **adaptive modeling** using adaptive filters (Widrow and Stearns, 1985) (see Chapter 3).

Finally, when the filter parameters are estimated, the last step is to evaluate the spectral density. This may be achieved by inserting

$$z = e^{-j2\pi(\omega/\omega_s)} \tag{5.54}$$

into the transfer function $H(z)$ of the model, e.g. equation (5.41) and then, using the relation (5.40), evaluate the spectral density. It should however be pointed out that the calculations needed may be tiring, and that considerable error between the estimated and the true spectral density may occur. One has to remember that the spectral density is an approximation, based on the model used. Choosing an improper model for a specific signal may hence result in a poor spectral estimate.

The parametric analysis methods have other interesting features. In some applications, the parameters themselves are sufficient information about the signal. In speech-coding equipment, for instance, the parameters are transmitted to the receiver (RX) instead of the speech signal itself. Using a model filter and a noise source at the receiving site, the speech signal can be recreated. This technique is a type of data compression algorithm (see also Chapter 7), since transmitting the filter parameters requires less capacity then transmitting the signal itself.

5.2.5 Wavelet analysis

Wavelet analysis, also called **wavelet theory**, or just **wavelets** (Lee and Yamamoto, 1994; Bergh *et al.*, 1999), has attracted much attention recently. It has been used in transient analysis, image analysis and communication systems, and has been shown to be very useful for processing **non-stationary** signals.

Wavelet analysis deals with expansion of functions in terms of a set of **basis functions**, like Fourier analysis. Instead of trigonometric functions being used in Fourier analysis, wavelets are used. The objective of wavelet analysis is to define these wavelet basis functions. It can be shown that every application using the FFT can be reformulated using wavelets. In the latter case, more localized temporal and frequency information can be provided. Thus, instead of a conventional frequency spectrum, a "wavelet spectrum" may be obtained.

Wavelets are created by **translations** and **dilations** of a fixed function called the **mother wavelet**. Assume that $\Psi(t)$ is a complex function. If this function satisfies the following two conditions, it may be used as a mother wavelet

$$\int_{-\infty}^{\infty} |\Psi(t)|^2 \, dt < \infty \tag{5.55}$$

This expression implies finite energy of the function, the second condition is

$$c_\Psi = 2\pi \int_{-\infty}^{\infty} \frac{|\Psi(\omega)|^2}{\omega} \, d\omega < \infty \tag{5.56}$$

where $\Psi(\omega)$ is the Fourier transform of $\Psi(t)$. This condition implies that if $\Psi(\omega)$ is smooth, then $\Psi(0) = 0$. One example of a mother wavelet is the

Haar wavelet, which is historically the original and simplest wavelet

$$\Psi(t) = \begin{cases} 1 & \text{for } 0 < t \le 1/2 \\ -1 & \text{for } 1/2 < t \le 1 \\ 0 & \text{else} \end{cases} \tag{5.57}$$

Other examples of smooth wavelets having better frequency localization properties are the **Meyer** Wavelet, the **Morlet** Wavelet and the **Daubechies** Wavelet (Lee and Yamamoto, 1994; Bergh *et al.*, 1999). Figure 5.2 shows an example of a Morlet wavelet. The wavelets are obtained by translations and dilations of the mother wavelet

$$\Psi_{a,b}(t) = a^{-1/2}\,\Psi\!\left(\frac{t-b}{a}\right) \tag{5.58}$$

which means rescaling the mother wavelet by a and shifting in time by b, where $a > 0$ and $-\infty < b < \infty$.

The **wavelet transform** of the real signal $x(t)$ is the scalar product

$$X(b,a) = \int_{-\infty}^{\infty} \Psi_{a,b}^{*}(t)\,x(t)\,\mathrm{d}t \tag{5.59}$$

where * denotes the complex conjugate. There is also an inverse transform

$$x(t) = \frac{1}{c_\Psi} \int_{-\infty}^{\infty} \int_{0}^{\infty} X(b,a)\,\Psi_{a,b}(t)\,\frac{\mathrm{d}a\,\mathrm{d}b}{a^2} \tag{5.60}$$

where c_Ψ is obtained from equation (5.56). For the **discrete wavelet transform**, equations (5.58)–(5.60) are replaced by equations (5.61)–(5.63), respectively

$$\Psi_{m,n}(t) = a_0^{-m/2}\Psi\!\left(\frac{t-nb_0}{a_0^m}\right) \tag{5.61}$$

where m and n are integers and a_0 and b_0 are constants. Quite often a_0 is chosen as

$$a_0 = 2^{1/v}$$

where v is an integer and v pieces of $\Psi_{m,n}(t)$ are processed as one group called a **voice**

$$X_{m,n} = \int_{-\infty}^{\infty} \Psi_{m,n}^{*}(t)\,x(t)\,\mathrm{d}t \tag{5.62}$$

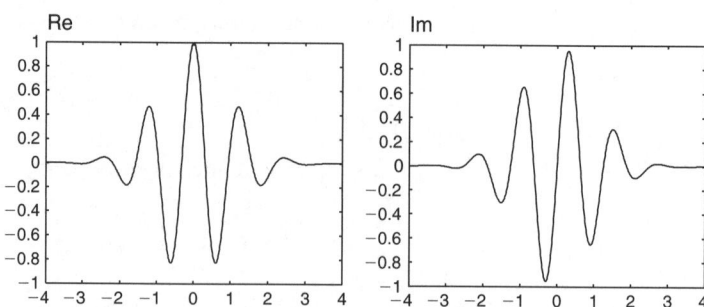

Figure 5.2 *Real and imaginary part of a **Morlet** wavelet*

$$x(t) = k_\Psi \sum_m \sum_n X_{m,n} \Psi_{m,n}(t) \tag{5.63}$$

Graphical representation of the complex functions obtained from equations (5.63) and (5.66) is not always entirely easy. Commonly, three-dimensional graphics, or gray-scale graphics showing the magnitude and phase as a function of a, are used. One example is Figure 5.3, showing the magnitude of a wavelet transform (Morlet) of a sinusoid with linearly decreasing frequency. Figure 5.4

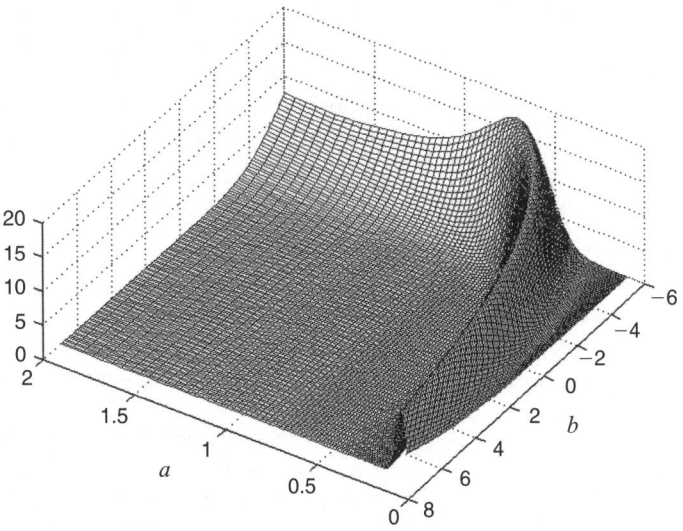

Figure 5.3 *Magnitude of a wavelet transform (Morlet) of a sinusoid with linearly decreasing frequency*

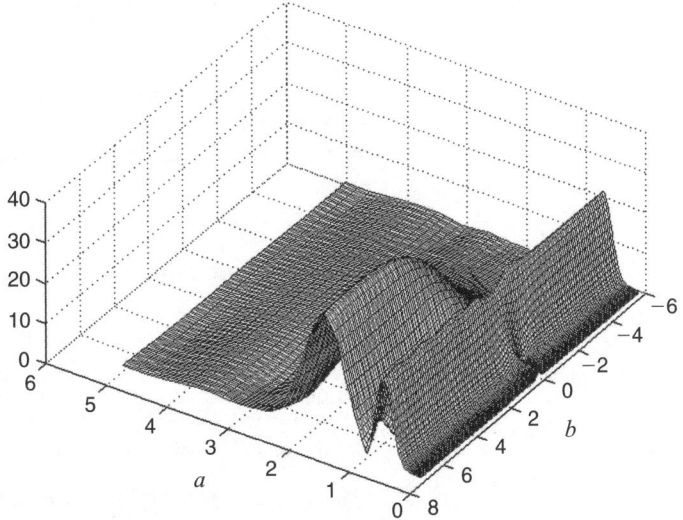

Figure 5.4 *A signal comprising of two sinusoids with different frequencies, where one frequency component is switched on with a time delay*

shows a signal comprising of two sinusoids with different frequencies, where one frequency component is switched on with a time delay.

5.3 Modulation

In many communication systems, especially in radio systems, the information ("digital" or "analog"), or the **baseband** signal to be transmitted is **modulated**, i.e. "encoded" onto a **carrier** signal. The carrier is chosen depending on the type of media available for the transmission, for instance, a limited band in the radio frequency spectrum, or an appropriate frequency for transmission over cables or fibers. If we assume that the carrier signal is a plain **cosine** signal there are three parameters of the carrier that can be modulated and used for information transfer. These parameters are the **amplitude** $a(t)$, the **frequency** $f(t)$ and the **phase** $\phi(t)$. The modulated carrier is then

$$s(t) = a(t) \cos\left(2\pi \int_0^t f(\tau)\,\mathrm{d}\tau + \phi(t) \right) \tag{5.64}$$

The corresponding modulation types are denoted **amplitude modulation (AM)**, **frequency modulation (FM)** and **phase modulation (PM)**. FM and PM are also referred to as **angular modulation** and are related in that the frequency is the phase changing speed; in other words, the derivative of the phase function is the frequency. Hence, frequency modulation can be achieved by first integrating the baseband signal and then feeding it into a phase modulator. This method is called Armstrong's indirect FM (Miller and Beasley, 2002).

In a **digital** radio communication system, it is common to modulate either by changing $f(t)$ between a number of discrete frequencies, **frequency shift keying (FSK)** or by changing $\phi(t)$ between a number of discrete phases, **phase shift keying (PSK)**. In some systems, the amplitude is also changed between discrete levels, **amplitude shift keying (ASK)**. So, if we are to design a modulation scheme which is able to transmit M different discrete symbols (for binary signals, $M = 2$), we hence have to define M unique combinations of amplitude, frequency and/or phase values of the carrier signal. The trick is to define these **signal points**, $a_n(t)$, $f_n(t)$ or $\phi_n(t)$, where $n \in \{0, \ldots, M-1\}$, in a way that communication speed can be high, influence of interference low, spectral occupancy low, and modulation and demodulation equipment can be made fairly simple.

5.3.1 Amplitude shift keying (ASK)

The concept of ASK is rather straightforward. Using this modulation method, the amplitude of the carrier can be one out of M given baseband amplitude functions $a_n(t)$, corresponding to the M information symbols used in the system.

$$s_n(t) = a_n(t) \cos(2\pi f_c t + \phi) \tag{5.65}$$

The carrier frequency $f(t) = f_c$ and the phase shift $\phi(t) = \phi$ are constant. The amplitude functions $a_n(t)$ are defined over a finite period of time, i.e. $t_0 \leq t < t_0 + T$, where T is the **symbol time**. A simple example would be a system using binary symbols, i.e. $M = 2$ and square pulses for amplitude functions

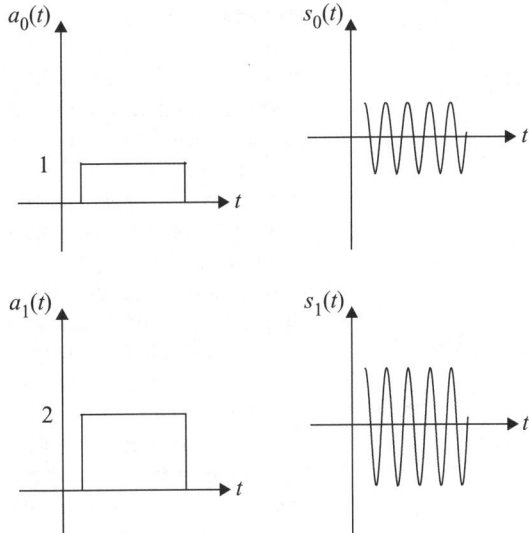

Figure 5.5 *Baseband signal and modulated signal for ASK, M = 2*

according to the below (see Figure 5.5)

symbol "0":

$$a_0(t) = \begin{cases} 1 & t_0 \le t < t_0 + T \\ 0 & \text{else} \end{cases}$$

$$\Rightarrow s_0(t) = \begin{cases} \cos(2\pi f_c t + \phi) & t_0 \le t < t_0 + T \\ 0 & \text{else} \end{cases} \tag{5.66a}$$

symbol "1":

$$a_1(t) = \begin{cases} 2 & t_0 \le t < t_0 + T \\ 0 & \text{else} \end{cases}$$

$$\Rightarrow s_1(t) = \begin{cases} 2\cos(2\pi f_c t + \phi) & t_0 \le t < t_0 + T \\ 0 & \text{else} \end{cases} \tag{5.66b}$$

The above can be regarded as a kind of pulse amplitude modulation (PAM) system. The data transfer capacity D of such a system is determined by equation (5.67)

$$D = \frac{1}{T}\,\text{lb}(M) = R\,\text{lb}(M)\,\text{bits/s} \tag{5.67}$$

where $\text{lb}(x) = \log_2(x)$ is the logarithm base 2 and R is the symbol rate. In the example above binary symbols were used, i.e. $M = 2$ which implies that $\text{lb}(2) = 1$, and that the data transfer capacity becomes $D = 1/T = R$. In other words, the symbol time equals the bit time. So one obvious recipe for achieving high data transfer capacity is to use a high bit rate R, i.e. a short bit (symbol) time T. Unfortunately, in a real world application, we also have to take the background noise and interference into account. Using a too short symbol time implies that the energy in each symbol will be small compared to the energy in the background noise. This, in turn, means that the receiver (demodulator), retrieving the encoded information, will be disturbed by the background noise in such a way that it makes frequent misinterpretations of the received signal. In other words, errors in the transmission will be common.

From equation (5.67) we find that another way of increasing the data transfer capacity is to increase the number of symbols M. In all real world applications, we only have a limited transmitter power (or voltage) available. Increasing the number of levels M means that the voltage (or power) **difference** between adjacent levels will decrease. Taking the background noise into account again, the probability of errors in the receiving process will thus increase. Hence, there is a limit to how many symbols M can be used in a transmission system, depending on the level of the background noise. These issues will be elaborated on further in Chapter 8.

Another problem with the amplitude functions in the example above is the shape of the waveform. A square pulse like this will have a quite wide-frequency spectrum, increasing the risk of causing interference to other users in adjacent frequency bands. For this reason, smoother waveforms than square pulses are commonly used.

5.3.2 Frequency shift keying (FSK)

In this case, the frequency of the signal is controlled by the baseband signal, i.e.

$$s(t) = a \cos\left(2\pi \int_0^t f_n(\tau)\, d\tau + \phi\right) \tag{5.68}$$

The amplitude $a(t) = a$ and the phase shift $\phi(t) = \phi$ are constant. The frequency functions $f_n(t)$ are defined over a finite period of time, i.e. $t_0 \leq t < t_0 + T$, where T is the symbol time, in a similar way to ASK above. A simple example would be a system using binary symbols, i.e. $M = 2$ and square pulses for frequency functions according to the below (see Figure 5.6). An FSK modulation scheme with $M = 2$ is sometimes referred to as a **binary frequency shift keying (BFSK)** system. Commonly, the shape of the frequency functions is chosen to be smoother than square pulses, for the spectral occupancy reasons mentioned above

symbol "0":

$$f_0(t) = \begin{cases} f_0 & t_0 \leq t < t_0 + T \\ 0 & \text{else} \end{cases}$$

$$\Rightarrow s_0(t) = \begin{cases} a \cos(2\pi f_0 t + \phi) & t_0 \leq t < t_0 + T \\ 0 & \text{else} \end{cases} \tag{5.69a}$$

symbol "1":

$$f_1(t) = \begin{cases} f_1 & t_0 \leq t < t_0 + T \\ 0 & \text{else} \end{cases}$$

$$\Rightarrow s_1(t) = \begin{cases} a \cos(2\pi f_1 t + \phi) & t_0 \leq t < t_0 + T \\ 0 & \text{else} \end{cases} \tag{5.69b}$$

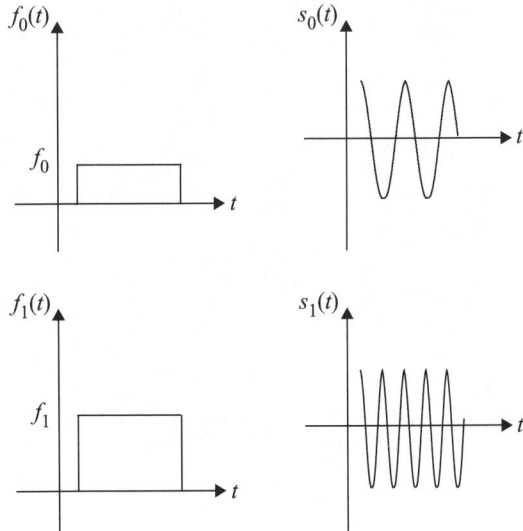

Figure 5.6 *Baseband signal and modulated signal for FSK, M = 2*

For the receiver to determine what symbol was transmitted, the frequency of the carrier $s_n(t)$ needs to be determined. This can be accomplished in two ways: non-coherent and coherent detection.

In **non-coherent** detection, the input signal enters a bank of bandpass filters, where there is a filter for each frequency used, or in other words, one filter for each symbol used in the system. The outputs of the filters are compared to each other, and the received symbol is supposed to be the one that corresponds to the filter having the strongest output signal magnitude. The phase of the received signal in relation to the phase of the transmitted signal does not matter; the method is non-coherent. This approach is straightforward and quite easy to implement.

In the **coherent** detection system, the phase of the signal does matter. Hence, phase synchronization between transmitter and receiver is needed, which complicates the system. On the other hand, a coherent system has a greater resistance against noise and interference than a non-coherent system. In a coherent system, detection (demodulation) is performed using a cross-correlation technique (see Figure 5.7). The received signal is cross-correlated with locally generated prototype signals $s'_n(t)$ corresponding to all the M symbols used in the system. Since coherence is needed, the phase of $s'_n(t)$ and the received signal $s_n(t)$ have to agree. The prototype signals for our example above would be

symbol "0":

$$s'_0(t) = \begin{cases} \cos(2\pi f_0 t + \phi) & t_0 \leq t < t_0 + T \\ 0 & \text{else} \end{cases} \tag{5.70a}$$

symbol "1":

$$s'_1(t) = \begin{cases} \cos(2\pi f_1 t + \phi) & t_0 \leq t < t_0 + T \\ 0 & \text{else} \end{cases} \tag{5.70b}$$

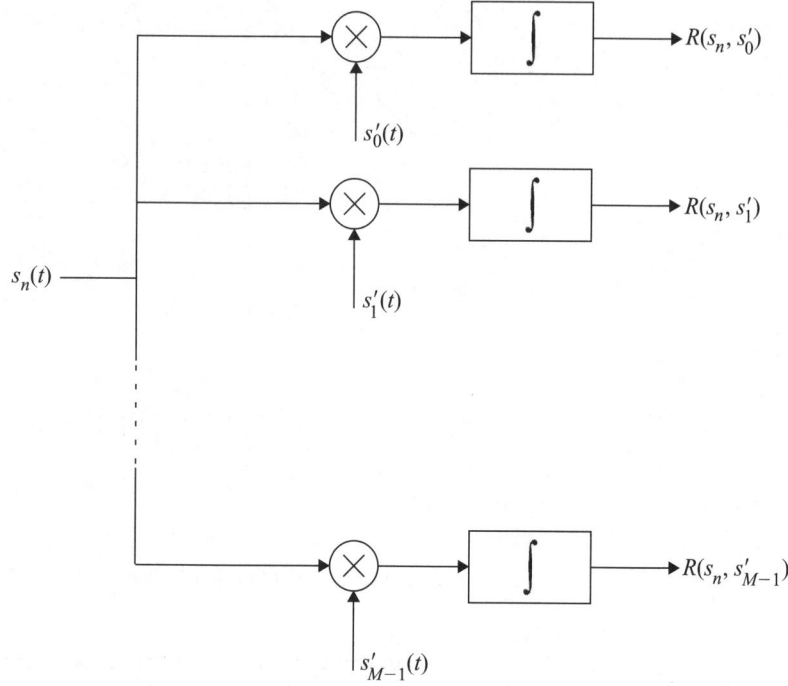

Figure 5.7 *Coherent detection of FSK signals using cross-correlation*

The cross-correlation is performed as $R_{s_n, s'_n}(\tau) = \int_0^T s_n(t)\, s'_n(t + \tau)\, dt$ where τ is the delay, i.e. phase shift between $s_n(t)$ and $s'_n(t)$. Since we have a coherent system, the phase shift between the two signals should ideally be equal to zero, i.e. the delay should be an integer number of periods. Therefore, in this application our main interest is the variable $R_{s_n, s'_n}(0)$ where we expect to find maximum correlation. For clarity, the correlation operation in this application will be written as

$$R(s_n, s'_n) = \int_0^T s_n(t)\, s'_n(t)\, dt \qquad (5.71)$$

The complete demodulator works as follows: the input signal enters a bank of cross-correlators. The outputs of these correlators are compared to each other, and the received symbol is supposed to be the one that corresponds to the correlator having the strongest output signal.

We have been discussing a frequency shift system. One question that needs to be answered is how large the shift in frequency Δf preferably should be. Obviously, the larger the shift, the more spectrum bandwidth will be occupied, which is commonly a bad thing, but how small can the shift be for the system to work well? We may argue like this: if the received signal is $s_n(t)$, the correlation to $s'_n(t)$ should be as large as possible, while the (undesired) correlations between $s_n(t)$ and $s'_m(t)$ where $m \neq n$ should be at a minimum. Using the binary example above the situation can be formulated as

$$\left| R(s_0, s'_0) \right| = \text{max}, \quad \left| R(s_1, s'_1) \right| = \text{max}, \quad R(s_0, s'_1) = R(s_1, s'_0) = 0$$

From the last condition above, we can find the minimum frequency shift required

$$R(s_0, s'_1) = \int_0^T s_0(t) s'_1(t) \, \mathrm{d}t = a \int_0^T \cos(2\pi f_0 t) \cos(2\pi f_1 t) \, \mathrm{d}t = 0$$

$$(5.72)$$

where we have assumed $\phi = 0$. Solving equation (5.72) (Proakis, 1989) above, we find that the frequency shift $|f_0 - f_1| = \Delta f$ needs to be

$$\Delta f = \frac{k}{2T}, \quad k = 1, 2, 3, \ldots \tag{5.73}$$

If we set $k = 1$ the minimum frequency shift is obtained $\Delta f = 1/2T$. An FSK system using this frequency shift is called a **minimum shift keying (MSK)** system.

If two signals $s_n(t)$ and $s_m(t)$, where $m \neq n$, has the property

$$R(s_n(t), s_m(t)) = 0,$$

the signals are said to be **orthogonal**. If all signals representing different symbols in a modulation scheme are orthogonal to each other, the risk of misinterpretation and transmission errors is minimized.

In the calculations above, continuous-time signals have been assumed, since integrals have been used. If dealing with a "digital" discrete-time system, the integrals can of course be approximated by summations, i.e.

$$\int g(t) \, \mathrm{d}t \approx (1/f_s) \sum_n g\left(\frac{n}{f_s}\right).$$

5.3.3 Phase shift keying (PSK)

In a PSK system, the phase shift of the signal is controlled by the baseband signal, i.e.

$$s_n(t) = a \cos(2\pi f_c t + \phi_n(t)) \tag{5.74}$$

The carrier frequency $f(t) = f_c$ and the amplitude $a(t) = a$ are constant. The phase functions $\phi_n(t)$ are defined over a finite period of time, i.e. $t_0 \leq t < t_0 + T$, where T is the symbol time as before. A simple example would be a system using binary symbols, i.e. $M = 2$ and square pulses for phase functions according to the below (see Figure 5.8). Commonly, the shape of the phase functions is chosen to be smoother than square pulses, for spectral occupancy reasons

symbol "0":

$$\phi_0(t) = \begin{cases} 0 & t_0 \leq t < t_0 + T \\ 0 & \text{else} \end{cases}$$

$$\Rightarrow s_0(t) = \begin{cases} a \cos(2\pi f_c t) & t_0 \leq t < t_0 + T \\ 0 & \text{else} \end{cases} \tag{5.75a}$$

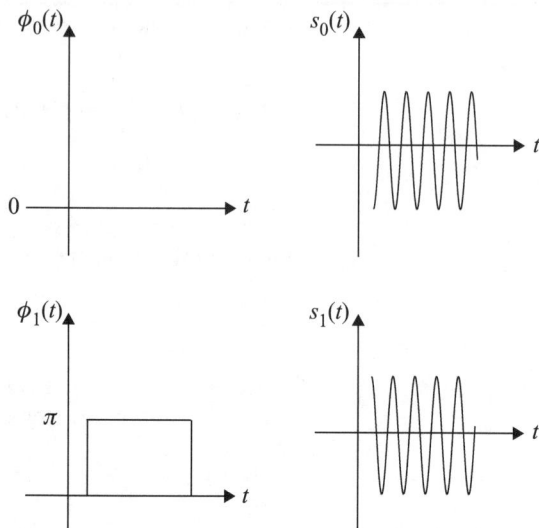

Figure 5.8 *Baseband signal and modulated signal for PSK,* M $= 2$

symbol "1":

$$\phi_1(t) = \begin{cases} \pi & t_0 \leq t < t_0 + T \\ 0 & \text{else} \end{cases}$$

$$\Rightarrow s_1(t) = \begin{cases} a\cos(2\pi f_c t + \pi) & t_0 \leq t < t_0 + T \\ 0 & \text{else} \end{cases} \tag{5.75b}$$

Often, the use of **phasors** (Denbigh, 1998) can be beneficial when describing modulation processes. Starting out with a general cosine signal of a given frequency f_n, amplitude a_n and phase shift ϕ_n we have

$$s_n(t) = a_n \cos(2\pi f_n t + \phi_n) \tag{5.76}$$

using Euler's formula, equation (5.76) can be rewritten as

$$\begin{aligned} s_n(t) &= a_n \cos(2\pi f_n t + \phi_n) \\ &= \text{Re}[a_n(\cos(2\pi f_n t + \phi_n) + j\sin(2\pi f_n t + \phi_n))] \\ &= \text{Re}\left[a_n\, e^{j(2\pi f_n t + \phi_n)}\right] = \text{Re}\left[a_n\, e^{j\phi_n}\, e^{j2\pi f_n t}\right] = \text{Re}\left[z_n\, e^{j2\pi f_n t}\right] \end{aligned} \tag{5.77}$$

where the complex number z_n is denoted the phasor of the signal. If the carrier frequency $f_n = f_c$ is given and constant, the phasor will represent the signal without any ambiguity, since

$$a_n = |z_n| \tag{5.78a}$$

$$\phi_n = \angle z_n \tag{5.78b}$$

Further, the phasor, being a complex number, can be drawn as a vector starting from the origin and having magnitude a_n and angle ϕ_n in a complex xy-plane.

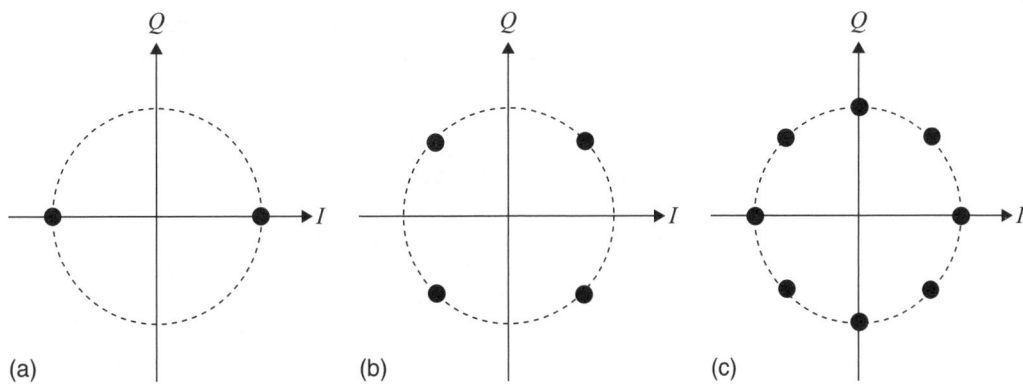

Figure 5.9 *Signal constellations. (a) BPSK* M = *2, (b) QPSK* M = *4 and (c) 8PSK* M = *8*

Phasors are a handy tool, used in many applications, not only in signal theory. However, a caution should be issued: when dealing with signals and modulation the magnitude of the phasor, denoted a_n, commonly refers to the amplitude of the signal, i.e. the peak value of the cosine signal. In literature dealing with electrical power applications the magnitude of the phasor may refer to the RMS value of the cosine (Denbigh, 1998).

Now, when using phasors in a PSK context, the phasors for all M symbols in the system can be drawn in a polar diagram, but for clarity, only the endpoint of the vectors (phasors) are drawn as a dot. The diagram thus obtained is sometimes called a **signal constellation**.

In Figure 5.9 some signal constellations are shown. Our PSK example in equations (5.75a) and (5.75b) above is shown as example in Figure 5.9(a). Since $M = 2$, i.e. that we have a binary system, this modulation type is called **binary phase shift keying (BPSK)** or **2PSK**. In Figure 5.9(b), we have used $M = 4$, giving four signal points and a system called **quadruple phase shift keying (QPSK)** or **4PSK**. In Figure 5.9(c) $M = 8$, i.e. the signal constellation **8PSK**, etc. Note that in all cases, the signal points are situated on a circle with radius a_n, since we are only changing the phase shift of the signal, not the amplitude. An obvious question in this situation is how many symbols, i.e. how large, can M be? We can conclude that the larger number of symbols we are using, the smaller the difference in phase shift. Considering the background noise and interference, the smaller the phase difference between adjacent symbols, the larger the risk for misinterpretations and transmission errors. A good rule of thumb is that the larger the distance between the signal points, the more resistance against noise and interference is obtained. Hence, given a limited transmitter power and background noise level, a BSPK system performs better than a QPSK system regarding transmission errors. On the other hand, the QPSK system has the potential of being twice as fast as the BPSK system. As expected, designing optimum modulation methods is a matter of compromises.

Obviously, demodulation of PSK signals can only be carried out coherently. The receiver needs to have some synchronized reference of the phase in order to be able to determine the phase shift ϕ_n of the received signals. This problem can however be circumvented by using a **differential phase shift keying (DPSK)** method. In such a system, the absolute phase shift is not of interest, but rather the **difference** in phase shift between successive symbols.

5.3.4 Complex modulation

A good approach when working with modulation is to utilize the concept of **complex modulation**, a general method being able to achieve all the modulation types mentioned above. Assume that the bandwidth of the baseband signal, the information modulated onto the carrier, is small relative to the frequency of the carrier signal. This implies that the modulated carrier, the signal being transmitted over the communications media, can be regarded as a **narrowband passband signal**, or simply a **bandpass signal** (Proakis, 1989; Ahlin and Zander, 1998)

$$
\begin{aligned}
s(t) &= a(t)\cos(2\pi f_c t + \phi(t)) \\
&= a(t)\cos(\phi(t))\cos(2\pi f_c t) - a(t)\sin(\phi(t))\sin(2\pi f_c t) \\
&= x(t)\cos(2\pi f_c t) - y(t)\sin(2\pi f_c t)
\end{aligned}
\tag{5.79}
$$

where $x(t)$ and $y(t)$ are the **quadrature components** of the signal $s(t)$, defined as

$$
x(t) = a(t)\cos(\phi(t)) \quad \text{is the \textbf{in phase} or } I \text{ component}
\tag{5.80a}
$$

and

$$
y(t) = a(t)\sin(\phi(t)) \quad \text{is the \textbf{quadrature phase} or } Q \text{ component}
\tag{5.80b}
$$

The two components above can be assembled into one handy entity by defining the **complex envelope**

$$
z(t) = x(t) + jy(t) = a(t)\cos(\phi(t)) + ja(t)\sin(\phi(t)) = a(t)\,e^{j\phi(t)}
\tag{5.81}
$$

Hence, the modulated carrier can now be written (see also equation (5.79)) as

$$
s(t) = \mathrm{Re}\!\left[z(t)\,e^{j2\pi f_c t}\right]
\tag{5.82}
$$

(compare to the phasor concept in equation (5.77)). The modulation process is performed by simply making a **complex multiplication** of the carrier and the complex envelope. This can be shown on component level

$$
s(t) = x(t)\cos(2\pi f_c t) - y(t)\sin(2\pi f_c t)
\tag{5.83}
$$

Figure 5.10 shows a simple, digital quadrature modulator, implementing expression (equation (5.83)) above. The M-ary symbol sequence to be transmitted enters the symbol mapper. This device determines the quadrature components $x(t)$ and $y(t)$ according to the modulation scheme in use.

For example, let us assume that the incoming data stream $A(n)$ consists of binary digits (BITs), i.e. ones and zeros. We have decided to use a QPSK, in other words, we are using $M = 4$ symbols represented by four equally spaced phase shifts of the carrier signal. So, two binary digits $A(n)$ and $A(n-1)$ are transmitted simultaneously. If we assume that $a(t) = 1$ (constant amplitude), the mapper then implements a table like Table 5.1 (many schemes exist).

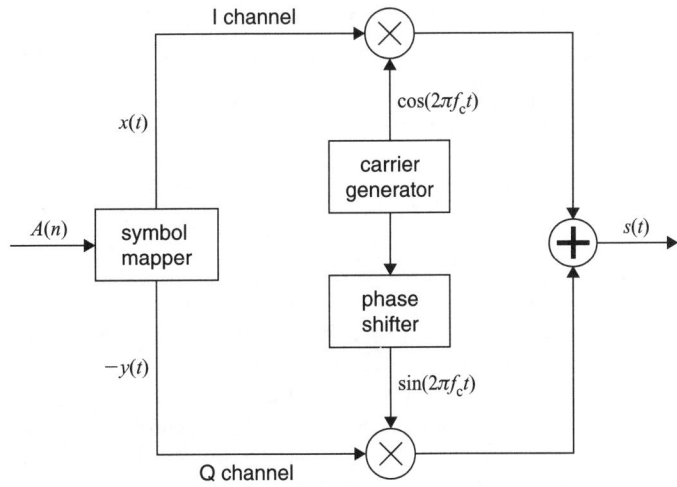

Figure 5.10 *Simple* M-*ary quadrature modulator*

Table 5.1 *Example,* I *and* Q *components and phase angle as a function of input bits*

$A(n)$	$A(n-1)$	$\phi(t)$	$x(t)$	$y(t)$
0	0	$45°$	$\frac{1}{\sqrt{2}}$	$\frac{1}{\sqrt{2}}$
0	1	$135°$	$\frac{-1}{\sqrt{2}}$	$\frac{1}{\sqrt{2}}$
1	0	$-45°$	$\frac{1}{\sqrt{2}}$	$\frac{-1}{\sqrt{2}}$
1	1	$-135°$	$\frac{-1}{\sqrt{2}}$	$\frac{-1}{\sqrt{2}}$.

The quadrature components coming from the mapper, having half the data rate of the binary symbols, is then multiplied by the carrier and a $-90°$ phase-shifted version (the quadrature channel) of the carrier, since $\cos(\alpha - 90) = \sin(\alpha)$. The *I* and *Q* channels are finally added and the modulated carrier $s(t)$ is obtained.

Demodulation of the signal at the receiving site can easily be done in a similar way, by performing a complex multiplication

$$\hat{z}(t) = s(t)\,e^{-j2\pi f_c t} \tag{5.84}$$

and removing high-frequency components using low-pass filters. From the received estimate $\hat{z}(t)$ the quadrature components can be obtained and from these, the information symbols. In practice, the situation is more complicated, because the received signal is not $s(t)$, but rather a distorted version with noise and interference added (see Chapter 8).

An advantage with this kind of modulator is that almost any modulation type can be achieved. From equations (5.80a) and (5.80b) we realize that

$$a(t) = \sqrt{x^2(t) + y^2(t)} \tag{5.85a}$$

$$\phi(t) = \arcsin\left(\frac{y(t)}{x(t)}\right) \tag{5.85b}$$

Since we are free to chose the I channel signal $x(t)$ and the Q channel signal $y(t)$ as we please, we can obviously perform amplitude modulation (or ASK) and phase modulation (or PSK) simultaneously. This implies that we can locate our signal points arbitrary in the complex plane. We are not limited to locating the signal points on a circle as in, for instance, the case of QPSK. In Figure 5.11 the signal points are placed in an array which can serve as an example of **quadrature amplitude modulation (QAM)**.

Another way of viewing QAM is to make a cross-correlation analysis (equation (5.71)) of the carriers used: $\cos(2\pi f_c t)$ and $\sin(2\pi f_c t)$. Such an analysis shows that the carriers are orthogonal, i.e. ideally there is no "cross-talk" in between them. This means that the I and Q channels can be viewed as two independent amplitude modulated transmission channels. Therefore, in some systems, different information is transmitted over these channels.

Yet, another possibility is to achieve frequency modulation (or FSK), since indirect frequency modulation (Armstrong modulation) can be performed by integrating the baseband signal and feeding it to a phase modulator (Miller and Beasley, 2002). An example of such a system is the cellular mobile telephone system GSM, using **Gaussian minimum shift keying (GMSK)**. To obtain good spectral properties, the baseband signal is first processed by a low-pass filter having an approximation of a Gaussian impulse response (equation (5.86)) before entering the integrator and the quadrature modulator

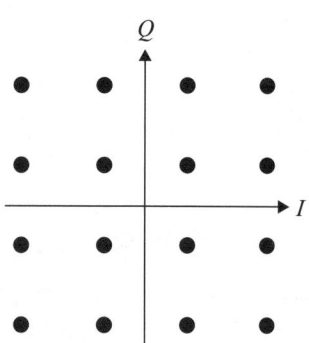

Figure 5.11 *An example signal constellation for QAM*

$$h(t) = \frac{1}{2T}\left(Q\left(2\pi B_b \frac{t - (T/2)}{\sqrt{2\ln(2)}}\right) - Q\left(2\pi B_b \frac{t + (T/2)}{\sqrt{2\ln(2)}}\right)\right) \tag{5.86}$$

where B_b is the bandwidth of the filter, T is the bit time and $Q(t)$ is the Q-function

$$Q(t) = \int_t \frac{1}{\sqrt{2}} e^{-x^2/2} \, dx$$

5.3.5 The Hilbert transformer

A special kind of "ideal" filter is the **quadrature filter**. Since this filter is non-causal, only approximations (Mitra and Kaiser, 1993) of the filter can be implemented in practice. This filter model is often used when dealing with **single sideband (SSB)** signals. The ideal frequency response is

$$H(\omega) = \begin{cases} -j & \omega > 0 \\ 0 & \omega = 0 \\ j & \omega < 0 \end{cases} \tag{5.87}$$

This can be interpreted as an **all-pass** filter, providing a phase shift of $\pi/2$ radians at all frequencies. Hence, if the input signal is $\cos(\omega t)$, the output will be $\cos(\omega t - 90) = \sin(\omega t)$.

The corresponding impulse response is

$$h(t) = \frac{1}{\pi t} \quad \text{for } t \neq 0 \tag{5.88}$$

The response of the quadrature filter to a real input signal is denoted the **Hilbert transform** and can be expressed as the convolution of the input signal $x(t)$ and the impulse response of the filter

$$y(t) = x(t) * h(t) = \frac{1}{\pi} \int_{-\infty}^{\infty} \frac{x(\tau)}{t - \tau} \, d\tau \tag{5.89}$$

where $*$ denotes convolution. Now, the complex **analytic signal** $z(t)$ associated with $x(t)$ can be formed as

$$z(t) = x(t) + jy(t) = x(t) + jx(t) * h(t) \tag{5.90}$$

It is clear that $z(t)$ above is the response of the system

$$G(\omega) = 1 + jH(\omega) \tag{5.91}$$

The frequency response above implies attenuation of "negative" frequency components while passing "positive" frequency components. In an SSB situation, this relates to the lower and upper sideband (Proakis, 1989), respectively

$$G(\omega) = \begin{cases} 1 + j(-j) = 2 & \text{for } \omega > 0 \\ 1 + j(j) = 0 & \text{for } \omega < 0 \end{cases} \tag{5.92}$$

Hence, if we want to transmit the analog signal $x(t)$ using SSB modulation, equation (5.90) replaces equation (5.81). Further, the symbol mapper in Figure 5.10 is replaced by a direct connection of $x(t)$ to the I channel and a connection via a quadrature filter (Hilbert transformer) to the Q channel. (The phase shifter connected to the carrier generator in Figure 5.10 can, of course also, be implemented as a Hilbert transformer.)

When dealing with discrete-time systems, the impulse response (equation (5.88)) is replaced by

$$h(n) = \begin{cases} 1/\pi n & \text{for odd } n \\ 0 & \text{for even } n \end{cases} \tag{5.93}$$

This function can, for instance, be approximated by an FIR filter. There are different ways of approximating the time-discrete Hilbert transformer (Proakis, 1989).

Summary

In this chapter the following topics have been treated:

- DFT, FFT, twiddle factors and zero padding
- Windowing techniques, spectrum estimation, Welch's and Bartlett's methods
- Correlation and covariance, the Wiener–Khintchine theorem
- AR, MA and ARMA models, the Yule–Walker equations
- Wavelets

- ASK, FSK, PSK, BPSK, QPSK, MSK, GMSK and QAM
- Phasors, the quadrature modulator, Armstrong's indirect FM method and the Hilbert transform.

Review questions

R5-1 What is the relation between DFT and FFT? What tricks are typically used by FFT?

R5-2 Why are windowing techniques used? Pros and cons?

R5-3 Explain how auto-correlation, auto-covariance, cross-correlation and cross-covariance are related.

R5-4 What is the basic idea behind parametric spectrum analysis? What do AR, MA and ARMA stand for?

R5-5 Under what circumstances is wavelet analysis better than using the Fourier transform?

R5-6 What is the trade-off when determining the number of symbols M in a modulator?

R5-7 What is Armstrong's indirect FM method? How can it be used in practice?

R5-8 What are the properties of the Hilbert transformer? Give an application example.

Solved problems

P5-1 Draw a block diagram of a quadrature modulator and demodulator, including proper signal names and equations.

P5-2 Show that equations (5.85a) and (5.85b) can be obtained from equations (5.80a) and (5.80b).

P5-3 Solve equation (5.72) and prove that equation (5.73) is the solution for Δf in an MSK system.

P5-4 Write a MATLAB^{TM} program making an FFT and plotting the magnitude of the spectrum of the digital AM signal $s(n) = (1 + \cos(\Omega_i n)) \cos(\Omega_c n)$, where $\Omega_i = 0.01$, $\Omega_c = 2$ and $N = 1024$. Try the rectangular window and the Hamming window. What conclusions can be made?

6 Introduction to Kalman filters

Background

In earlier chapters, we have mainly been dealing with deterministic signals having fairly simple frequency spectra. Such signals can often be processed successfully using classical filters. When it comes to filtering of stochastic (random) signals, things get worse. Since the frequency spectra of a stochastic signal commonly is quite complex, it will be difficult to extract or reject the desired parts of the spectra to obtain the required filtering action. In such a situation, a Kalman filter may come in handy. Using a Kalman filter, signals are filtered according to their statistical properties, rather than their frequency contents.

The Kalman filter has other interesting properties. The filter contains a signal model, a type of "simulator" that produces the output signal. When the quality of the input signal is good (for instance, a small amount of noise or interference), the signal is used to generate the output signal and the internal model is adjusted to follow the input signal. When, on the other hand, the input signal is poor, it is ignored and the output signal from the filter is mainly produced by the model. In this way, the filter can produce a reasonable output signal even during drop out of the input signal. Further, once the model has converged well to the input signal, the filter can be used to simulate future output signals, i.e. the filter can be used for prediction.

Kalman filters are often used to condition transducer signals and in control systems for satellites, aircraft and missiles. The filter is also used in applications dealing with examples, such as economics, medicine, chemistry and sociology.

Objectives

In this chapter we will discuss:

- Recursive least square (RLS) estimation and the underlying idea of Kalman filters
- The pseudo-inverse and how it is related to RLS and Kalman filters
- The signal model, a dynamic system as a state–space model
- The measurement-update equations and the time-update equations
- The innovation, the Kalman gain matrix and the Riccatti equation
- A simple example application, estimating position and velocity while cruising down main street
- Properties of Kalman filters.

6.1 An intuitive approach

Filters were originally viewed as systems or algorithms with frequency selective properties. They could discriminate between unwanted and desired signals found in different frequency bands. Hence, by using a filter, these unwanted signals could be rejected.

In the 1940s, theoretical work was performed by N. Wiener and N.A. Kolomogorov on statistical approaches to filtering. In these cases, signals could be filtered according to their **statistical properties**, rather than their frequency contents. At this time, the Wiener and Kolomogorov theory required the signals to be **stationary**, which means that the statistical properties of the signals were not allowed to change with time.

In late 1950s and early 1960s a new theory was developed capable of coping with **non-stationary** signals as well. This theory came to be known as the **Kalman filter theory**, named after R.E. Kalman.

The Kalman filter is a **discrete-time**, **linear filter** and it is also an **optimal filter** under conditions that shall be described later in the text. The underlying theory is quite advanced and is based on statistical results and estimation theory. This presentation of the Kalman filter (Anderson and Moore, 1979; Åström and Wittenmark, 1984; Bozic, 1994) will begin with a brief discussion of estimation.

6.1.1 Recursive least square estimation

Let us start with a simple example to illustrate the idea of **recursive least square (RLS) estimation**. Assume that we would like to measure a constant signal level (DC level) x. Unfortunately, our measurements $z(n)$ are disturbed by noise $v(n)$. Hence, based on the observations

$$z(n) = x + v(n) \tag{6.1}$$

Our task is to filter out the noise and make the best possible estimation of x. This estimate is called $\hat{x}$. Our quality criteria is finding the estimate $\hat{x}$ that minimizes the least square criteria

$$J(\hat{x}) = \sum_{n=1}^{N} (z(n) - \hat{x})^2 \tag{6.2}$$

The minimum of equation (6.2) can be found by setting the first derivative to 0

$$\frac{\partial J(\hat{x})}{\partial \hat{x}} = \sum_{n=1}^{N} 2(\hat{x} - z(n)) = 0 \tag{6.3}$$

Solving for $\hat{x}(N)$, i.e. the estimate $\hat{x}$ used during N observations of $z(n)$, we obtain

$$\sum_{n=1}^{N} \hat{x} = \sum_{n=1}^{N} z(n) \tag{6.4}$$

where

$$\sum_{n=1}^{N} \hat{x} = N \hat{x}(N) = \sum_{n=1}^{N} z(n) \tag{6.5}$$

Hence, the best way to filter our observations in the sense of minimizing our criteria (6.2) is

$$\hat{x}(N) = \frac{1}{N} \sum_{n=1}^{N} z(n) \tag{6.6}$$

That is, taking the average of N observations. Now, we do not want to wait for N observations before obtaining an estimate. We want to have an estimate almost at once, when starting to measure. Of course, this estimate will be quite poor, but we expect it to grow better and better as more observations are gathered. In other words, we require a **recursive** estimation (filtering) method. This can be achieved in the following way.

From equation (6.5) we can express the estimate after $N + 1$ observations

$$(N + 1)\hat{x}(N + 1) = \sum_{n=1}^{N+1} \hat{x} = \sum_{n=1}^{N+1} z(n) \tag{6.7}$$

Now, we want to know how the estimate changes from N to $N + 1$ observations, in order to find a recursive algorithm. Taking equation (6.7) minus equation (6.5) gives

$$(N + 1)\hat{x}(N + 1) - N\hat{x}(N) = \sum_{n=1}^{N+1} z(n) - \sum_{n=1}^{N} z(n) = z(N + 1) \tag{6.8}$$

Rearranging in a way that $\hat{x}(N + 1)$ (the "new" estimate) is expressed in terms of $\hat{x}(N)$ (the "old" estimate), from equation (6.8) we get

$$\hat{x}(N + 1) = \frac{1}{N + 1}(z(N + 1) + N\hat{x}(N))$$

$$= \frac{1}{N + 1}(z(N + 1) + (N + 1)\hat{x}(N) - \hat{x}(N))$$

$$= \hat{x}(N) + \frac{1}{N + 1}(z(N + 1) - \hat{x}(N)) \tag{6.9}$$

where we assume the initial condition $\hat{x}(0) = 0$. The filter can be drawn as in Figure 6.1. Note that the filter now contains a **model** (the delay z^{-1} or "memory") of the signal x. Presently, the model now holds the best estimate $\hat{x}(N)$ which is compared to the new measured value $z(N + 1)$. The difference $z(N + 1) - \hat{x}(N)$, sometimes called the **innovation**, is amplified by the **gain factor** $1/(N + 1)$ before it is used to update our estimate, i.e. our model to $\hat{x}(N + 1)$.

Two facts need to be noted. Firstly, if our estimate $\hat{x}(N)$ is close to the measured quantity $z(N + 1)$, our model is not adjusted significantly and the output is quite good as is. Secondly, the gain factor $1/(N + 1)$ will grow smaller and smaller as time goes by. The filter will hence pay less and less attention to the measured noisy values, making the output level stabilize to a fixed value of $\hat{x}$.

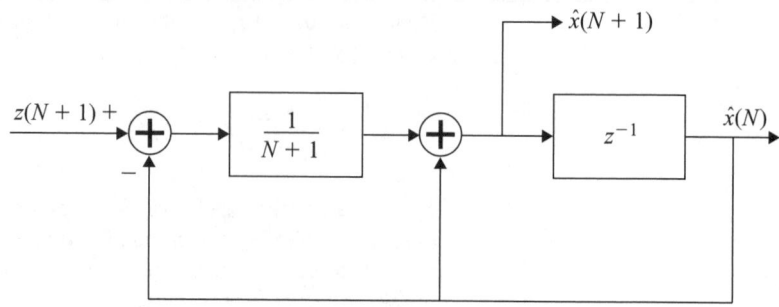

Figure 6.1 *An example of an RLS estimator*

In the discussion above, we have assumed that the impact of the noise has been constant. The "quality" of the observed variable $z(n)$ has been the same for all n. If, on the other hand, we know that the noise is very strong at certain times, or even experience a disruption in the measuring process, then we should, of course, pay less attention to the input signal $z(n)$. One way to solve this is to implement a kind of quality weight $q(n)$. For instance, this weight could be related to the magnitude of the noise as

$$q(n) \propto \frac{1}{v^2(n)} \tag{6.10}$$

Inserting this into equation (6.2), we obtain a **weighted least square** criteria

$$J(\hat{x}) = \sum_{n=1}^{N} q(n)(z(n) - \hat{x})^2 \tag{6.11}$$

Using equation (6.11) and going through the same calculations as before, equations (6.3)–(6.9) we obtain the expression for the gain factor in this case (compare to equation (6.9))

$$\hat{x}(N+1) = \hat{x}(N) + \frac{q(N+1)}{\sum_{n=1}^{N+1} q(n)}(z(N+1) - \hat{x}(N))$$

$$= \hat{x}(N) + Q(N+1)(z(N+1) - \hat{x}(N)) \tag{6.12}$$

It can be shown that the gain factor $Q(N+1)$ can also be calculated recursively

$$Q(N+1) = \frac{q(N+1)}{\sum_{n=1}^{N+1} q(n)} = \frac{q(N+1)}{\sum_{n=1}^{N} q(n) + q(N+1)}$$

$$= \frac{q(N+1)}{\frac{q(N)}{Q(N)} + q(N+1)} = \frac{Q(N)q(N+1)}{q(N) + Q(N)q(N+1)} \tag{6.13}$$

where the starting conditions are: $q(0) = 0$ and $Q(0) = 1$. We can now draw some interesting conclusions about the behavior of the filter. If the input signal quality is extremely low (or if no input signal is available) $q(n) \to 0$ implies that

$Q(n + 1) \to 0$ and the output from the filter equation (6.12) is

$$\hat{x}(N + 1) = \hat{x}(N) \tag{6.14}$$

In other words, we are running "dead reckoning" using the model in the filter only. If, on the other hand, the input signal quality is excellent (no noise present), $q(n) \to \infty$ and $Q(n + 1) \to 1$ then

$$\hat{x}(N + 1) = z(N + 1) \tag{6.15}$$

In this case, the input signal is fed directly to the output, and the model in the filter is updated.

For intermediate quality levels of the input signal, a mix of measured signal and modeled signal is presented at the output of the filter. This mix thus represents the best estimate of x, according to the weighted least square criteria.

So far, we have assumed x to be a **constant**. If $x(n)$ is allowed to change over time, but considerably slower than the noise, we must introduce a "forgetting factor" into equation (6.2), or else all "old" values of $x(n)$ will counteract changes of $\hat{x}(n)$. The forgetting factor $w(n)$ should have the following property

$$w(n) > w(n - 1) > \cdots > w(1) \tag{6.16}$$

In other words, the "oldest" values should be forgotten the most. One example is

$$w(n) = \alpha^{N-n} \tag{6.17}$$

where $0 < \alpha < 1$. Inserting equation (6.17) into equation (6.2) and going through the calculations equations (6.3)–(6.9) again, we obtain

$$\hat{x}(N + 1) = \hat{x}(N) + \frac{1}{\sum_{n=1}^{N+1} \alpha^{N+1-n}}(z(N + 1) - \hat{x}(N))$$
$$= \hat{x}(N) + P(N + 1)(z(N + 1) - \hat{x}(N)) \tag{6.18}$$

The gain factor $P(n)$ can be calculated recursively by

$$P(N + 1) = \frac{P(N)}{\alpha + P(N)} \tag{6.19}$$

Note, the gain factors can be calculated off-line, in advance.

6.1.2 The pseudo-inverse

If we now generalize the scalar measurement problem outlined above and go into a multidimensional problem, we can reformulate equation (6.1) using vector notation

$$\mathbf{z}(n) = \mathbf{H}^{\mathrm{T}}(n)\,\mathbf{x} + \mathbf{v}(n) \tag{6.20}$$

where the dimensions of the entities are

$$\mathbf{z}(n): (K \times 1) \quad \mathbf{H}(n): (L \times K)$$
$$\mathbf{x}: (L \times 1) \quad \mathbf{v}(n): (K \times 1)$$

In a simple case, where there is no noise present, i.e. $\mathbf{v}(n) = \mathbf{0}$ (or when the noise is known) solving $\mathbf{x}$ from equation (6.20) could be done simply by inverting the matrix $\mathbf{H}(n)$

$$\mathbf{x} = (\mathbf{H}^{\mathrm{T}}(n))^{-1}\,(\mathbf{z}(n) - \mathbf{v}(n)) \tag{6.21}$$

This corresponds to solving a system of L equations. Inverting the matrix $\mathbf{H}(n)$ may not be possible in all cases for two reasons, either the inverse does not exist or we do not have access to a sufficient amount of information. The latter is the case after each observation of $\mathbf{z}(n)$ when noise $\mathbf{v}(n)$ is present or if $K < L$, i.e. we have too few observations. Hence, a straightforward matrix inversion is not possible. In such a case, where the inverse of the matrix cannot be found, a **pseudo-inverse** (Anderson and Moore, 1979; Åström and Wittenmark, 1984), also called the *Moore–Penrose inverse*, may be used instead. Using the pseudo-inverse, equation (6.21) can be solved "approximately".

For a true inverse: $(\mathbf{H}^{\mathrm{T}})^{-1}\mathbf{H}^{\mathrm{T}} = \mathbf{I}$, where $\mathbf{I}$ is the identity matrix. In a similar way, using the pseudo-inverse $\mathbf{H}^{\#}$ we try to make the matrix product $(\mathbf{H}^{\mathrm{T}})^{\#}\,\mathbf{H}^{\mathrm{T}}$ as close to the identity matrix as possible in a least squares sense. Note, this is similar to fitting a straight line to a number of measured and plotted points in a diagram using the least square method.

The pseudo-inverse of $\mathbf{H}(n)$ can be expressed as

$$\mathbf{H}^{\#} = (\mathbf{H}^{\mathrm{T}}\mathbf{H})^{-1}\,\mathbf{H}^{\mathrm{T}} \tag{6.22}$$

Now, finding the pseudo-inverse corresponds to minimizing the criteria (compare to equation (6.2))

$$J(\hat{\mathbf{x}}) = \sum_{n=1}^{N} \left\| \mathbf{z}(n) - \mathbf{H}^{\mathrm{T}}(n)\,\hat{\mathbf{x}} \right\|^2 \tag{6.23}$$

where the Euclidean matrix norm is used, i.e.

$$\|\mathbf{A}\|^2 = \sum_{i}\sum_{j} a_{ij}^2$$

Minimizing the above is finding the vector estimate $\hat{\mathbf{x}}$ that results in the least square error. Taking the derivative of equation (6.23) for all components, we obtain the gradient. Similar to equations (6.3)–(6.5) we solve for $\hat{\mathbf{x}}$ as the gradient is set equal to 0, thus obtaining the following (compare to equation (6.6))

$$\hat{\mathbf{x}}(N) = \left(\sum_{n=1}^{N} \mathbf{H}(n)\,\mathbf{H}^{\mathrm{T}}(n) \right)^{-1} \sum_{n=1}^{N} \mathbf{H}(n)\,\mathbf{z}(n) \tag{6.24}$$

Note, the sums above constitute the pseudo-inverse. In the same way as before, we like to find a recursive expression for the best estimate. This can be achieved by going through calculations similar to equations (6.7) and (6.8), but using vector and matrix algebra. These calculations will yield the result (compare to equation (6.9))

$$\hat{\mathbf{x}}(N+1) = \hat{\mathbf{x}}(N) + \left(\sum_{n=1}^{N+1} \mathbf{H}(n)\,\mathbf{H}^{\mathrm{T}}(n) \right)^{-1} \mathbf{H}(N+1)(\mathbf{z}(N+1)$$

$$- \mathbf{H}^{\mathrm{T}}(N+1)\,\hat{\mathbf{x}}(N))$$

$$= \hat{\mathbf{x}}(N) + \mathbf{P}(N+1)\,\mathbf{H}(N+1)(\mathbf{z}(N+1) - \mathbf{H}^{\mathrm{T}}(N+1)\,\hat{\mathbf{x}}(N))$$

$$(6.25)$$

where $\mathbf{P}(n)$ is the gain factor as before. Reasoning in a similar way to the calculations of equation (6.13), we find that this gain factor can also be calculated recursively

$$\mathbf{P}(N+1) = \mathbf{P}(N) - \mathbf{P}(N)\,\mathbf{H}(N+1)$$

$$(\mathbf{I} + \mathbf{H}^{\mathrm{T}}(N+1)\,\mathbf{P}(N)\,\mathbf{H}(N+1))^{-1}\,\mathbf{H}^{\mathrm{T}}(N+1)\,\mathbf{P}(N)$$

$$(6.26)$$

The equations (6.25) and (6.26) are a recursive method of obtaining the pseudo-inverse $\mathbf{H}^{\#}$ using a filter model as in Figure 6.1. The ideas presented above constitute the underlying ideas of Kalman filters.

6.2 The Kalman filter

6.2.1 The signal model

The **signal model**, sometimes also called the **process model** or the **plant**, is a model of the "reality" which we would like to measure. This "reality" also generates the signals we are observing. In this context, dynamic systems are commonly described using a state–space model (see Chapter 1). A simple example may be the following.

Assume that our friend Bill is cruising down Main Street in his brand new Corvette. Main Street can be approximated by a straight line, and since Bill has engaged the cruise control, we can assume that he is traveling at a constant speed (no traffic lights). Using a simple radar device, we try to measure Bill's position along Main Street (starting from the Burger King) at every second.

Now, let us formulate a discrete-time, **state–space model**. Let Bill's position at time n be represented by the discrete-time variable $x_1(n)$ and his speed by $x_2(n)$. Expressing this in terms of a recursive scheme we can write

$$x_1(n+1) = x_1(n) + x_2(n)$$
$$x_2(n+1) = x_2(n) \qquad (6.27)$$

The second equation simply tells us that the speed is constant. The equations above can also be written using vector notation by defining a **state vector** $\mathbf{x}(n)$ and a **transition matrix** $\mathbf{F}(n)$, as

$$\mathbf{x}(n) = \begin{bmatrix} x_1(n) \\ x_2(n) \end{bmatrix} \quad \text{and} \quad \mathbf{F}(n) = \begin{bmatrix} 1 & 1 \\ 0 & 1 \end{bmatrix} \qquad (6.28)$$

The equations (6.27) representing the system can now be nicely formulated as a simple state–space model

$$\mathbf{x}(n+1) = \mathbf{F}(n)\,\mathbf{x}(n) \tag{6.29}$$

This is, of course, a quite trivial situation and an ideal model, but Bill is certainly not. Now and then, he brakes a little when there is something interesting at the sidewalk. This repeated braking and putting the cruise control back into gear, changes the speed of the car. If we assume that the braking positions are randomly distributed along Main Street, we can hence take this into account by adding a white Gaussian noise signal $w(n)$ to the speed variable in our model

$$\mathbf{x}(n+1) = \mathbf{F}(n)\,\mathbf{x}(n) + \mathbf{G}(n)\,w(n) \tag{6.30}$$

where

$$\mathbf{G}(n) = \begin{bmatrix} 0 \\ 1 \end{bmatrix}$$

The noise, sometimes called "process noise" is supposed to be scalar in this example, having the variance: $Q = \sigma_w^2$ and a mean equal to 0. Note, $\mathbf{x}(n)$ is now a stochastic vector.

So far, we have not considered the errors of the measuring equipment. What entities in the process (Bill and Corvette) can we observe using our simple radar equipment? To start with, since the equipment is old and not of Doppler type, it will only give a number representing the **distance**. Speed is not measured. Hence, we can only get information about the state variable $x_1(n)$. This is represented by the observation matrix $\mathbf{H}(n)$. Further, there are, of course, random errors present in the distance measurements obtained. On some occasions, no sensible readings are obtained, when cars are crossing the street. This uncertainty can be modeled by adding another white Gaussian noise signal $v(n)$. This so-called "measurement noise" is scalar in this example, and is assumed to have zero mean and a variance $R = \sigma_v^2$. Hence, the measured signal $z(n)$ can be expressed as

$$z(n) = \mathbf{H}^{\mathrm{T}}(n)\,\mathbf{x}(n) + v(n) \tag{6.31}$$

Equations (6.30) and (6.31) now constitute our basic signal model, which can be drawn as in Figure 6.2.

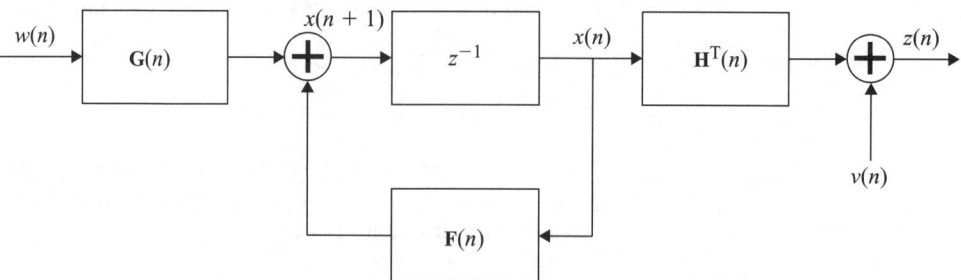

Figure 6.2 *Basic signal model as expressed by equations (6.30) and (6.31)*

6.2.2 The filter

The task of the filter, given the observed signal $\mathbf{z}(n)$ (a vector in the general case), is to find the best possible estimate of the state vector $\mathbf{x}(n)$ in the sense of the criteria given below. We should however remember, $\mathbf{x}(n)$ is now a stochastic signal rather than a constant, as in the previous section.

For our convenience, we will introduce the following notation: the estimate of $\mathbf{x}(n)$ at time n, based on the $n-1$ observations $\mathbf{z}(0)$, $\mathbf{z}(1)$, ..., $\mathbf{z}(n-1)$, will be denoted as $\hat{\mathbf{x}}(n\,|\,n-1)$ and the set of observations $\mathbf{z}(0)$, $\mathbf{z}(1)$, ..., $\mathbf{z}(n-1)$ itself will be denoted $\mathbf{Z}(n-1)$.

Our quality criteria is finding the estimate that minimizes the conditional error covariance matrix (Anderson and Moore, 1979)

$$\mathbf{C}(n\,|\,n-1) = \mathrm{E}\left[(\mathbf{x}(n) - \hat{\mathbf{x}}(n\,|\,n-1))(\mathbf{x}(n) - \hat{\mathbf{x}}(n\,|\,n-1))^{\mathrm{T}}\,|\,\mathbf{Z}(n-1)\right]$$

(6.32)

This is a **minimum variance** criteria and can be regarded as a kind of "stochastic version" of the least square criteria used in the previous section. Finding the minimum is a bit more complicated in this case than in the previous section. The best estimate, according to our criteria, is found using the **conditional mean** (Anderson and Moore, 1979), i.e.

$$\hat{\mathbf{x}}(n\,|\,n) = \mathrm{E}[\mathbf{x}(n)\,|\,\mathbf{Z}(n)]$$

(6.33)

The underlying idea is as follows: $\mathbf{x}(n)$ and $\mathbf{z}(n)$ are both random vector variables of which $\mathbf{x}(n)$ can be viewed as being a "part" of $\mathbf{z}(n)$ (see equation (6.31)). The statistical properties of $\mathbf{x}(n)$ will be "buried" in the statistical properties of $\mathbf{z}(n)$. For instance, if we now want to have a better knowledge of the mean of $\mathbf{x}(n)$, uncertainty can be reduced by considering the actual values of the measurements $\mathbf{Z}(n)$. This is called **conditioning**. Hence, equation (6.33) the conditional mean of $\mathbf{x}(n)$ is the **most probable** mean of $\mathbf{x}(n)$, given the measured values $\mathbf{Z}(n)$.

We will show how this conditional mean equation (6.33) can be calculated recursively, which is exactly what the **Kalman filter** does. (Later we shall return to the example of Bill and his Corvette.)

Since we are going for a recursive procedure, let us start at time $n = 0$. When there are no measurements made, we have from equation (6.32)

$$\mathbf{C}(0\,|-1) = \mathrm{E}[\mathbf{x}(0)\,\mathbf{x}^{\mathrm{T}}(0)\,|\,\mathbf{Z}(-1)] = \mathbf{P}(0)$$

(6.34)

It can be shown (Anderson and Moore, 1979), that the conditional mean of $\mathbf{x}(0)$ can be obtained from the cross-covariance of $\mathbf{x}(0)$ and $\mathbf{z}(0)$, the auto-covariance and the mean of $\mathbf{z}(0)$ and the measured value $\mathbf{z}(0)$ itself

$$\hat{\mathbf{x}}(0\,|\,0) = \mathrm{E}[\mathbf{x}(0)] + \mathbf{C}_{xz}(0)\,\mathbf{C}_{zz}^{-1}(0)(\mathbf{z}(0) - \mathrm{E}[\mathbf{z}(0)])$$

(6.35)

The covariance matrix at time n, based on n observations is denoted $\mathbf{C}(n\,|\,n)$. It can be obtained from (for $n = 0$)

$$\mathbf{C}(0\,|\,0) = \mathbf{C}_{xx}(0) - \mathbf{C}_{xz}(0)\,\mathbf{C}_{zz}^{-1}(0)\,\mathbf{C}_{zx}(0)$$

(6.36)

Next, we need to find the mean vectors and covariance matrices to plug into equations (6.35) and (6.36). Both the process noise and the measurement noise are Gaussian, and we assume that they are uncorrelated.

The mean of $\mathbf{x}(0)$ is denoted $E[\mathbf{x}(0)]$. Using equation (6.31), the mean of $\mathbf{z}(0)$ is

$$E[\mathbf{z}(0)] = E[\mathbf{H}^T(0)\,\mathbf{x}(0) + \mathbf{v}(0)] = \mathbf{H}^T(0)\,E[\mathbf{x}(0)] \tag{6.37}$$

where, we have used the fact that the mean of the measurement noise $\mathbf{v}(0)$ is 0.

The auto-covariance of $\mathbf{x}(0)$

$$\mathbf{C}_{xx}(0) = E[\mathbf{x}(0)\,\mathbf{x}^T(0)] = \mathbf{P}(0) \tag{6.38}$$

Using equation (6.31), the auto-covariance of $\mathbf{z}(0)$ can be expressed as

$$\begin{aligned}
\mathbf{C}_{zz}(0) &= E[\mathbf{z}(0)\,\mathbf{z}^T(0)] \\
&= E[(\mathbf{H}^T(0)\,\mathbf{x}(0) + \mathbf{v}(0))(\mathbf{H}^T(0)\,\mathbf{x}(0) + \mathbf{v}(0))^T] \\
&= E\left[\mathbf{H}^T(0)\,\mathbf{x}(0)\,\mathbf{x}^T(0)\,\mathbf{H}(0) + \mathbf{H}^T(0)\,\mathbf{x}(0)\,\mathbf{v}^T(0) \right. \\
&\qquad \left. + \mathbf{v}(0)\,\mathbf{x}^T(0)\,\mathbf{H}(0) + \mathbf{v}(0)\,\mathbf{v}^T(0)\right] \\
&= \mathbf{H}^T(0)\,E[\mathbf{x}(0)\,\mathbf{x}^T(0)]\,\mathbf{H}(0) + E[\mathbf{v}(0)\,\mathbf{v}^T(0)] \\
&= \mathbf{H}^T(0)\,\mathbf{P}(0)\,\mathbf{H}(0) + \mathbf{R}(0)
\end{aligned} \tag{6.39}$$

where the cross-covariance between $\mathbf{x}(0)$ and $\mathbf{v}(0)$ is 0, since the measurement noise was assumed to be uncorrelated to the process noise. $\mathbf{R}(0)$ is the auto-correlation matrix of the measurement noise $\mathbf{v}(0)$. In a similar way, the cross-covariance between $\mathbf{x}(0)$ and $\mathbf{z}(0)$ can be expressed as

$$\begin{aligned}
\mathbf{C}_{zx}(0) &= E[\mathbf{z}(0)\,\mathbf{x}^T(0)] \\
&= E[(\mathbf{H}^T(0)\,\mathbf{x}(0) + \mathbf{v}(0))\,\mathbf{x}^T(0)] \\
&= E[\mathbf{H}^T(0)\,\mathbf{x}(0)\,\mathbf{x}^T(0) + \mathbf{v}(0)\,\mathbf{x}^T(0)] \\
&= \mathbf{H}^T(0)\,E[\mathbf{x}(0)\,\mathbf{x}^T(0)] = \mathbf{H}^T(0)\,\mathbf{P}(0)
\end{aligned} \tag{6.40}$$

and

$$\mathbf{C}_{xz}(0) = E[\mathbf{x}(0)\,\mathbf{z}^T(0)] = \mathbf{P}(0)\,\mathbf{H}(0) \tag{6.41}$$

Inserting these results into equations (6.35) and (6.36), respectively, we obtain the conditioned mean and the covariance

$$\hat{\mathbf{x}}(0\,|\,0) = E[\mathbf{x}(0)] + \mathbf{P}(0)\,\mathbf{H}(0)(\mathbf{H}^T(0)\,\mathbf{P}(0)\,\mathbf{H}(0) + \mathbf{R}(0))^{-1}$$
$$(\mathbf{z}(0) - \mathbf{H}^T(0)\,E[\mathbf{x}(0)]) \tag{6.42}$$

$$\mathbf{C}(0\,|\,0) = \mathbf{P}(0) - \mathbf{P}(0)\,\mathbf{H}(0)(\mathbf{H}^T(0)\,\mathbf{P}(0)\,\mathbf{H}(0) + \mathbf{R}(0))^{-1}\,\mathbf{H}^T(0)\,\mathbf{P}(0)$$
$$\tag{6.43}$$

Let us now take a step forward in time, i.e. for $n = 1$, before we have taken the measurement $\mathbf{z}(1)$ into account, from equation (6.30) and various independence assumptions (Anderson and Moore, 1979) we have the mean

$$\hat{\mathbf{x}}(1 \mid 0) = \mathbf{F}(0)\,\hat{\mathbf{x}}(0 \mid 0) \tag{6.44}$$

and the covariance

$$\mathbf{C}(1 \mid 0) = \mathbf{F}(0)\,\mathbf{C}(0 \mid 0)\,\mathbf{F}^{\mathrm{T}}(0) + \mathbf{G}(0)\,\mathbf{Q}(0)\,\mathbf{G}^{\mathrm{T}}(0) \tag{6.45}$$

where $\mathbf{Q}(0)$ is the auto-correlation matrix of the process noise $\mathbf{w}(0)$. We are now back in the situation where we started for $n = 0$, but now with $n = 1$. The calculations starting with equation (6.35) can be repeated to take the measurement $\mathbf{z}(1)$ into account, i.e. to do the conditioning. For convenience, only the last two equations in the sequel will be shown (compare to equations (6.42) and (6.43))

$$\hat{\mathbf{x}}(1 \mid 1) = \hat{\mathbf{x}}(1 \mid 0) + \mathbf{C}(1 \mid 0)\,\mathbf{H}(1)(\mathbf{H}^{\mathrm{T}}(1)\,\mathbf{C}(1 \mid 0)\,\mathbf{H}(1) + \mathbf{R}(1))^{-1}$$
$$(\mathbf{z}(1) - \mathbf{H}^{\mathrm{T}}(1)\,\hat{\mathbf{x}}(1 \mid 0)) \tag{6.46}$$

$$\mathbf{C}(1 \mid 1) = \mathbf{C}(1 \mid 0) - \mathbf{C}(1 \mid 0)\,\mathbf{H}(1)(\mathbf{H}^{\mathrm{T}}(1)\,\mathbf{C}(1 \mid 0)\,\mathbf{H}(1) + \mathbf{R}(1))^{-1}$$
$$\mathbf{H}^{\mathrm{T}}(1)\,\mathbf{C}(1 \mid 0) \tag{6.47}$$

Repeating the steps as outlined above, we can now formulate a general, recursive algorithm that constitutes the **Kalman filter equations** in their basic form. These equations are commonly divided into two groups, the **measurement-update equations** and the **time-update equations**.

Measurement-update equations

$$\hat{\mathbf{x}}(n \mid n) = \hat{\mathbf{x}}(n \mid n - 1) + \mathbf{C}(n \mid n - 1)\,\mathbf{H}(n)(\mathbf{H}^{\mathrm{T}}(n)\,\mathbf{C}(n \mid n - 1)\mathbf{H}(n)$$
$$+ \mathbf{R}(n))^{-1}(\mathbf{z}(n) - \mathbf{H}^{\mathrm{T}}(n)\,\hat{\mathbf{x}}(n \mid n - 1)) \tag{6.48}$$

$$\mathbf{C}(n \mid n) = \mathbf{C}(n \mid n - 1) - \mathbf{C}(n \mid n - 1)\,\mathbf{H}(n)(\mathbf{H}^{\mathrm{T}}(n)\,\mathbf{C}(n \mid n - 1)\,\mathbf{H}(n)$$
$$+ \mathbf{R}(n))^{-1}\,\mathbf{H}^{\mathrm{T}}(n)\,\mathbf{C}(n \mid n - 1) \tag{6.49}$$

Time-update equations

$$\hat{\mathbf{x}}(n + 1 \mid n) = \mathbf{F}(n)\,\hat{\mathbf{x}}(n \mid n) \tag{6.50}$$

$$\mathbf{C}(n + 1 \mid n) = \mathbf{F}(n)\,\mathbf{C}(n \mid n)\,\mathbf{F}^{\mathrm{T}}(n) + \mathbf{G}(n)\,\mathbf{Q}(n)\,\mathbf{G}^{\mathrm{T}}(n) \tag{6.51}$$

The equations above are straightforward to implement in software. An alternative way of writing them, making it easier to draw the filter in diagram form, is

$$\hat{\mathbf{x}}(n + 1 \mid n) = \mathbf{F}(n)\,\hat{\mathbf{x}}(n \mid n - 1) + \mathbf{K}(n)(\mathbf{z}(n) - \mathbf{H}^{\mathrm{T}}(n)\,\hat{\mathbf{x}}(n \mid n - 1)) \tag{6.52}$$

where $\mathbf{K}(n)$ is the **Kalman gain** matrix

$$\mathbf{K}(n) = \mathbf{F}(n)\,\mathbf{C}(n \mid n - 1)\,\mathbf{H}(n)(\mathbf{H}^{\mathrm{T}}(n)\,\mathbf{C}(n \mid n - 1)\,\mathbf{H}(n) + \mathbf{R}(n))^{-1} \tag{6.53}$$

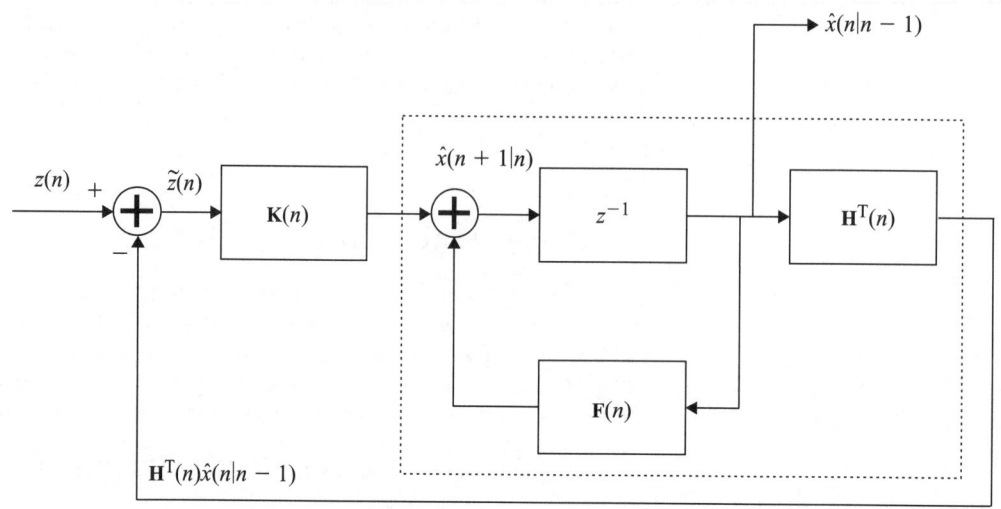

Figure 6.3 *A basic Kalman filter. Note the copy of the signal model in the dotted area*

and the conditional error covariance matrix is given recursively by a discrete-time **Riccati equation**

$$\mathbf{C}(n+1 \mid n) = \mathbf{F}(n)(\mathbf{C}(n \mid n-1) - \mathbf{C}(n \mid n-1)\,\mathbf{H}(n)$$

$$(\mathbf{H}^{\mathrm{T}}(n)\,\mathbf{C}(n \mid n-1)\,\mathbf{H}(n) + \mathbf{R}(n))^{-1}\,\mathbf{H}^{\mathrm{T}}(n)$$

$$\mathbf{C}(n \mid n-1))\,\mathbf{F}^{\mathrm{T}}(n) + \mathbf{G}(n)\,\mathbf{Q}(n)\,\mathbf{G}^{\mathrm{T}}(n) \qquad (6.54)$$

The structure of the corresponding Kalman Filter is shown in Figure 6.3. The filter can be regarded as a copy of the signal model (see Figure 6.2) put in a feedback loop. The input to the model is the difference between the actual measured signal $\mathbf{z}(n)$ and our estimate of the measured signal $\mathbf{H}^{\mathrm{T}}(n)\,\hat{\mathbf{x}}(n \mid n-1)$ multiplied by the Kalman gain $\mathbf{K}(n)$. This is mainly the same approach as the recursive least square estimator in Figure 6.1. The difference

$$\tilde{\mathbf{z}}(n) = \mathbf{z}(n) - \mathbf{H}^{\mathrm{T}}(n)\,\hat{\mathbf{x}}(n \mid n-1) \qquad (6.55)$$

is sometimes referred to as the **innovation**.

Now, to conclude this somewhat simplified presentation of the Kalman filter, let us go back to Bill and the Corvette. As an example, we will design a Kalman filter to estimate Bill's speed and position from the not-too-good measurements available from the radar equipment.

We have earlier defined the signal model, i.e. we know $\mathbf{x}(n)$, $\mathbf{F}$ and $\mathbf{G}$. We are now writing the matrices without indices, since they are constants. Further, we recall that the process noise is a Gaussian random variable $w(n)$, which in this example is a scalar resulting in the auto-correlation matrix $\mathbf{Q}$ turning into a scalar as well. This is simply the variance $Q = \sigma_w^2$ which is assumed to be constant over time.

Regarding the measurement procedure, since we can only measure position and not speed, the observation matrix $\mathbf{H}$ (constant) will be (see also equation (6.31))

$$\mathbf{H} = \begin{bmatrix} 1 \\ 0 \end{bmatrix} \tag{6.56}$$

Since the measured signal $z(n)$ is a scalar in this example, the measurement noise $v(n)$ will also be a scalar, and the auto-correlation matrix $\mathbf{R}$ will simply consist of the constant variance $R = \sigma_v^2$. Now, all the details regarding the signal model are determined, but we also need some vectors and matrices to implement the Kalman filter.

First, we need vectors for the estimates of $\mathbf{x}(n)$

$$\hat{\mathbf{x}}(n \mid n-1) = \begin{bmatrix} \hat{x}_1(n \mid n-1) \\ \hat{x}_2(n \mid n-1) \end{bmatrix}$$

Then, we will need matrices for the error covariance

$$\mathbf{C}(n \mid n-1) = \begin{bmatrix} C_{11}(n \mid n-1) & C_{12}(n \mid n-1) \\ C_{21}(n \mid n-1) & C_{22}(n \mid n-1) \end{bmatrix}$$

If we insert all the known facts above into equations (6.48)–(6.51), the Kalman filter for this example is defined. From equation (6.48), the measurement-update yields

$$\hat{\mathbf{x}}(n \mid n) = \hat{\mathbf{x}}(n \mid n-1) + \begin{bmatrix} C_{11}(n \mid n-1) \\ C_{21}(n \mid n-1) \end{bmatrix} \frac{z(n) - \hat{x}_1(n \mid n-1)}{C_{11}(n \mid n-1) + R} \tag{6.57}$$

which can be expressed in component form

$$\hat{x}_1(n \mid n) = \hat{x}_1(n \mid n-1) + \frac{C_{11}(n \mid n-1)(z(n) - \hat{x}_1(n \mid n-1))}{C_{11}(n \mid n-1) + R} \tag{6.58a}$$

$$\hat{x}_2(n \mid n) = \hat{x}_2(n \mid n-1) + \frac{C_{21}(n \mid n-1)(z(n) - \hat{x}_1(n \mid n-1))}{C_{11}(n \mid n-1) + R} \tag{6.58b}$$

Using equation (6.49) the covariance can be updated

$$\mathbf{C}(n \mid n) = \mathbf{C}(n \mid n-1)$$
$$- \begin{bmatrix} C_{11}(n \mid n-1)\,C_{11}(n \mid n-1) & C_{11}(n \mid n-1)\,C_{12}(n \mid n-1) \\ C_{21}(n \mid n-1)\,C_{11}(n \mid n-1) & C_{21}(n \mid n-1)\,C_{12}(n \mid n-1) \end{bmatrix}$$
$$\cdot \frac{1}{C_{11}(n \mid n-1) + R} \tag{6.59}$$

or expressed in component form

$$C_{11}(n \mid n) = C_{11}(n \mid n-1) - \frac{(C_{11}(n \mid n-1))^2}{C_{11}(n \mid n-1) + R} \tag{6.60a}$$

$$C_{12}(n \mid n) = C_{12}(n \mid n-1) - \frac{C_{11}(n \mid n-1) \, C_{12}(n \mid n-1)}{C_{11}(n \mid n-1) + R} \tag{6.60b}$$

$$C_{21}(n \mid n) = C_{21}(n \mid n-1) - \frac{C_{21}(n \mid n-1) \, C_{11}(n \mid n-1)}{C_{11}(n \mid n-1) + R} \tag{6.60c}$$

$$C_{22}(n \mid n) = C_{22}(n \mid n-1) - \frac{C_{21}(n \mid n-1) \, C_{12}(n \mid n-1)}{C_{11}(n \mid n-1) + R} \tag{6.60d}$$

and then, the time-update equations, starting with equation (6.50)

$$\hat{\mathbf{x}}(n+1 \mid n) = \begin{bmatrix} \hat{x}_1(n \mid n) + \hat{x}_2(n \mid n) \\ \hat{x}_2(n \mid n) \end{bmatrix} \tag{6.61}$$

This can be compared to the basic state–space signal model (6.27). If equation (6.61) is expressed in component form, we get

$$\hat{x}_1(n+1 \mid n) = \hat{x}_1(n \mid n) + \hat{x}_2(n \mid n) \tag{6.62a}$$

$$\hat{x}_2(n+1 \mid n) = \hat{x}_2(n \mid n) \tag{6.62b}$$

Finally, equation (6.51) gives

$$\mathbf{C}(n+1 \mid n)$$
$$= \begin{bmatrix} C_{11}(n \mid n) + C_{21}(n \mid n) + C_{12}(n \mid n) + C_{22}(n \mid n) & C_{12}(n \mid n) + C_{22}(n \mid n) \\ C_{21}(n \mid n) + C_{22}(n \mid n) & C_{22}(n \mid n) \end{bmatrix}$$
$$+ \begin{bmatrix} 0 & 0 \\ 0 & Q \end{bmatrix} \tag{6.63}$$

In component form

$$C_{11}(n+1 \mid n) = C_{11}(n \mid n) + C_{21}(n \mid n) + C_{12}(n \mid n) + C_{22}(n \mid n) \tag{6.64a}$$

$$C_{12}(n+1 \mid n) = C_{12}(n \mid n) + C_{22}(n \mid n) \tag{6.64b}$$

$$C_{21}(n+1 \mid n) = C_{21}(n \mid n) + C_{22}(n \mid n) \tag{6.64c}$$

$$C_{22}(n+1 \mid n) = C_{22}(n \mid n) + Q \tag{6.64d}$$

Hence, the Kalman filter for estimating Bill's position and speed is readily implemented by repetitively calculating the 12 equations (6.58a), (6.58b), (6.60a)–(6.60d), (6.62a), (6.62b) and (6.64a)–(6.64d) above.

For this example, Figure 6.4 shows the true velocity and position of the car and the noisy, measured position, i.e. the input to the Kalman filter. Figure 6.5

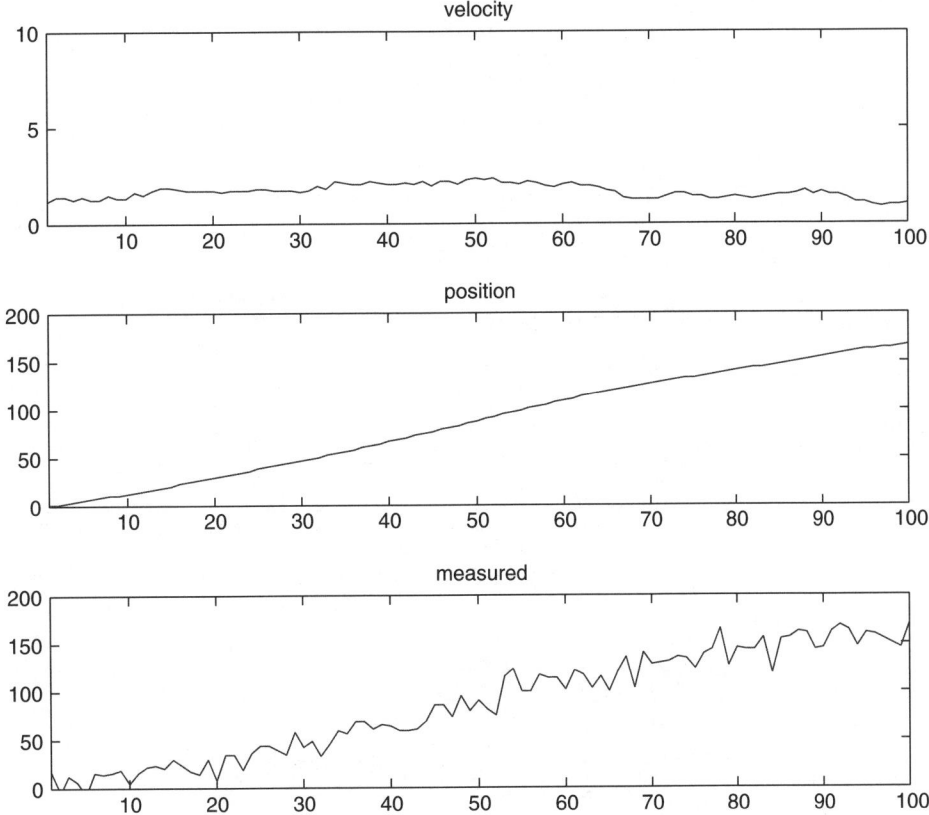

Figure 6.4 *True velocity, position and noisy measured position*

shows the output from the filter, the estimated velocity and position. An overshot can be seen in the beginning of the filtering process, before the filter is tracking. Figure 6.6 shows the two components of the decreasing Kalman gain as a function of time.

6.2.3 Kalman filter properties

At first, it should be stressed that the brief presentation of the Kalman filter in the previous section is simplified. For instance, the assumption about Gaussian noise is not necessary in the general case (Anderson and Moore, 1979). Nor is the assumption that the process and measurement noise is uncorrelated. There are also a number of extensions (Anderson and Moore, 1979) of the Kalman filter which have not been described here. Below, we will however discuss some interesting properties of the general Kalman filter.

The Kalman filter is linear. This is obvious from the preceding calculations. The filter is also a **discrete-time** system and has **finite dimensionality**.

The Kalman filter is an optimal filter in the sense of achieving minimum variance estimates. It can be shown that in Gaussian noise situations, the Kalman

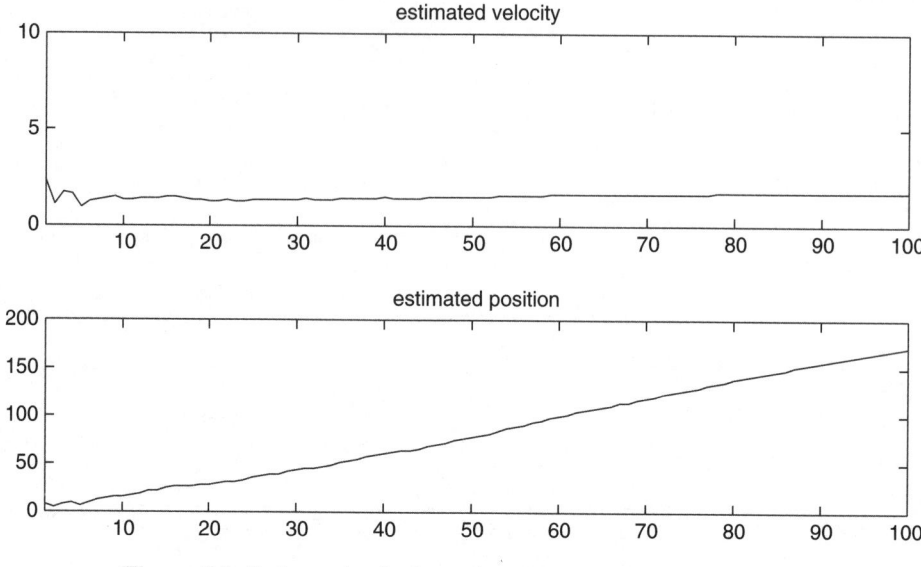

Figure 6.5 *Estimated velocity and position*

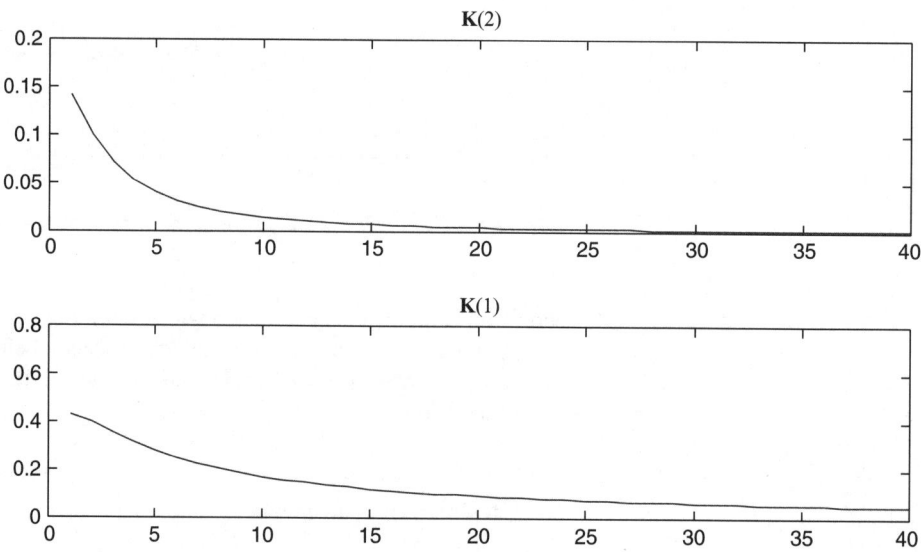

Figure 6.6 *Kalman gain as a function of time*

filter is the **best** possible filter, and in non-Gaussian cases, the **best linear** filter. The optimal filter in the latter case is non-linear and may therefore be very hard to find and analyze in the general case.

The gain matrix $\mathbf{K}(n)$ **can be calculated off-line, in advance, before the Kalman filter is actually run**. The output $\hat{\mathbf{x}}(n \mid n - 1)$ of the filter is obviously dependent on the input $\mathbf{z}(n)$, but the covariance matrix $\mathbf{C}(n \mid n - 1)$ and hence the Kalman gain $\mathbf{K}(n)$ is **not**. $\mathbf{K}(n)$ can be regarded as the smartest way of taking the measurements into account given the signal model and statistical properties.

From the above, we can also conclude that since $\mathbf{C}(n \mid n - 1)$ is independent of the measurements $\mathbf{z}(n)$, no one set of measurements helps more than any other to eliminate the uncertainty about $\mathbf{x}(n)$.

Another conclusion that can be drawn is that the filter is only optimal given the signal model and statistical assumptions made at design time. If there is a poor match between the real world signals and the assumed signals of the model, the filter will, of course, not perform optimally in reality. This problem is, however, common to all filters.

Further, even if the signal model is time invariant and the noise processes are stationary, i.e. $\mathbf{F}(n)$, $\mathbf{G}(n)$, $\mathbf{H}(n)$, $\mathbf{Q}(n)$ and $\mathbf{R}(n)$ are constant, in general $\mathbf{C}(n \mid n - 1)$ and hence $\mathbf{K}(n)$ will **not** be constant. This implies that in the general case, the **Kalman filter will be time varying**.

The Kalman filter contains a **model**, which **tracks** the true system we are observing. So, from the model, we can obtain estimates of state variables that we are only measuring in an indirect way. We could, for instance, get an estimate of the speed of Bill's Corvette in our example above, despite only measuring the position of the car.

Another useful property of the built-in model is that in case of missing measurements during a limited period of time, the filter can "**interpolate**" the state variables. In some applications, when the filter model has "stabilized", it can be "sped up" and even be used for **prediction**.

Viewing the Kalman filter in the frequency domain, it can be regarded as a low-pass filter with varying cut-off frequency (Bozic, 1994). Take for instance a scalar version of the Kalman filter, e.g. the RLS algorithm equation (6.9), repeated here for convenience

$$\hat{x}(N + 1) = \hat{x}(N) + \frac{1}{N + 1}(z(N + 1) - \hat{x}(N)) \tag{6.65}$$

To avoid confusion we shall rename the variables so that $u(n)$ is the input signal and $v(n)$ is the output signal of the filter, resulting in

$$v(N + 1) = v(N) + \frac{1}{N + 1}(u(N + 1) - v(N)) \tag{6.66}$$

Denoting the gain factor $k = 1/(N + 1)$ and taking the z-transform of equation (6.66) we get

$$zV(z) = V(z) + k(zU(z) - V(z)) \tag{6.67}$$

Of course, the z in equation (6.67) is the z-transform parameter, while in equation (6.65) it is the input signal to the filter. They are certainly **not** the same entity. Rewriting equation (6.67) we obtain the transfer function of the filter

$$H(z) = \frac{V(z)}{U(z)} = \frac{kz}{z - 1 + k} \tag{6.68}$$

Now, when the filter is just started, and $N = 0$ we get $k = 1$ and the transfer function will be

$$H(z) = \frac{z}{z} = 1 \tag{6.69}$$

In this case, the filter has a pole and a zero in the center of the unit circle in the z-plane and the magnitude of the amplitude function is unity. This is nothing but an **all-pass filter** with gain one.

Later, when the filter has been running for a while and $N \to \infty$, $k \to 0$, the transfer function (6.68) turns into

$$H(z) = \frac{0 \cdot z}{z - 1} \tag{6.70}$$

This is a low-pass filter with a pole on the unit circle and a gain tending towards zero. Due to the placement of the pole, we are now dealing with a highly narrow-band low-pass filter, actually a pure digital integrator.

6.2.4 Applications

The Kalman filter is a very useful device that has found many applications in diverse areas. Since the filter is a discrete-time system, the advent of powerful and not too costly digital signal processing (DSP) circuits has been crucial to the possibilities of using Kalman filters in commercial applications.

The Kalman filter is used for **filtering** and **smoothing** measured signals, not only in electronic applications, but also in processing of data in the areas, e.g. economy, medicine, chemistry and sociology. Kalman filters are also to some extent used in **digital image processing**, when enhancing the quality of digitized pictures. Further detection of signals in radar and sonar systems and in telecommunication systems often requires filters for equalization. The Kalman filter, belonging to the class of adaptive filters, performs well in such contexts (see Chapter 3).

Other areas where Kalman filters are used are **process identification**, **modeling** and **control**. Much of the early developments of the Kalman filter theory came from applications in the aerospace industry. One control system example is keeping satellites or missiles on a desired trajectory. This task is often solved using some optimal control algorithm, taking estimated state variables as inputs. The estimation is, of course, done with a Kalman filter. The estimated state vector may be of dimension 12 (or more) consisting of position (x, y, z), yaw, roll, pitch and the first derivatives of these, i.e. speed (x, y, z) and speed of yaw, roll and pitch movements. Designing such a system requires a considerable amount of computer simulations.

An example of process modeling using Kalman filters is to analyze the behavior of the stock market, and/or to find parameters for a model of the underlying economic processes. Modeling of meteorological and hydrological processes as well as chemical reactions in the manufacturing industry are other examples.

Kalman filters can also be used for **forecasting**, for such as prediction of air pollution levels, air traffic congestion, etc.

Summary   In this chapter we have covered:

- Recursive least square (RLS) estimation
- The pseudo-inverse and how to obtain it in an iterative way
- The measurement-update equations and the time-update equations of the Kalman filter

- The innovation, the Kalman gain matrix and the Riccatti equation
- Properties and applications of the Kalman filter.

Review questions **R6-1** The gas tank gauge of a car is sampled every second by an analog-to-digtal (A/D) converter and an embedded computer. Unfortunately, the transducer signal is very noisy, resulting in a fluctuating reading on the digital display. To "clean" the transducer signal and obtain a stable reading, a suggestion is to employ an RLS algorithm as in equation (6.9). Why is this not a good idea? How could the algorithm be modified to perform better?

R6-2 What is the pseudo-inverse? When is it used?

R6-3 Draw a block diagram of a basic Kalman filter. Explain the functioning of the different blocks and state the signal names (compare to Figure 6.3).

R6-4 In the "Bill and Corvette example", we have assumed an almost constant speed and the small variations were represented by the process noise. If we extend the state–space model assuming a constant acceleration $x_3(n)$ (Bill has got a gas turbine for his car?), how will this be reflected in the state vector and the transition matrix?

R6-5 Explain briefly how a Kalman filter can be used for prediction of a future signal. Which are the dangers?

Solved problems **P6-1** Examine the measurement-update and time-update equations (6.48)–(6.51), and explain the significance of the different terms in the equations. How will the filter perform if the signal quality is perfect (no measurement noise)? What happens if the quality is extremely bad? Try setting the measurement noise to zero and infinity, respectively, in the measurement-update and time-update equations and explain the behavior of the filter.

P6-2 Derive the Kalman filter function (6.52), the Kalman gain equation (6.53) and the Riccatti equation (6.54) starting from the measurement-update and time-update equations. What happens to the Kalman gain and the Riccatti equation if the signal quality is perfect (no measurement noise)? What happens if the quality is extremely bad?

P6-3 Write a MATLAB™ program to simulate Bill and his Corvette. The program should calculate and plot the true velocity, position and the measured position as in Figure 6.4. Use the MATLAB™ expressions `w = 0.15*randn` to generate the process noise and `v = 10*randn` for the measurement noise. Further, assume the following starting conditions (at time 0): position equal to 0 and speed equal to 1.

P6-4 Write a MATLAB™ program implementing the Kalman filter as used in the "Bill and Corvette example". The program should use the measured position generated by the program in P6-3 above as the input signal and plot the estimated speed and estimated position as in Figure 6.5. Assume the variance of the process noise to be 0.2 and the variance of the measurement noise to be equal to 4. Start the Kalman gain vector and the covariance matrix with all ones.

7 Data compression

Background What is information? Certainly a relevant question, but is there a unique answer? If you ask a journalist, a politician or an engineer, you will probably get quite different answers. For people working in the telecommunications and computing businesses, however, the answer is simple. Transporting and storing information is money.

In this chapter, we will look deeper into the nature of information, and we will do it from a mathematical and statistical point of view. We will also consider the problem of data compression, i.e. different ways of minimizing the amount of digital data symbols, e.g. bits, need to represent a given amount of information in, for instance, a computer file. Fewer bits to store per file means more files and more users on a hard disk. Fewer bits to transport means faster communications, better transmission capacity and more customers in our networks and communication satellite links. Hence, data compression is a good business.

Objectives In this chapter we will deal with:

- General ideas of information theory, how to quantize information, the "bit"
- Mutual information, entropy and redundancy, source coding
- Huffman algorithm and prefix-free variable length codes
- Different versions of delta modulation and differential pulse code modulation (PCM) algorithms
- Speech coding techniques and vocoders
- Image coding, joint photographics expert group (JPEG) and moving pictures expert group (MPEG)
- Other common data compression methods, layer-3 of MPEG-1 (MP3) and Lempel–Ziv–Welch (LZW).

7.1 An information theory primer

7.1.1 Historic notes

The concept of **information theory** (Cover and Thomas, 1991) in a strict mathematical sense was born in 1948, when C.E. Shannon published his celebrated work "A Mathematical Theory of Communication", later published using the title "The Mathematical Theory of Communication". Obviously, Shannon was ahead of his time and many of his contemporary communication specialists did not understand his results. Gradually it became apparent, however, that Shannon had indeed created a new scientific discipline.

Besides finding a way of quantizing information in a mathematical sense, Shannon formulated three important fundamental theorems: **the source coding theorem**, **the channel coding theorem** and **the rate distortion theorem**. In this chapter, we will concentrate on the source coding theorem, while implications of the channel coding theorem will be discussed in Chapter 8.

The works of Shannon are an important foundation of modern information theory. It has proven useful not only in circuit design, computer design and communications technology, but it is also being applied to biology and psychology, to phonetics and even to semantics and literature.

7.1.2 Information and entropy

There are many different definitions of the term "information"; however, in this case we will define it as receiving **information** implies a **reduction of uncertainty** about a certain condition. Now, if we assume that this condition or variable can take a finite number of values or states, these states can be represented using a finite set of **symbols**, an "**alphabet**".

Consider the following example: in a group of 10 persons, we need to know in which country in Scandinavia each person is born. There are three possibilities; Denmark, Norway and Sweden. We decide to use a subset of the digits 0, 1, 2, ..., 9 as our symbol alphabet as follows

$$1 = \text{Denmark}, \quad 2 = \text{Norway}, \quad 3 = \text{Sweden}$$

The complete **message** will be a 10 digit long string of symbols, where each digit represents the birthplace of a person. Assuming that the probability of being born in any of the three countries is equal, $p_1 = 1/3$ (born in Denmark), $p_2 = 1/3$ (born in Norway) and $p_3 = 1/3$ (born in Sweden), the total uncertainty about the group (in this respect) before the message is received in one out of $3^{10} = 59\,049$ (the number of possible combinations). For every symbol of the message we receive, the uncertainty is reduced by a factor of three, and when the entire message is received, there is only one possible combination left. Our uncertainty has hence been reduced and the information has been received.

Extending our example above, we now need to know in which country in Europe the persons are born. We realize that our symbol alphabet must be modified, since there are 34 countries in Europe today (changes rather quickly these days). We extend our alphabet in the following way. First we use 0, 1, 2, ..., 9 as before, and after that we continue using the normal letters, A, B, C, ..., Z. This will give us an alphabet consisting of $10 + 26 = 36$ different possible symbols, out of which we use 34. If all the countries are equally probable birthplaces, the uncertainty before we get the message is one out of $34^{10} \approx 2.06 \cdot 10^{15}$ possible combinations. Every symbol reduces the information by a factor of 34 and when the entire 10 symbols string is received, no uncertainty exists.

From the above example, we draw the conclusion that in the latter case, the message contained a larger amount of information than in the first case. This is because of the fact that in the latter case there were more countries to choose from than in the former. The uncertainty of "which country in Europe?" is of course larger than the uncertainty of "which country in Scandinavia?". If we know in advance that there are only persons born in Scandinavia in the group, this represents a certain amount of information that we already have and that need not be included in the message. From this discussion, we are now aware of the fact that when dealing with information measures, a very relevant question is "information about **what**?".

In the latter case above, dealing with extending the alphabet, we could have chosen another approach. Instead of using **one** character per symbol, we could have used **two** characters, e.g. 01, 02, 03, ..., 33, 34. In this way, only the digits 0, 1, 2, 3, ..., 9 would have been used as in the earlier case, but on the other hand, the message would have been twice as long, i.e. 20 characters. Actually, we could have used any set of characters to define our alphabet. There is, however, a trade-off in that the smaller the set of characters, the longer the message. The number of characters per symbol L needed can be expressed as

$$L = \left\lceil \frac{\log(N_S)}{\log(N_C)} \right\rceil = \lceil \log_{N_C}(N_S) \rceil \tag{7.1}$$

where N_S is the number of symbols needed, in other words, the number of discrete states, N_C is the number of different characters used and $\lceil\ \rceil$ is the "ceiling" operator (the first integer greater or equal to the argument). For instance, if we use the popular **binary digits** (**BITs** for short) 0 and 1, this implies that $N_C = 2$. Using binary digits in our examples above implies

$$\text{"Scandinavia"}: \quad N_S = 3 \quad L = \lceil \log_2(3) \rceil = \lceil \text{lb}(3) \rceil = \lceil 1.58 \rceil$$
$$= 2 \text{ char/symbol}$$
$$\text{"Europe"}: \quad N_S = 34 \quad L = \lceil \text{lb}(34) \rceil = \lceil 5.09 \rceil = 6 \text{ char/symbol}$$

From above, it can be seen that lb( ) means logarithm base 2, i.e. $\log_2($).

A binary message for the "Scandinavia" example would be 20-bit long, while for the "Europe" example 60 bits. It is important to realize that the "shorter" version (using many different characters) of the respective strings contains the same amount of information as the "longer", binary version.

Now in an information theory context, we can see that the choice of characters or the preferred entries used to represent a certain amount of information does not matter. Using a smaller set of characters would, however, result in longer messages for a given amount of information. Quite often, the choice of character set is made upon the way a system is implemented. When analyzing communication systems, binary characters (bits) are often assumed for convenience, even if another representation is used in the actual implementation.

Let us look at another situation and assume that we also want to know the gender of the persons in the group. We choose to use symbols consisting of normal letters. The two symbols will be the abbreviations "fe" for female and "ma" for male. Hence, information about the entire group with respect to sex will be a 20 characters long string, e.g.

fefemafemamamafefema

Taking a closer look at this string of information, we realize that if every second letter is removed, we can still obtain all the information we need

ffmfmmmffm

or

eeaeaaaeea

would be OK. In this case, the longer string obviously contains "unnecessary information" or so-called **redundant information**, i.e. information that could be deducted from other parts of the message and thus not reduce our uncertainty (no "news"). If the redundant information is removed, we can still reduce the uncertainty to a desired level and obtain the required information. In this simple example, the redundancy in the longer string is due to the **dependency** between the characters. If we receive, for instance, the character "f", we can be 100% sure that the next character will be "e", so there is no uncertainty. The shorter versions of the string cannot be made any more shorter, because there is no dependency between the characters. The sex of a person is independent of the sex of another person. The appearance of the characters "f" and "m" seems to be completely random. In this case, no redundancy is present. The example above is of course a very simple and obvious one. Redundancy is in many cases "hidden" in more sophisticated ways.

Removing redundancy is mainly what **data compression** is all about. A very interesting question in this context is how much information is left in a message when all redundancy is removed? What maximum data compression factor is possible? This question is answered by the source coding theorem of Shannon. Unfortunately, the theorem does not tell us the smartest way of finding this maximum data compression method; it only tells us that there is a minimum message size that contains exactly all the information we need.

A drawback when removing redundancy is that our message becomes more vulnerable to errors. In the example above, assume that the letters are written by hand in haste, and we find a distorted character looking like an "o". Since we have only defined "e" for female and "a" for male, this leads to an uncertainty. If we had used the system with redundant characters, we would have seen either "fo" or "mo", giving us a chance to resolve the ambiguity. This is the basic idea behind error-correcting codes, dealt with in Chapter 8.

Now, let us be a bit more formal about the above. To start with, we shall define the concept of **mutual information**. Assuming that we want to gain information about an event A by observing the event B, the mutual information is defined as

$$I(A,B) = \log_b\left(\frac{P(A\mid B)}{P(A)}\right) \tag{7.2}$$

where we have assumed that the probability of the event A is non-zero, i.e. $P(A) \neq 0$ and that $P(B) \neq 0$. $P(A|B)$ is the **conditional probability**, i.e. the probability of A given that B is observed. The unit will be "bits" if logarithm base 2 (lb( )) is used ($b = 2$) and "nats" if the natural logarithm (ln( )) is used ($b = e$).

Using Bayes' theorem (Papoulis and Pillai, 2001) it is straightforward to show that $I(A,B)$ is symmetric with respect to A and B

$$I(A,B) = \log\left(\frac{P(A\mid B)}{P(A)}\right) = \log\left(\frac{P(AB)}{P(A)P(B)}\right) = \log\left(\frac{P(B\mid A)}{P(B)}\right) = I(B,A) \tag{7.3}$$

Hence, it does not matter if B is observed to gain information about A, or vice versa. The same amount of information is obtained. This is why $I(A,B)$ is called the **mutual** information between the events A and B.

If we now assume that the events A and B are completely independent, there is no "coupling" in between them what-so-ever. In this case, $P(A|B) = P(A)$ and the knowledge of A is not increased by the fact that B is observed. The mutual information in this case is

$$I(A,B) = \log\left(\frac{P(A)}{P(A)}\right) = \log(1) = 0 \tag{7.4}$$

On the other hand, if there is a complete dependence between A and B or, in other words, if when observing B we are 100% sure that A has happened, then $P(A|B) = 1$ and

$$I(A,B) = \log\left(\frac{1}{P(A)}\right) = -\log(P(A)) \tag{7.5}$$

The 100% dependence assumed above is equivalent to observing A itself instead of B, hence

$$I(A,A) = -\log(P(A)) \tag{7.6}$$

which represents the maximum information we can obtain about the event A or, in other words, the maximum amount of information inherent in A.

So far we have discussed single events, but if we now turn to the case of having a discrete random variable X that can take one of the values $x_1, x_2, \ldots, x_K$, we can define the K events as A_i when $X = x_i$. Using equation (7.6) above and taking the average of the information for all events, the **entropy** of the stochastic, discrete variable X can be calculated as

$$H(X) = E[I(A_i, A_i)] = -\sum_{i=1}^{K} f_X(x_i) \log(f_X(x_i)) \tag{7.7}$$

where $f_X(x_i)$ is the probability that $X = x_i$, i.e. that event A_i has happened.

The entropy can be regarded as the information inherent in the variable X. As stated in the source coding theorem, this is the minimum amount of information we must keep to be able to reconstruct the behavior of X without errors. In other words, this is the redundancy-free amount of information that should preferably be produced by an ideal data compression algorithm.

An interesting thing about the entropy is that the maximum information (entropy) is obtained when all the outcomes, i.e. all the possible values x_i, are equally probable. In other words, the signal X is completely random. This seems intuitively right, since when all the alternatives are equally probable the uncertainty is the greatest.

Going back to our examples in the beginning of this section, let us calculate the entropy. For the case of "birthplace in Scandinavia" the entropy is (assuming equal probabilities)

$$H(X) = -\sum_{i=1}^{3} f_X(x_i) \, \text{lb}(f_X(x_i)) = -\frac{1}{3}\sum_{i=1}^{3} \text{lb}\left(\frac{1}{3}\right) = 1.58 \text{ bits/person}$$

Hence, the minimum amount of information for 10 independent persons is 15.8 bits (we used 20 bits in our message). Data compression is therefore possible. We only have to find the "smartest" way.

For the "birthplace in Europe" example, the entropy can be calculated in a similar way

$$H(X) = -\sum_{i=1}^{34} f_X(x_i)\,\mathrm{lb}(f_X(x_i)) = -\frac{1}{34}\sum_{i=1}^{34}\mathrm{lb}\left(\frac{1}{34}\right) = 5.09 \text{ bits/person}$$

The minimum amount of information for 10 independent persons is 50.9 bits and we used 60 bits in our message. Data compression is possible in this case too.

7.1.2.1 Some concluding remarks about the concept of entropy

The term entropy is commonly used in thermodynamics, where it refers to the disorder of molecules. In a gas, molecules move around in random patterns, while in a crystal lattice (solid state) they are lined up in a very ordered way; in other words, there is a certain degree of dependency. The gas is said to have higher entropy than the crystal state. Now, this has a direct coupling to the information concept. If we had to describe the position of a number of molecules, it would require more information for the gas than for the crystal. For the molecules in the gas we would have to give three coordinates (x, y, z) in space for every molecule. For the crystal, the positions could be represented by giving the coordinates for the reference corner of the lattice and the spacing of the lattice in x, y and z directions.

If you take, for instance, water and turn it into ice (a crystal with lower entropy), you have to put the water in the freezer and add energy. Hence, **reduction of entropy requires energy**. This is also why you become exhausted by cleaning your room (reduction of entropy), while the "messing up" process seems to be quite relaxing. In an information theory context, freezing or cleaning translates into getting rid of information, i.e. "forgetting" requires energy.

7.2 Source coding

In the previous section, we concluded that getting rid of redundancy was the task of data compression. This process is also called **source coding**. The underlying idea is based on the fact that we have an **information source** generating information (discrete or continuous) at a rate of H bits/s (the entropy). If we need to store this information, we want to use as small storage as possible per time unit. If we are required to transmit the information, it is desirable to use as low data rate R as possible. From the source coding theorem, we however know that $R \geq H$ to avoid loss of information.

There are numerous examples of information sources, such as a keyboard (+human) generating text, a microphone generating a speech signal, a television camera, an image scanner, transducers in a measurement logging system, etc.

The output information flow (signals) from these sources is commonly the subject of source coding (data compression) performed using general purpose digital computers or digital signal processors (DSPs). There are two classes of data compression methods. In the first class, we find **general data compression algorithms**, i.e. methods that work fairly well for any type of input data (text, images and speech signals). These methods are commonly able to "decompress", in other words, restore the original information data sequence without any errors ("lossless"). File compression programs for computers (e.g. "double space", "pkzip" and "arc") are typical examples. This class of data compression methods therefore adhere to the source coding theorem.

In the other class of data compression methods, we find **specialized algorithms**, for instance, speech coding algorithms in cellular mobile telephone systems. This kind of data compression algorithms makes use of prior information about the data sequences and may, for instance, contain a model of the information source. For a special type of information, these specialized methods are more efficient than the general type of algorithms. On the other hand, since they are specialized, they may exhibit poor performance if used with other types of input data. A speech coding device designed for Nordic language speech signals may, for instance, be unusable for Arabian users. Another common property of the algorithms in this class is the inability to restore exactly the original information ("lossy"). They can only restore it "sufficiently" well. Examples are speech coding systems (your mother almost sounds like your girlfriend) or image compression algorithms, producing more or less crude pictures with poor resolution and a very limited number of color shades.

7.2.1 Huffman algorithm

In the following discussion, we will mainly use binary digits, i.e. 0 and 1. Other character sets can, of course, be used. The basic idea of the **Huffman coding** algorithm is to use variable length, prefix-free symbols and to minimize the average length of the symbols (in number of characters), taking the statistics of the symbols into account. This algorithm belongs to the class of general data compression algorithms.

Variable length means that a certain symbol can be one, two, three and so on characters long, e.g. 0, 10, 1111, etc. To be able to decode a string of variable length symbols, no symbol is, however, allowed to be a prefix of a longer symbol. We must be able to identify the symbol as soon as the last character of the symbol is received. The code must be **prefix free**. A simple example of a prefix-free code is a code consisting of the symbols

$$0, 10, 11$$

while the code using the symbols

$$1, 11, 101$$

is **not** prefix free. If we, for instance, receive "111011" we cannot tell if it is "1", "1", "101", "1" or "11", "101", "1" or …?

The average length of a symbol can be expressed as

$$\mathrm{E}[L] = \sum_{i=1}^{K} l_i f_U(u_i) \tag{7.8}$$

where l_i is the number of characters (the length) of symbol u_i and $f_U(u_i)$ is the probability of symbol u_i. The way to minimize the average symbol length (equation (7.8)) is obviously to assign the most probable symbols the shortest codes and the least probable symbols the longest codes. This idea is not new. It is, for instance, used in telegraphy in form of the **Morse code** (Carron, 1991), invented by Samuel Morse in the 1840s. The most probable letters like "e" and "t" have the shortest Morse codes, only one dot or one dash, respectively, while uncommon characters like question mark are coded as long sequences of dots and dashes.

Designing a code having the desired properties can be done using Huffman algorithm, which is a tree algorithm. It will be demonstrated for binary codes, which result in binary trees, i.e. trees that branch into **two** branches, 0 and 1. Huffman algorithm can be expressed using "pseudo-code" in the following way:

```
huffman:   assign every symbol a node
           assign every node the probability of the symbol
           make all the nodes active.

           while (active nodes left) do
           {
                take the two least probable active nodes
                join these nodes into a binary tree
                deactivate the nodes
                add the probabilities of the nodes
                assign the root node this probability
                activate the root node.
           }
```

The following example will be used to illustrate the way the algorithm works. Assume that we have five symbols from our information source with the following probabilities

$u_1 \quad f_U(u_1) = 0.40$
$u_2 \quad f_U(u_2) = 0.20$
$u_3 \quad f_U(u_3) = 0.18$
$u_4 \quad f_U(u_4) = 0.17$
$u_5 \quad f_U(u_5) = 0.05$

The resulting Huffman tree is shown in Figure 7.1 (there are many possibilities). Probabilities are shown in parentheses.

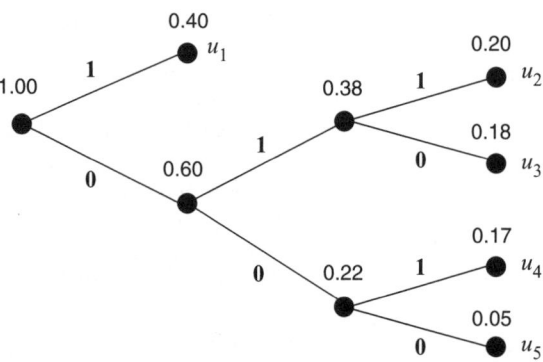

Figure 7.1 *A tree diagram used to find the optimum binary Huffman code of the example*

Building the tree is as follows. Firstly, we take the symbols having the smallest probability, i.e. u_5 and u_4. These two symbols now constitute the first little sub-tree down to the right. The sum of the probabilities is $0.05 + 0.17 = 0.22$. This probability is assigned to the root node of this first little sub-tree. The nodes u_5 and u_4 are now deactivated, i.e. dismissed from the further process, but the root node is activated.

Looking at our table of active symbols, we now realize that the two smallest active nodes (symbols) are u_2 and u_3. We form another binary sub-tree above the first one in a similar way. The total probability of this sub-tree is $0.20 + 0.18 = 0.38$, which is assigned to the root node.

During the next iteration of our algorithm, we find that the two active root nodes of the sub-trees are those possessing the smallest probabilities, 0.22 and 0.38, respectively. These root nodes are hence combined into a new sub-tree, having the total probability of $0.22 + 0.38 = 0.60$.

Finally, there are only two active nodes left, the root node of the sub-tree just formed and the symbol node u_1. These two nodes are joined by a last binary sub-tree and the tree diagram for the optimum Huffman code is completed. Using the definition that "up-going" (from left to right) branches are represented by a binary "1" and "down-going" branches by a "0" (this can, of course, be done in other ways as well), we can now write down the binary, variable length, prefix-free codes for the respective symbols

$$
\begin{array}{ll}
u_1 & 1 \\
u_2 & 011 \\
u_3 & 010 \\
u_4 & 001 \\
u_5 & 000
\end{array}
$$

Using equation (7.8) the average symbol length can be calculated as $E[L] = 2.2$ bits/symbol. The entropy can be found to be $H = 2.09$ bits/symbol using equation (7.7). As can be seen, we are quite close to the maximum data compression possible. Using fixed length, binary coding would have resulted in $E[L] = 3$ bits/symbol.

Using this algorithm in the "born in Scandinavia" example, an average symbol length of $E[L] = 1.67$ bits/symbol can be achieved. The entropy is $H = 1.58$ bits/symbol. For the "born in Europe" example the corresponding figures are $E[L] = 5.11$ bits/symbol and $H = 5.09$ bits/symbol.

The **coding efficiency** can be defined as

$$\eta = \frac{H(X)}{E[L]} \tag{7.9}$$

and the **redundancy** can be calculated using

$$r = 1 - \eta = 1 - \frac{H(X)}{E[L]} \tag{7.10}$$

If the efficiency is equal to one, the redundancy is zero and we have found the most compact, "lossless" way of coding the information. For our examples above, we can calculate the efficiency and the redundancy

$$\text{"Born in Scandinavia":} \quad \eta = \frac{1.58}{1.67} = 0.946 \Rightarrow r = 1 - 0.946 = 0.054$$

$$\text{"Born in Europe":} \quad \eta = \frac{5.09}{5.11} = 0.996 \Rightarrow r = 1 - 0.996 = 0.004$$

In principle, the Huffman algorithm assumes that the information source is "without memory", which means that there is no dependency between successive source symbols. The algorithm can, however, be used even for information sources with "memory", but better algorithms can often be found for these cases. One such case is the coding of facsimile signals, described at the end of this chapter.

7.2.2 Delta modulation, adaptive delta modulation and continuously variable slope delta modulation

Delta modulation (DM) was briefly discussed in Chapter 2 in conjunction with sigma–delta analog-to-digital (A/D) converters. The delta modulation technique can also be viewed as a data compression method. In this case, redundancy caused by dependency between successive samples in, for instance, a pulse code modulation (PCM) data stream can be removed. The idea behind delta modulation is to use the **difference** between two successive data samples, rather than the samples themselves. Hopefully, the difference will be small and can be represented by fewer bits than the value of the samples themselves. In the simplest form of delta modulation, the difference is fixed to only ± 1 PCM step and can hence be represented by 1 bit. Assuming the PCM word has a word length of N bits, and that it is transmitted using a bit rate of R bits/s, the transmission rate of a DM signal would be only R/N bits/s. Figure 7.2 shows a block diagram of a simple delta modulator (data compressing algorithm) and delta demodulator ("decompressing" algorithm).

Starting with the modulator, the incoming data sequence $x(n)$ is compared to a predicted value $\hat{x}(n)$, based on earlier samples. The difference $d(n) = x(n) - \hat{x}(n)$ ("prediction error") is fed to a **1-bit quantizer**, in this case basically a sign detector. The output from this detector is $s(n)$

$$s(n) = \begin{cases} +\delta & \text{if } d(n) \geq 0 \\ -\delta & \text{if } d(n) < 0 \end{cases} \tag{7.11}$$

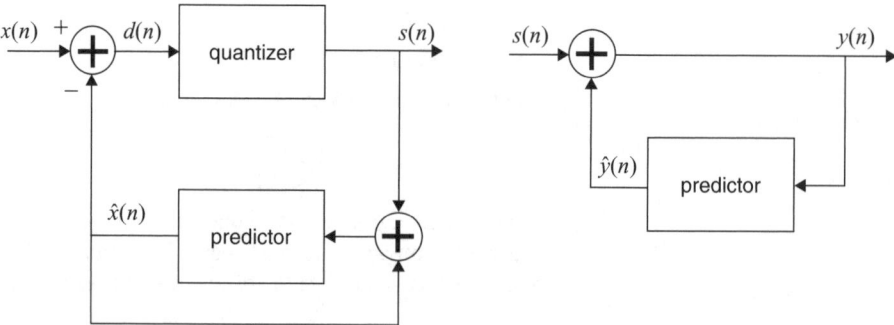

Figure 7.2 *A generic delta modulator and demodulator*

where δ is referred to as the **step size**, in this generic system $\delta = 1$. Further, in this basic-type delta modulation system, the predictor is a **first-order predictor**, basically an accumulator, in other words, a one sample delay (memory), hence

$$\hat{x}(n) = \hat{x}(n-1) + s(n-1) \tag{7.12}$$

The signal $x(n)$ is reconstructed in the demodulator using a predictor similar to the one used in the modulator. The demodulator is fed the delta-modulated signal $s(n)$, which is a sequence of $\pm\delta$ values. The reconstructed output signal is denoted $y(n)$ in Figure 7.2

$$y(n) = \hat{y}(n) + s(n) = y(n-1) + s(n) \tag{7.13}$$

A simple DM system like above, using a fixed step size δ, is called a **linear delta modulation (LDM)** system. There are two problems associated with this type of system, **slope overload** and **granularity**. The system is only capable of tracking a signal with a derivative (slope) smaller than $\pm\delta/\tau$ where τ is the time between two consecutive samples, i.e. the sampling period. If the slope of the signal is too steep, slope overload will occur. This can, of course, be cured by increasing the sampling rate (for instance, oversampling which is used in the sigma–delta A/D converter, see Chapter 2) or increasing the step size. By increasing the sampling rate, the perceived advantage using delta modulation may be lost. If the step size is increased, the LDM system may exhibit poor signal-to-noise ratio (SNR) performance when a weak input signal is present, due to the ripple caused by the relatively large step size. This is referred to as the granularity error.

Choosing a value of the step size δ is hence crucial to the performance of an LDM system. This kind of data compression algorithm works best for slowly varying signals having a moderate dynamic range. LDM is often inappropriate for speech signals, since in these signals there are large variations in amplitude between silent sounds (e.g. "ph" or "sh") and loud sounds (like "t" or "a").

A way to reduce the impact of the problems above is to use **adaptive delta modulation (ADM)**. In the ADM system, the step size is adapted as a function of the output signal $s(n)$. If, for instance, $s(n)$ has been $+\delta$ for a certain number of consecutive samples, the step size is increased. This simple approach may improve the DM system quite a bit. However, a new problem is now present. If there are errors when transmitting the data sequence $s(n)$, the modulator and demodulator may disagree about the present step size. This may cause considerable errors that will be present for a large number of succeeding samples. A way to recover from these errors is to introduce "leakage" into the step size adaptation algorithm. Such a system is the **continuously variable slope delta (CVSD) modulation** system.

In CVSD, the step size $\delta(n)$ depends on the two previous values of the output $s(n)$. If there has been consecutive runs of ones or zeros, there is a risk for slope overload and the step size is increased to be able to "track" the signal faster. If, on the other hand, the previous values of $s(n)$ have been alternating, the signal is obviously not changing very fast, and the step size should be reduced (by "leakage") to minimize the granulation noise. The step size in a CVSD system is adapted as

$$\delta(n) = \begin{cases} \gamma\,\delta(n-1) + C_2 & \text{if } s(n) = s(n-1) = s(n-2) \\ \gamma\,\delta(n-1) + C_1 & \text{else} \end{cases} \tag{7.14}$$

where $0 < \gamma < 1$ is the "leakage" constant and $C_2 \geq C_1 > 0$. The constants C_1 and C_2 are used to define the minimum and maximum step size as

$$\frac{C_1}{1 - \gamma} < \delta(n) < \frac{C_2}{1 - \gamma} \tag{7.15}$$

Still, the system may suffer from slope overload and granularity problems. CVSD, e.g. being less sensitive to transmission errors than ADM, was however widely popular until the advent of adaptive differential pulse code modulation (ADPCM) algorithms (see below) standardized by the Comité Consultatif International Télegraphique et Télephonique (CCITT).

7.2.3 Differential pulse code modulation and adaptive differential pulse code modulation

The **differential pulse code modulation (DPCM)** scheme can be viewed as an extension of delta modulation presented above. The same block diagram Figure 7.2 applies, but in the DPCM case, we have a quantizer using p values, not a 1-bit quantizer as in the former case, but rather an $\mathrm{lb}(p)$-bit quantizer. Further, the predictor in this case is commonly of higher order, and can be described by its impulse response $h(n)$, hence

$$\hat{x}(n) = \sum_{k=1}^{N} h(k)(\hat{x}(n - k) + s(n - k)) \tag{7.16}$$

Note, if $h(1) = 1$ and $h(k) = 0$ for $k > 1$, then we are back in the first-order predictor used by the delta modulation algorithm (7.12). The predictor is commonly implemented as a finite impulse response (FIR) filter (see Chapter 1) which can be readily implemented using a DSP. The length of the filter is often quite moderate, $N = 2$ up to $N = 6$ is common.

The advantage of DPCM over straightforward PCM (see Chapter 2) is of course the data compression achieved by utilizing the dependency between samples in, for instance, an analog speech signal. If data compression is not the primary goal, DPCM also offers the possibility of enhancing the SNR. As DPCM is only quantizing the difference between consecutive samples and not the sample values themselves, a DPCM system has less quantization error and noise than a pure PCM system using the same word length.

To be able to find good parameters $h(n)$ for the predictor and a proper number of quantization levels p, knowledge of the signal statistics is necessary. In many cases, although the long-term statistics is known, the signal may depart significantly from these during shorter periods of time. In such applications, an adaptive algorithm may be advantageous.

Adaptive differential pulse code modulation (ADPCM) is a term used for two different methods, namely adaptation of the **quantizer** and adaptation of the **predictor** (Marwen and Ewers, 1993). Adaptation of the quantizer involves estimation of the step size, based on the level of the input signal. This estimation can be done in two ways, **forward** estimation (**DPCM-AQF**) and **backward** estimation (**DPCM-AQB**) (see Figure 7.3). In the forward case AQF, a number of input samples are first buffered in the modulator and used to estimate the

AQF

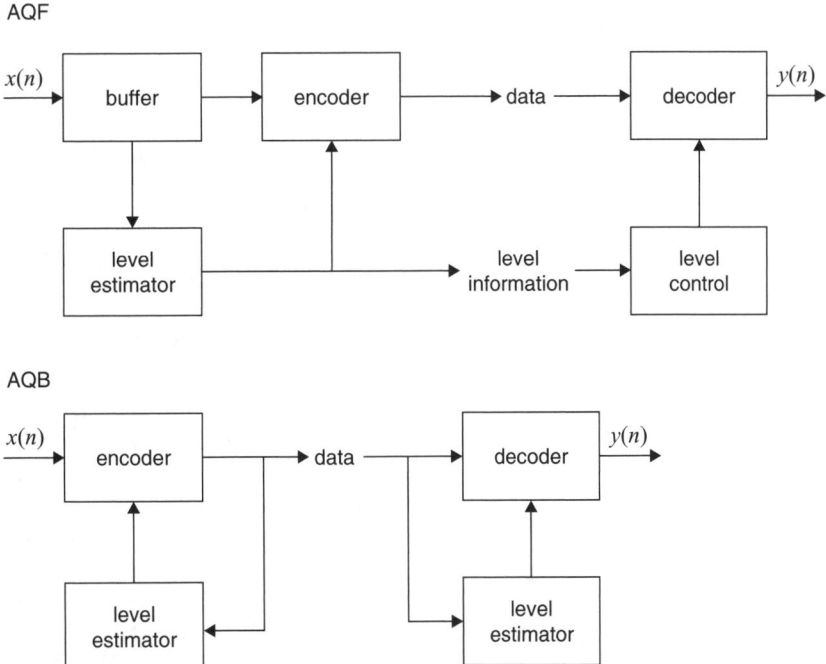

AQB

Figure 7.3 *Forward estimation (DPCM-AQF) and backward estimation (DPCM-AQB)*

signal level. This level information is sent to the demodulator, together with the normal DPCM data stream. The need to send the level information is a drawback of AQF and for this reason, AQB is more common. Another drawback of AQF is the delay introduced by the buffering process. This delay may render AQF unusable in certain applications. AQF, however, has the potential of performing better than AQB.

In backward level estimation AQB, estimation is based on the out-going DPCM data stream. A similar estimation is made in the demodulator, using the incoming DPCM data. This scheme has some resemblance to CVSD, discussed above. Adaptive quantizers commonly offer an improvement in SNR of roughly 3–7 dB compared to fixed quantizers. The quality of the adaptation depends to a large extent on the quality of the level estimator. There is of course a trade-off between the complexity and the cost of a real world implementation.

When using the acronym ADPCM, adaptation of the **predictor** or adaptation of both the predictor and the quantizer is often assumed. The predictor is commonly adapted using methods based on gradient descent type algorithms used in adaptive filters (see Chapter 3). Adaptive predictors can increase the SNR of the system significantly compared to fixed predictors.

As an example, in the standard CCITT G.721, the input is a CCITT standard 64-kbits/s PCM-coded speech signal and the output is a 32-kbits/s ADPCM data stream. In this standard, feedback adaptation of both the quantizer and predictor is used. In a typical CCITT G.721 application (Marwen and Ewers, 1993), the adaptive quantizer uses 4 bits and the predictor is made up of a sixth-order FIR

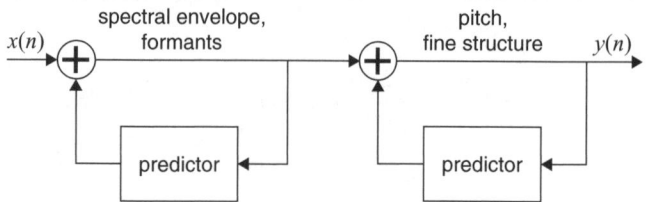

Figure 7.4 *Encoder for adaptive, predictive coding of speech signals. The decoder is mainly a mirrored version of the encoder*

filter and a second-order infinite impulse response (IIR) filter. A **"transcoder"** of this kind can be implemented as a mask programmed DSP at a competitive cost.

7.2.4 Speech coding, adaptive predictive coding and sub-band coding

Adaptive predictive coding (APC) is a technique used for speech coding, i.e. data compression of speech signals. APC assumes that the input speech signal is repetitive with a period significantly longer than the average frequency content. Two predictors are used in APC. The high-frequency components (up to 4 kHz) are estimated using a "spectral" or "formant" predictor, and the low-frequency components (50–200 Hz) by a "pitch" or "fine structure" predictor (see Figure 7.4). The spectral estimator may be of order 1–4, and the pitch estimator about order 10. The low-frequency components of the speech signal are due to the movement of the tongue, chin and lips. The high-frequency components originate from the vocal cords and the noise-like sounds (like in "s") produced in the front of the mouth.

The output signal $y(n)$ together with the predictor parameters, obtained adaptively in the encoder, are transmitted to the decoder where the speech signal is reconstructed. The decoder has the same structure as the encoder, but the predictors are not adaptive and are invoked in the reverse order. The prediction parameters are adapted for blocks of data corresponding to, for instance, 20 ms time periods.

APC is used for coding speech at 9.6 and 16 kbits/s. The algorithm works well in noisy environments, but unfortunately the quality of the processed speech is not as good as for other methods like code excited linear prediction (CELP) described below.

Another coding method is **sub-band coding (SBC)** (see Figure 7.5) which belongs to the class of **waveform coding** methods, in which the frequency domain properties of the input signal are utilized to achieve data compression.

The basic idea is that the input speech signal is split into **sub-bands** using band-pass filters. The sub-band signals are then encoded using ADPCM or other techniques. In this way, the available data transmission capacity can be allocated between bands according to perceptual criteria, enhancing the speech quality as perceived by listeners. A sub-band that is more "important" from a human listening point of view can be allocated more bits in the data stream, while less important sub-bands will use fewer bits.

A typical setup for a sub-band coder would be a bank of N (digital) band-pass (BP) filters followed by decimators, encoders (for instance ADPCM) and a multiplexer combining the data bits coming from the sub-band channels. The

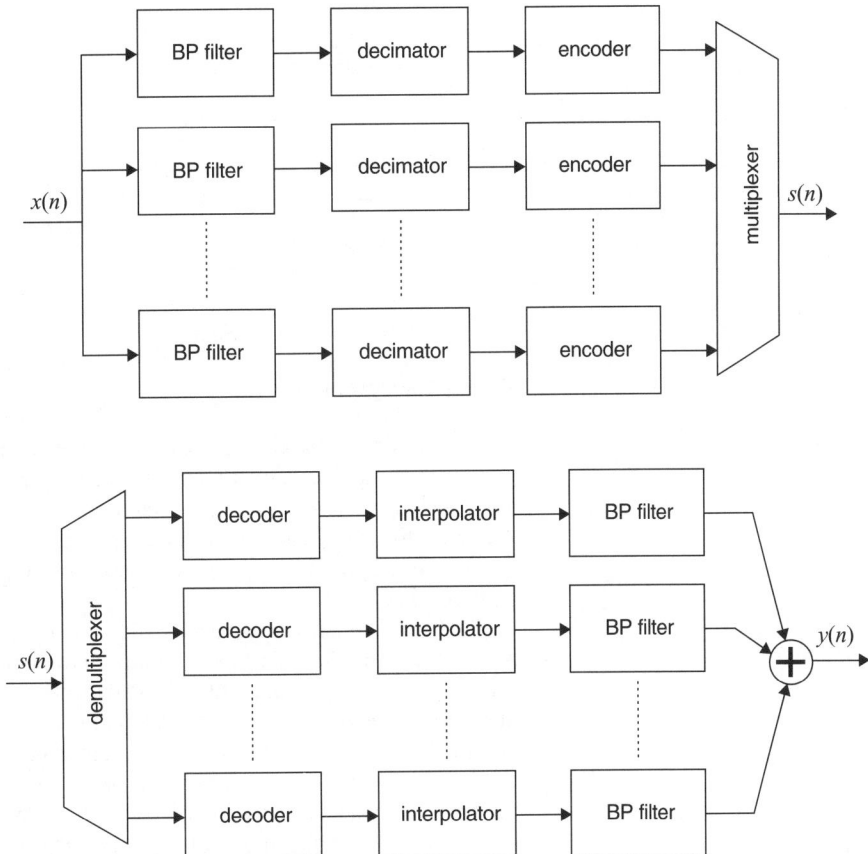

Figure 7.5 *An example of a sub-band coding system*

output of the multiplexer is then transmitted to the sub-band decoder having a demultiplexer splitting the multiplexed data stream back into N sub-band channels. Every sub-band channel has a decoder (for instance ADPCM), followed by an interpolator and a band-pass filter. Finally, the outputs of the band-pass filters are summed and result in a reconstructed output signal.

Sub-band coding is commonly used at bit rates between 9.6 and 32 kbits/s and performs quite well. The complexity of the system may, however, be considerable if the number of sub-bands is large. The design of the band-pass filters is also a critical topic working with sub-band coding systems.

7.2.5 Vocoders and linear predictive coding

In the methods described above, APC, SBC and ADPCM, speech signal applications have been used as examples. Modifying the structure and parameters of the predictors and filters, the algorithms may as well be used for other signal types. The main objective was to achieve a reproduction that was as faithful as possible to the original signal. Data compression was possible by removing redundancy in the time or frequency domain.

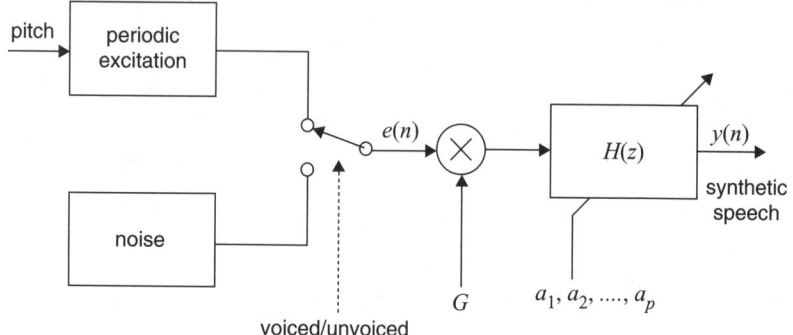

Figure 7.6 *The LPC model*

The class of **vocoders** (also called source coders) is a special class of data compression devices aimed only for speech signals. The input signal is analyzed and described in terms of speech model parameters. These parameters are then used to synthesize a voice pattern having an acceptable level of perceptual quality. Hence, waveform accuracy is not the main goal as it was in the previous methods discussed.

The first vocoder was designed by H. Dudley in the 1930s and demonstrated at the "New York Fair" in 1939. Vocoders have become popular as they achieve reasonably good speech quality at low data rates, from 2.4 to 9.6 kbits/s. There are many types of vocoders (Marven and Ewers, 1993), and some of the most common techniques will be briefly presented below.

Most vocoders rely on a few basic principles. Firstly, the characteristics of the speech signal is assumed to be fairly constant over a time of approximately 20 ms, hence most signal processing is performed on (overlapping) data blocks of 20–40 ms length. Secondly, the speech model consists of a time varying filter corresponding to the acoustic properties of the mouth and an excitation signal. The excitation signal is either a periodic waveform, as being created by the vocal cords, or a random noise signal for production of "unvoiced" sounds, e.g. "s" and "f". The filter parameters and excitation parameters are assumed to be independent of each other and are commonly coded separately.

Linear predictive coding (LPC) is a popular method, which has, however, been replaced by newer approaches in many applications. LPC works exceedingly well at low-bit rates and the LPC parameters contain sufficient information of the speech signal to be used in speech recognition applications. The LPC model is shown in Figure 7.6.

LPC is basically an auto-regressive model (see Chapter 5) and the vocal tract is modeled as a time varying all-pole filter (IIR filter) having the transfer function $H(z)$

$$H(z) = \frac{1}{1 + \sum_{k=1}^{p} a_k z^{-k}} \tag{7.17}$$

where p is the order of the filter. The excitation signal $e(n)$, being either noise or a periodic waveform, is fed to the filter via a variable gain factor G. The output signal can be expressed in the time domain as

$$y(n) = Ge(n) - a_1 y(n-1) - a_2 y(n-2) - \cdots - a_p y(n-p) \tag{7.18}$$

The output sample at time n is a linear combination of p previous samples and the excitation signal (linear predictive coding). The filter coefficients a_k are time varying.

The model above describes how to **synthesize** the speech given the pitch information (if noise or periodic excitation should be used), the gain and the filter parameters. These parameters must be determined by the encoder or the **analyzer**, taking the original speech signal $x(n)$ as input.

The analyzer windows the speech signal in blocks of 20–40 ms, usually with a Hamming window (see Chapter 5). These blocks or "frames" are repeated every 10–30 ms, hence there is a certain overlap in time. Every frame is then analyzed with respect to the parameters mentioned above.

Firstly, the pitch frequency is determined. This also tells if we are dealing with a voiced or unvoiced speech signal. This is a crucial part of the system, and many pitch detection algorithms have been proposed. If the segment of the speech signal is voiced and has a clear periodicity or if it is unvoiced and not periodic, things are quite easy. Segments having properties in between these two extremes are difficult to analyze. No algorithm has been found thus far i.e. "perfect" for all listeners.

Now, the second step of the analyzer is to determine the gain and the filter parameters. This is done by estimating the speech signal using an adaptive predictor. The predictor has the same structure and order as the filter in the synthesizer. Hence, the output of the predictor is

$$\hat{x}(n) = -a_1 x(n-1) - a_2 x(n-2) - \cdots - a_p x(n-p) \tag{7.19}$$

where $\hat{x}(n)$ is the predicted input speech signal and $x(n)$ is the actual input signal. The filter coefficients a_k are determined by minimizing the square error

$$\sum_n (x(n) - \hat{x}(n))^2 = \sum_n r^2(n) \tag{7.20}$$

This can be done in different ways, either by calculating the auto-correlation coefficients and solving the Yule–Walker equations (see Chapter 5) or by using some recursive, adaptive filter approach (see Chapter 3).

So, for every frame, all the parameters above are determined and transmitted to the synthesizer, where a synthetic copy of the speech is generated.

An improved version of LPC is **residual excited linear prediction (RELP)**. Let us take a closer look of the error or residual signal $r(n)$ resulting from the prediction in the analyzer equation (7.19). The residual signal (we are trying to minimize) can be expressed

$$r(n) = x(n) - \hat{x}(n) = x(n) + a_1 x(n-1) + a_2 x(n-2) + \cdots + a_p x(n-p) \tag{7.21}$$

From this it is straightforward to find out that the corresponding expression using the z-transforms is

$$R(z) = \frac{X(z)}{H(z)} = X(z) H^{-1}(z) \tag{7.22}$$

Hence, the predictor can be regarded as an "inverse" filter to the LPC model filter. If we now pass this residual signal to the synthesizer and use it to excite the LPC filter, i.e. $E(z) = R(z)$ instead of using the noise or periodic waveform sources, we get

$$Y(z) = E(z)H(z) = R(z)H(z) = X(z)H^{-1}(z)H(z) = X(z) \qquad (7.23)$$

In the ideal case, we would hence get the original speech signal back. When minimizing the variance of the residual signal (7.20), we gathered as much information about the speech signal as possible using this model in the filter coefficients a_k. The residual signal contains the remaining information. If the model is well suited for the signal type (speech signal), the residual signal is close to white noise, having a flat spectrum. In such a case, we can get away with coding only a small range of frequencies, for instance 0–1 kHz of the residual signal. At the synthesizer, this baseband is then repeated to generate higher frequencies. This signal is used to excite the LPC filter.

Vocoders using RELP are used with transmission rates of 9.6 kbits/s. The advantage of RELP is better speech quality compared to LPC for the same bit rate. However, the implementation is more computationally demanding.

Another possible extension of the original LPC approach is to use **multipulse excited linear predictive coding (MLPC)**. This extension is an attempt to make the synthesized speech less "mechanical" by using a number of different pitches of the excitation pulses rather than only the two (periodic and noise) used by standard LPC.

The MLPC algorithm sequentially detects k pitches in a speech signal. As soon as one pitch is found it is subtracted from the signal and detection starts over again, looking for the next pitch. Pitch information detection is a hard task and the complexity of the required algorithms is often considerable. However, MLPC offers better speech quality than LPC for a given bit rate, and is used in systems working with 4.8–9.6 kbits/s.

Yet another extension of LPC is the code excited linear prediction (CELP). The main feature of the CELP compared to LPC is the way in which the filter coefficients are handled. Assume that we have a standard LPC system, with a filter of the order p. If every coefficient a_k requires N bits, we need to transmit $N \cdot p$ bits per frame for the filter parameters only. This approach is all right if all combinations of filter coefficients are equally probable. However, this is not the case. Some combinations of coefficients are very probable, while others may never occur. In CELP, the coefficient combinations are represented by p-dimensional vectors. Using vector quantization techniques, the most probable vectors are determined. Each of these vectors is assigned an **index** and stored in a **code book**. Both the analyzer and synthesizer of course have identical copies of the code book, typically containing 256–512 vectors. Hence, instead of transmitting $N \cdot p$ bits per frame for the filter parameters, only 8–9 bits are needed.

This method offers high-quality speech at low-bit rates but requires considerable computing power to be able to store and match the incoming speech to the "standard" sounds stored in the code book. This is, of course, especially true if the code book is large. Speech quality degrades as the code book size decreases.

Most CELP systems do not perform well with respect to higher-frequency components of the speech signal at low-bit rates. This is counteracted in newer systems using a combination of CELP and MLPC.

There is also a variant of CELP called **vector sum excited linear prediction (VSELP)**. The main difference between CELP and VSELP is the way the code book is organized. Further, since VSELP uses fixed-point arithmetic algorithms, it is possible to implement using cheaper DSP chips than CELP, which commonly requires floating-point arithmetics.

7.2.6 Image coding, joint photographics expert group (JPEG), moving pictures expert group (MPEG)

Digitized images consist of large amounts of picture elements (pixels), hence transmitting or storing images often involving large amounts of data. For this reason, data compression of image data, or image coding, is a highly interesting topic.

The same general fundamental idea for data compression applies in that there is a reduction of redundancy-utilizing statistical properties of the data set, such as dependencies between pixels. In the case of image compression, many of the algorithms turn two dimensional, unless, for instance, compression is applied to a scanned signal. One such application is telefax (facsimile) systems, i.e. systems for transmission (or storage) of black-and-white drawings or maps, etc. In the simplest case, there are only two levels to transmit, black and white, which may be represented by binary "1" and "0," respectively, coming from the scanning photoelectric device.

Studying the string of binary digits (symbols) coming from the scanning device (the information source), it is easy to see that ones and zeros often come in long **bursts**, and not in a "random" fashion. For instance, 30 zeros, followed by 10 ones, followed by 80 zeros and so on may be common. There is hence a considerable dependency between successive pixels. In this case, a **run length code** will be efficient. The underlying idea is not to transmit every single bit, but rather a number telling how many consecutive ones or zeros there are in a burst.

For facsimile applications, CCITT has standardized a number of run length codes. A simple code is the **modified Huffman code (MHC)**. Black-and-white burst of length 0–63 pixels are assigned their own code words according to the Huffman algorithm outlined earlier in this chapter. If a burst is 64 pixels or longer, the code word is preceded by a prefix code word, telling how many multiples of 64 pixels there are in the burst. The fact that long white bursts are more probable than long black bursts is also taken into account when assigning code words. Since the letters are "hollow", a typical page of text contains approximately 10% black and 90% white areas.

During the scanning process, every scan line is assumed to start with a white pixel, otherwise the code word for white burst length zero is transmitted. Every scanned line is ended by a special "end-of-line" code word to reduce the impact of possible transmission errors. Compression factors between 6 and 16 times can be observed (i.e. 0.06–0.17 bits/pixel), depending on the appearance of the scanned document. A more elaborate run length code is the **modified Read code (MRC)** (Hunter and Robinson, 1980) which is capable of higher compression factors.

Image compression algorithms of the type briefly described above are often denoted **lossless** compression techniques, since they make a faithful reproduction of the image after decompression, i.e. no information is lost. If we can accept a certain degradation of, for instance, resolution or contrast in a picture, different schemes of predictive compression and transform compression can be used.

Predictive compression techniques work in similar ways as DPCM and ADPCM as described earlier. The predictors may, however, be more elaborate, working in two dimensions, utilizing dependencies in both the x and y direction of an image. For moving pictures, predictors may also consider dependencies in the z-axis, between consecutive picture frames. This is sometimes called three-dimensional prediction. For slowly moving objects in a picture, the correlation between picture frames may be considerable, offering great possibilities for data compression. One such example is the videophone. The most common image in such a system is a more or less static face of a human.

The basic idea of **transform compression** (Gonzales and Woods, 2002) is to extract appropriate statistical properties, for instance Fourier coefficients, of an image and let the most significant of these properties represent the image. The image is then reconstructed (decompressed) using an inverse transform.

A picture can be Fourier transformed in a similar way to a temporal one-dimensional signal, and a spectrum can be calculated. When dealing with images, we however have two dimensions and hence two frequencies, one in the x direction and one in the y direction. Further, we are now dealing with **spatial** frequencies. For instance, a coarse chessboard pattern would have a lower spatial frequency than a fine pattern. We are now talking about cycles per length unit, not cycles per time unit.

Often it is convenient to express the transform coefficients as a matrix. In doing this, it is commonly found that the high-frequency components have smaller amplitude than lower frequency components. Hence, only a subset of the coefficients needs to be used to reproduce a recognizable image, but fine structure details will be lost. Transform compression methods are commonly not lossless.

It has been found that the standard two-dimensional fast Fourier transform (FFT) is not the best choice for image compression. Alternative transforms, such as **Walsh**, **Hadamard** and **Karhunen–Loève** have been devised. One of the most popular is, however, the two-dimensional **discrete cosine transform (DCT)**

$$C(0,0) = \frac{1}{N} \sum_{x=0}^{N-1} \sum_{y=0}^{N-1} f(x,y) \tag{7.24a}$$

and for $u, v = 1, 2, \ldots, N - 1$

$$C(u,v) = \frac{1}{2N^3} \sum_{x=0}^{N-1} \sum_{y=0}^{N-1} f(x,y) \cos((2x + 1)u\pi) \cos((2y + 1)v\pi)$$

$$\tag{7.24b}$$

where $C(u,v)$ is the two-dimensional transform coefficients ("spectrum") and $C(0,0)$ is the "DC" component. $f(x,y)$ is the pixel value. The inverse transform is

$$f(x,y) = \frac{1}{N} C(0,0) + \frac{1}{2N^3} \sum_{u=1}^{N-1} \sum_{v=1}^{N-1} C(u,v) \cos((2x + 1)u\pi) \cos((2y + 1)v\pi)$$

$$\tag{7.25}$$

One nice feature of DCT is that the transform is real, unlike FFT which is a complex transform. Another advantage is that DCT is separable, implying that it can be implemented as two successive applications of a one-dimensional DCT algorithm.

A common technique is to split a picture into 8×8 pixel blocks and apply DCT. High-amplitude, low-frequency coefficients are transmitted first. In this way, a "rough" picture is first obtained with an increasing resolution as higher-frequency coefficients arrive. Transform coding can offer a data compression ratio of approximately 10 times.

The CCITT H.261 standard covers a class of image compression algorithms for transmission bit rates from 64 kbits/s to 2 Mbits/s. The lowest bit rate can be used for videophones on narrow-band integrated services digital network (ISDN) lines. The H.261 is a hybrid DPCM/DCT system with motion compensation. The luminance (black-and-white brightness) signal is sampled at 6.75 MHz, and the chrominance (color information) signal is sampled at 3.375 MHz. The difference between the present frame and the previous one is calculated and split into 8×8 pixel blocks. These blocks are transformed using DCT, and the resulting coefficients are coded using Huffman coding.

The motion detection algorithm takes each 8×8 pixel block of the present frame and searches the previous frame by moving the block ± 15 pixels in the x and y directions. The best match is represented by a displacement vector. The DCT coefficients and the displacement vector are transmitted to the decompressor, where the reverse action takes place.

There are commercial video conferencing systems today using the H.261 standard. The algorithms used are, however, very computationally demanding. Floating-point or multiple fixed-point DSPs are required.

The **joint photographics expert group (JPEG)** is a proposed standard for compression of still pictures. The color signals red, green and blue are sampled and each color component is transformed by DCT in 8×8 pixel blocks. The DCT coefficients are quantized and encoded in a way that the more important lower frequency components are represented by more bits than the higher-frequency coefficients. The coefficients are reordered by reading the DCT coefficient matrix in a zigzag fashion (Marven and Ewers, 1993), and the data stream is Huffman coded (the "DC" components are differentially encoded with the previous frame, if there are any).

The JPEG compressor is simpler than in the H.261 system. There is, for instance, no motion compensation, but many other elements are quite similar. The JPEG decompressor is, however, more complicated in JPEG than in H.261.

The **moving pictures expert group (MPEG)** proposed standard (MPEG-1) is aimed for compression of full-motion pictures on digital storage media, for instance CD-ROM and digital video disc (DVD), with a bit transfer rate of about 1.5 Mbits/s. It is to some extent similar to both H.261 and JPEG but does not have the motion compensation found in JPEG.

A sampled frame is split into blocks and transformed using DCT in the same way as for JPEG. The coefficients are then coded with either forward or backward prediction or a combination of both. The output from the predictive coding is then quantized using a matrix of quantization steps. Since MPEG is more complicated than JPEG it requires even more computing power.

The area of image and video compression algorithms is constantly evolving and there are many new methods and novel, dedicated signal processing application specific integrated circuits (ASICs) and DSPs to come. As of today, we have seen many new standards, e.g. JPEG 2000 (using wavelet techniques), MPEG-2 (ISO/IEC-13818) and MPEG-4 (ISO/IEC-14496). Information on current standards can be found on the World Wide Web.

7.2.7 The layer-3 of MPEG-1 algorithm (MP3)

The MPEG-1 standard, discussed in Section 7.2.6 above, does not only specify methods for image compression, but also for audio compression. One very popular audio compression method today is layer-3 of MPEG-1, known as MP3. It was originally developed and patented by the Fraunhofer Institute in Germany in 1989. The algorithm is very commonly used for compressing music files to be exchanged over the Internet. Typically, a compression ratio of 10:1 is achieved, implying that 1 min of stereophonic music will produce about 1 Mbyte of data, rather than approximately 11 Mbyte needed for 1 min standard "CD quality" hi-fi music. Better compression algorithms are available today, but one of the reasons for the success of MP3 is the relatively open nature of the format.

The MP3 algorithm is "lossy" and exploits the weaknesses of the human ear in order to "cheat" in a non-audible way, i.e. the music still sounds ok, despite the fact that most of the original information is lost and cannot be retrieved. The most important mechanism utilized is the concept of **auditory masking**. Masking occurs in the human ear and means that if a strong signal appears in a given frequency band, the ear cannot hear weaker signals below a given threshold in the same band at the same time. An example of this could be a loud orchestra masking the sound of a single instrument playing softly.

The ear has a different resolution for different frequency bands (typically 100 Hz–4 kHz bandwidth), depending on the amplitude of the signal in the respective bands. This can, of course, be used for audio compression by not using more resolution, i.e. data bits, than the ear can perceive in the different frequency bands at a given time. In other words, at every time instant, the algorithm discards information that the ear cannot hear, thus achieving compression.

Figure 7.7 shows a simplified block diagram of an MP3 encoder. The underlying principle is a smart adaptive, sub-band compression algorithm utilizing the auditory masking. The digital input data enters the filter bank consisting of 32 polyphase sub-band (SB) filters. The filter bank is followed by a modified discrete cosine transform (MDCT), giving totally 384 sub-bands. Since there is commonly a large correlation between the information in the left and right channels of stereophonic music, some initial data compression takes place at this stage. The joint stereo encoder is governed by the perceptual model. This model analyzes the input signal and, using the masking properties of the human ear, generates control signals for the encoder and the quantizer.

Further, the quantizer and scaling module operates on blocks consisting of 36 samples, and determines scale factors and allocation of bits for the blocks belonging to the different sub-bands. Each sub-band is allocated as few bits as possible, by the rule that the quantization noise should be below the threshold of the human ear. (Masking thresholds are obtained from the perceptual model). Taking the masking effect into account, considerable quantization noise (i.e. low resolution and few bits) can be accepted for some sub-bands, since the ear will not be able to hear it.

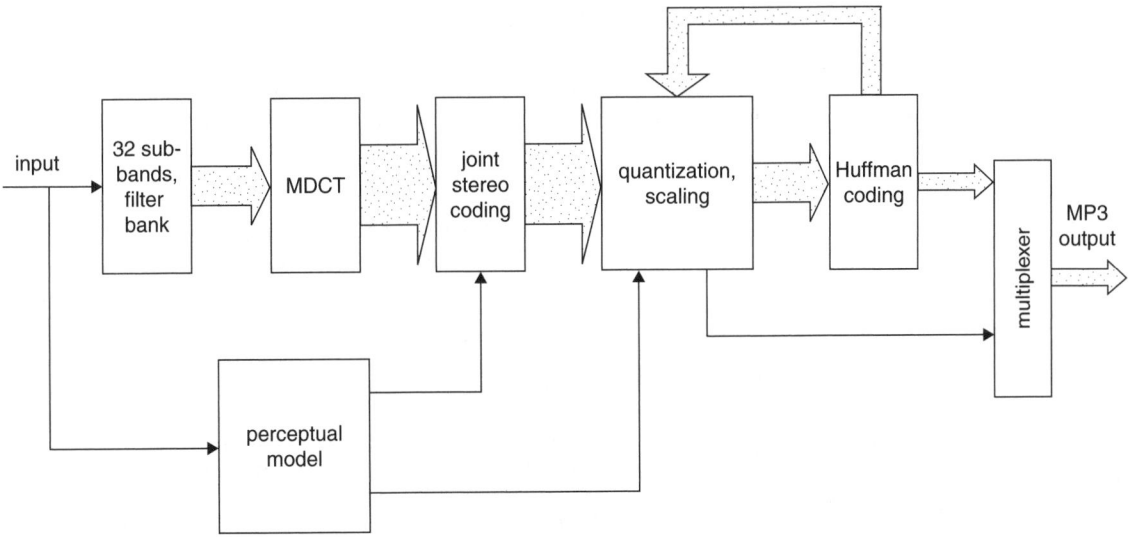

Figure 7.7 *A simplified block diagram of an MP3 encoder*

Finally, data compression using the Huffman algorithm is employed. Since optimum compression is a trade-off between allocation of bits and the statistics of the data, there is a feed back loop from the Huffman encoder to the quantizer. In this way, a good compromise can be found by iteration.

In the concluding multiplexer, the encoded sample data are merged with information about scale factors and resolution for the sample blocks.

7.2.8 The Lempel–Ziv algorithm

There are a number of data compression algorithms based on the **Lempel–Ziv (LZ)** method, named after A. Lempel and J. Ziv. The original ideas were published in 1977 and 1978, but later modified by Terry A. Welch, resulting in the **Lempel–Ziv–Welch (LZW)** algorithm (Smith, 2003). Only the basic method will be briefly described below.

The LZ data compression algorithm is a called a "universal" algorithm since it does not need any prior information about the input data statistics. Commonly, it works on a byte basis and is best suited for compression e.g. Amercian standard code for information interchange (ASCII) information. The algorithm creates a table or "dictionary" of frequently seen strings of bytes. When a string appears in the input data set, it is substituted by the index of a matching string in the dictionary. This dictionary-type algorithm is also referred to as a macro-replacement algorithm, because it replaces a string with a token. Decompression will simply be a table-lookup and string replacement operation. It is imperative, however, that no transmission errors occur, since these may add errors in the dictionary and jeopardize the entire decompression operation.

Let us illustrate the algorithm with a simple example. Assume that we are using an 8-bit ASCII code and want to transmit the following string

```
the_theme_theorem_theses
```

Since a space _ also represents an ASCII character, we have to transmit 24 characters, i.e. a total of $24 \times 8 = 192$ bits.

Now, let us use a simple version of the LZ method and limit the size of the dictionary to 16 entries, so that any entry in the dictionary can be pointed to by a 4-bit index. The compression and dictionary handling rule works like this:

```
lzcomp:    clear dictionary
           start from beginning of string
           while (string not traversed)
           {
             find the longest substring matching an entry
                in dictionary
             get the index of the entry in the dictionary
             get the next character in the string following
                the substring
             transmit index and next character
             append the substring + next character to
                the dictionary
             continue after next character in string
           }
```

The corresponding decompression and dictionary handling rule is:

```
lzdecomp:   clear dictionary
            clear buffer for decompressed string
            while (compressed characters received)
            {
              get index and next character from compressor
              use index to fetch substring from dictionary
              append substring to buffer
              append next character to buffer
              append the substring + next character to
                 the dictionary
            }
```

Below, the string, dictionary and transmitted symbols are shown. The indices are represented by the hexadecimal digits: 0, 1, ..., E, F.

The uncompressed string:

```
the_theme_theorem_theses
```

When the algorithm is started, the dictionary (Table 7.1) is empty. The first entry will be "nothing" which will be given index 0. The first substring in the input string, simply the letter "t", cannot be found in the dictionary. However, "nothing" can be found, hence we concatenate "nothing" and "t" (we do not include "nothing" in the dictionary from now on, since it is so small) and add it to the dictionary. The index of this new entry will be 1. We also submit the hexadecimal substring index 0 ("nothing") and the letter "t" to the output. These steps are repeated and the dictionary grows. Nothing interesting happens until we find the substring "t" for the second time (in "theme"). Since the substring "t" is already in the dictionary, it can be replaced by its index 1. So, following the rules, we send the index 1 and the next character "h", then we add the new substring "th" to the dictionary. This process is further repeated until the end of

Table 7.1 *Example LZ, 16-entry dictionary*

Index	Dictionary	Compressed output
0	nothing	
1	t	0t
2	h	0h
3	e	0e
4	_	0_
5	th	1h
6	em	3m
7	e_	3_
8	the	5e
9	o	0o
A	r	0r
B	em_	6_
C	thes	8s
D	es	3s
E		
F		

the uncompressed input string is reached. The total output of the compressed string will be:

```
0t0h0e0_1h3m3_5e0o0r6_8s3s
```

where every second character is an index (4 bits) and an ASCII character (8 bits), respectively. The total length of the compressed string will be: $13 \times 4 + 13 \times 8 = 156$ bits, implying a compression factor of $156/192 = 0.81$.

The decompressor receives the compressed string above and starts building the decompressed string and the dictionary in a similar way. To start with, "nothing" is assumed in position 0 of the dictionary. The first compressed information is index 0, corresponding to "nothing" in the dictionary and the character "t". Hence, "nothing" and "t" are placed in the first position of the decompressed string and the same information is added to the dictionary at position 1. The process is then repeated until the end of the compressed string. Now, the decompressed string is an exact copy of the uncompressed string and the dictionary is an exact copy of the dictionary used at compression time. It is worth noting that there are also many variations on this popular data compression algorithm.

7.3 Recognition techniques

7.3.1 A general problem formulation

In many applications, there is a need to recognize a given set of entities. In this book, without loss of generality, we will concentrate on situations where identification of a given sound or image fragment is required. In a way, recognition can be viewed as a kind of data compression technique. If we, for instance, can recognize the photo of a known person, we can transmit his or her name, typically a 100 bytes rather than the entire image of, say, 1 Mbyte. The receiver,

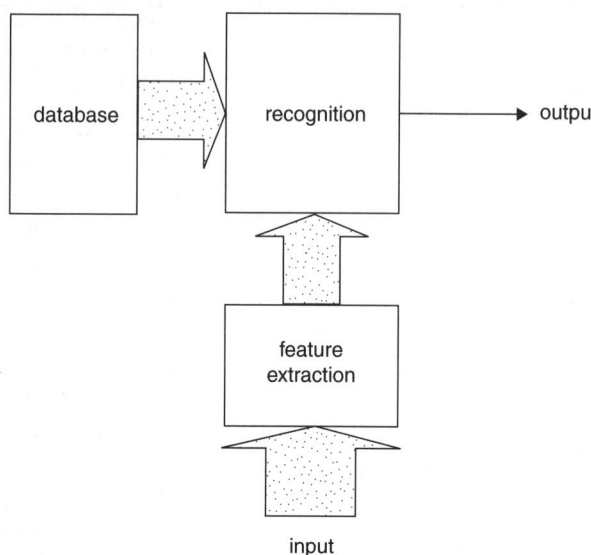

Figure 7.8 *A simplified block diagram of the recognition process*

of course, has to have a "standard" photo of the person in question to be able to decompress the message (i.e. translate the name back to an image). Obviously, this compression method is lossy, since there is no guarantee that the photo used by the transmitter and receiver is identical; they are, however, assumed to be photos of the same person for the system to work properly. Similar code book methods are used in speech coding systems, for instance, the CELP algorithm described in Section 7.2.5 above.

Recognition of speech or images can be regarded as a two-step operation (see Figure 7.8). The first stage is commonly a more or less complicated **mapping function**, transforming the input data (an image or a speech signal) into a number of significant **features**. These features are represented by a set of numbers, for instance, vectors or matrices. Stage two, is a **pattern-recognition** algorithm. The task of this algorithm is to compare the set of features received from the input to all known prototypes or templates in the database and to find the best match. There are many good pattern-recognition algorithms around, such as neural networks and fuzzy logic (see Chapter 4) and a number of statistical correlation-based methods, for instance, principal component analysis (PCA) (Esbensen *et al.*, 1994).

The main problem when designing an image or speech recognition system is that, commonly, there are considerable variations between individual samples of the given entity that we are expected to recognize. For instance, designing a system able to recognize photos of females would probably be quite difficult.

Two factors are very important for the system to work well. First, the mapping function or transformation mechanism producing the features needs to be designed carefully. It does not matter how good the pattern-recognition algorithm we are using is, if the features do not contain significant information about the input image or speech sample. So extracting relevant features is important. Second, the content of the database is also critical. It is important that the stored prototypes are good representatives, i.e. are "typical" (average) samples of the entities that should be recognized. For this reason, it is common that the database is adaptive and is updated every time a new set of features is received. In this way, the system "learns" good prototypes and improves its performance gradually. This is a common approach in neural network-based algorithms.

7.3.2 Speech recognition

Speech recognition research has been pursued for over 30 years, and the problems to be solved are difficult. Commonly, the aim of an automatic speech recognition (ASR) system is to be able to automatically produce written text from a spoken message. Many commercial products are on the market today, but the "perfect" solution has yet to come.

The feature extraction part commonly relies on some spectral analysis method, e.g. cepstrum analysis (Becchetti and Ricotti, 1999). The cepstrum of a digital speech signal $x(n)$ is obtained by first applying a discrete Fourier transform (DFT) (see Chapter 5) to get a complex frequency spectrum $X(\omega)$. The logarithm of the magnitude of the frequency spectrum is then processed by an inverse discrete Fourier transform (IDFT) to obtain the real part of the cepstrum $\hat{x}_r(n)$

$$\hat{x}_r(n) = \Im^{-1}\{\log |X(\omega)|\} = \Im^{-1}\{\log |\Im\{x(n)\}|\} \tag{7.26}$$

This algorithm belongs to the class of Homomorphic filters. From the cepstrum data, features are extracted. The features may be different "phonemes", i.e. sound fragments of spoken words, such as "th", "ye", "n" and so on. Using a pattern-recognition algorithm, groups of features (phonemes) are used to recognize the English prototype words stored in the database.

It all sounds quite straightforward, but there are many pitfalls on the way. One problem is that different persons pronounce words in different ways depending on dialect, emotional status, pitch of the voice and other physiological differences. The incoming speech signal is also affected by background noise, room acoustics and technical properties of the microphone. For this reason, an ASR system must be "trained" to a specific speaker before it can be used. During the training phase, the prototype database is updated to fit the specific speaker.

Another problem is that the speaker may use new, complicated words not found in the database. In such a case, either the system can present the closest match, resulting in rubbish, or it can mark the word as unknown. Yet another complication is the fact that some words, or parts of words, may be pronounced differently depending on the context and the position in a sentence. Hence, the recognition algorithm also needs to consider grammatical language rules.

To summarize, designing a reliable ASR system capable of understanding an arbitrary speaker is certainly not a simple task. Needless to say, every language has its own set of phonemes and grammatical rules.

7.3.3 Image recognition

In Section 7.3.2 above, we were dealing with speech signals (audio) being digitized as one-dimensional temporal information. When processing images it is about digital two-dimensional spatial information, often represented by a matrix of pixel values.

Image recognition can be used in different situations. One application is to find the best match of an entire image to a reference image, for instance, to find matching fingerprints in a register to determine the identity of a person. Another application may be to detect the presence of a specific object in an image or to determine the position (and velocity) of an object. The latter application is often referred to as "computer vision". A common example from the electronics manufacturing industry is the inspection of populated printed circuit boards. A video camera is used to obtain an image of the circuit board, and using image recognition algorithms, a computer will check the image to make sure that all electronic components are present and mounted in the right places. In both applications above there are some common problems, namely: illumination, contrast, shadows and imperfections in the camera, such as distortion introduced by lens limitations, etc.

In the first application above, an important issue is to make sure that all images are precisely aligned regarding translation in the x- and y-axes, and regarding rotation and scaling. If these conditions are met, a simple two-dimensional cross-correlation algorithm may perform well. The second application requires analysis of the image and conforms to the general recognition model outlined in Section 7.3.1 above. The features in such a system typically are contours, edges, corners and so on. A common way to find these features in an image is to use two-dimensional, discrete convolution.

−1	0	1
−1	0	1
−1	0	1

Figure 7.9 *The 3 × 3 Prewitt operator to estimate the horizontal (x) gradient*

$$R(x,y) = \sum_i \sum_j f(x - i, y - j)\, h(i,j) \tag{7.27}$$

The pixel matrix $f(x,y)$ of the image is convoluted with a convolution kernel, mask, template or operator $h(x,y)$. There are a number of operators to find gradients, do averaging or smoothing and to detect different features, such as an edge. For the x and y positions in the image where $R(x,y)$ shows a local maximum, we are likely to find a feature according to the operator used. Figure 7.9 shows an example operator, the 3×3 Prewitt operator to estimate the horizontal (x) gradient. This can be used to find vertical edges in an image (Nalwa, 1993).

In this way, we are able to find the location of different features in an image. These features are fed to the recognition unit, which can identify and position the objects stored in the database. For example, a square box can be represented by the features, four corners and four edges positioned in a given configuration relative to each other.

Describing more complicated objects take more features and have a tendency to become difficult, especially taking the perspective effect into account when objects are moved in the z-axis and may also be rotated. Advanced image recognition in the sense of computer vision often requires very powerful computers.

Summary In this chapter the following issues have been addressed:

- Basic information theory, mutual information, entropy and redundancy
- Lossless and lossy data compression
- Huffman algorithm and prefix-free variable length codes

- Delta modulation (DM), ADM and CVSD
- DPCM and ADPCM, DPCM-AQF and DPCM-AQB
- APC, SBC, LPC, MLPC, CELP, VSELP and vocoders
- Image coding, run length code, MHC, the DCT transform, JPEG and MPEG
- Other data compression methods, MP3 and LZW.

Review questions

R7-1 Explain the terms entropy and redundancy.

R7-2 What is the difference between "lossy" and "lossless" data compression algorithms? Give at least one example of an algorithm of each type.

R7-3 Explain the term prefix-free variable length code. What is the basic idea of the Huffman code algorithm?

R7-4 The algorithms DM, ADM, CVSD, DPCM, ADPCM and APC use some kind of predictor. What is the idea using a predictor? Give an example of how to implement a simple predictor. What problems can be anticipated using predictors? How can these be overcome?

R7-5 Sub-band coding (SBC) is used in, for instance, the MP3 algorithm. What is the basic idea of sub-band coding?

R7-6 What is the main idea behind the algorithms APC, LPC, MLPC, CELP and VSELP?

R7-7 Give a brief presentation of DCT used in, for instance, JPEG and MPEG.

Solved problems

P7-1 Decompress the string `0t0h0e0_1h3m3_5e0o0r6_8s3s` which was used as an example presenting the Lempel–Ziv algorithm in Section 7.2.8. Show that the original message can be recovered without any ambiguities.

P7-2 Make a binary Huffman code for a system using an alphabet of eight characters having the probabilities shown below

(1) $p_1 = 0.512$
(2) $p_2 = 0.128$
(3) $p_3 = 0.128$
(4) $p_4 = 0.032$
(5) $p_5 = 0.128$
(6) $p_6 = 0.032$
(7) $p_7 = 0.032$
(8) $p_8 = 0.008$

P7-3 For the system in P7-2 above, calculate the entropy, the average symbol length, the coding efficiency and the redundancy.

P7-4 Write a MATLAB program to simulate the delta modulator and demodulator as described by equations (7.11), (7.12) and (7.13). Generate an input signal using $x(n) = A \cos(\Omega n)$. Initially use amplitude equal to one, $\Omega = \pi/100$ and step size 0.08. Plot the input to the modulator and the output of the demodulator. Change the amplitude and frequency of the input signal and the step size of the modulator to demonstrate the problems of "slope overload" and "granularity". How are these problems related to the parameters above?

8 Error-correcting codes

Background All known communication mechanisms today employ some kind of signal. A signal is a detectable change of properties of some entity, a "carrier". The signal can consist of modulated light or radio waves, varying air pressure (sound) or changes in current or voltage levels and so on. Disregarding the type of carrier, a maximum communication speed determined by the channel capacity, is always imposed. The reason for this is the presence of noise and interference disturbing our signals and causing random misinterpretations, i.e. communication errors. If there were no such disturbing factors, we would in theory be able to communicate with an infinite speed. This in turn would imply that all known information in the universe would be accessible instantly, everywhere ... I am not sure that this would be an entirely good situation?

In any case, in this chapter we will look into the background of the channel capacity and we will explore a number of methods and algorithms which enable communication to be as close to the maximum information transfer rate as possible. The issue is to find smart methods to minimize the probability of communication errors, without slowing the communication process. Such methods are called error-correcting codes or error-control codes (ECC).

Objectives In this chapter we will cover:

- The binary symmetric channel (BSC) model
- Bit error probability and mutual information
- The additive white Gaussian noise channel (AWGN) and channel capacity
- Hamming distance, error detection and correction
- Linear block codes, matrix representation and syndromes
- Cyclic codes and Bose, Chaudhuri, Hocquenghem (BCH) codes and polynomial representation
- Convolution codes and the Viterbi decoder
- Interleaving, concatenated codes and turbo codes.

8.1 Channel coding

In Chapter 7, the source coding theorem by Shannon was briefly presented. In this chapter, some implications of his **channel coding theorem** will be discussed. The idea of source coding or data compression is to find a "minimum form" in which to represent a certain amount of information thus making transmission faster and/or data storing more compact. Data compression is based on removing redundancy, i.e. "excess" information. The problem, however, arises that the redundancy-free pieces of information will be extremely **vulnerable**. One example of this is the data compression algorithms used to "compress" data files on magnetic discs. If one single binary digit (BIT) in the "wrong" place happens to be corrupt, the entire disc (maybe several gigabytes) may be completely unreadable and almost all information is lost.

The same problem is even more likely to appear when transmitting data over wires or radio links. In such applications, transmission errors are common. Hence, to be able to transmit information error free, we must include mechanisms capable of detecting and preferably correcting transmission (or storage) errors. This requires **adding redundancy**. When adding redundancy, i.e. making a compressed piece of data large again, we take the properties of the transmission (or storage) process into account. If we know what kinds of errors are the most probable, for instance in a radio link, we can add redundant information tailored to correct these errors. **Channel coding** is mainly to "vaccinate" the information against expected transmission errors. If unexpected errors occur, we will of course be in trouble. If we, however, design our channel code properly, this will rarely occur.

As mentioned above, adding this "error-correcting" redundancy will increase the amount of data involved. Hence, we are making a trade-off between transmission time or storage space for reliability.

8.1.1 The channel model

In the following text, we will denote both a data transmission (in space) and a data storing (transmission in time) mechanism called a "**channel**". This is hence an abstract entity and will not necessarily relate to any physical media, device or system. It will only be used as a model, i.e. a description of how the received (or retrieved) data is related to the transmitted (or stored) data.

In Chapter 7, the concept of mutual information $I(A, B)$ was introduced (equation (7.2))

$$I(A, B) = \log\left(\frac{P(A \mid B)}{P(A)}\right) \tag{8.1}$$

where $P(A)$ is the probability that event A, generated by an information source, takes place. $P(A \mid B)$ is the conditional probability of A taking place given we have observed another event B. Now, in the channel context, the event A corresponds to the transmitted data symbol and event B to the received symbol, in other words what we observe is probably caused by A.

Now, for a simple example, let us assume that the binary digits 1 and 0 are used to communicate over a channel and that the input symbols of the channel are denoted X and the output symbols Y. Event A_i implies that $X = x_i$ and event B_j implies that $Y = y_j$. We can rewrite equation (8.1) as

$$I(X = x_i, Y = y_j) = \log\left(\frac{P(X = x_i \mid Y = y_j)}{P(X = x_i)}\right) \tag{8.2}$$

We now define the **average mutual information**

$$I(X, Y) = E[I(A_i, B_j)] = \sum_{i=1}^{K} \sum_{j=1}^{L} f_{XY}(x_i, y_j) \log\left(\frac{f_{X|Y}(x_i \mid y_j)}{f_X(x_i)}\right)$$

$$= \sum_{i=1}^{K} \sum_{j=1}^{L} f_{XY}(x_i, y_j) \log(f_{X|Y}(x_i \mid y_j)) - \sum_{i=1}^{K} f_X(x_i) \log(f_X(x_i))$$

$$= -H(X \mid Y) + H(X) \tag{8.3}$$

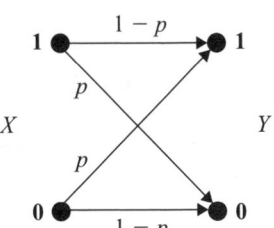

Figure 8.1 *The BSC, error probability* p

where $H(X)$ is the entropy as defined in the previous chapter (equation (7.7)) or the "input information". Further, we have defined the conditional entropy $H(X \mid Y)$ in a similar way, which can be viewed as the "uncertainty" caused by the channel.

In Figure 8.1 a simple but useful model is shown called the **binary symmetric channel (BSC)**. The symbol 1 or 0 is inputted from the left, and outputted to the right. The BSC is "memoryless", which means that there is no dependency between successive symbols. The numbers on the branches are the corresponding **transition probabilities**, i.e. the conditional probabilities $P(Y \mid X)$ which in this case can be expressed in terms of the error probability p

$$P(Y = 1 \mid X = 1) = (1 - p)$$
$$P(Y = 1 \mid X = 0) = p$$
$$P(Y = 0 \mid X = 1) = p$$
$$P(Y = 0 \mid X = 0) = (1 - p)$$

As shown in equation (7.3) the mutual information is symmetric with respect to X and Y, hence using Bayes' theorem equation (8.3) can be rewritten as

$$I(X, Y) = E[I(A_i, B_j)] = E[I(B_j, A_i)]$$

$$= \sum_{j=1}^{L} \sum_{i=1}^{K} f_{XY}(x_i, y_j) \log\left(\frac{f_{Y|X}(y_j \mid x_i)}{f_Y(y_j)}\right)$$

$$= \sum_{j=1}^{L} \sum_{i=1}^{K} f_{XY}(x_i, y_j) \log(f_{Y|X}(y_j \mid x_i)) - \sum_{j=1}^{L} f_Y(y_j) \log(f_Y(y_j))$$

$$= \sum_{j=1}^{L} \sum_{i=1}^{K} f_{Y|X}(y_j \mid x_i) f_X(x_i) \log(f_{Y|X}(y_j \mid x_i))$$

$$\quad - \sum_{j=1}^{L} f_Y(y_j) \log(f_Y(y_j))$$

$$= -H(Y \mid X) + H(Y) \tag{8.4}$$

where the probabilities

$$f_Y(y_j) = \sum_{i=1}^{K} f_{Y|X}(y_j \mid x_i) f_X(x_i).$$

Assume that the inputs $x_1 = 0$ and $x_2 = 1$ are equally probable, i.e. $f_X(0) = f_X(1) = 0.5$. This implies an entropy (equation (7.7)) of (using logarithm base 2)

$$H(X) = -\sum_{i=1}^{2} f_X(x_i) \operatorname{lb}(f_X(x_i)) = -\frac{1}{2} \operatorname{lb}\left(\frac{1}{2}\right) - \frac{1}{2} \operatorname{lb}\left(\frac{1}{2}\right) = 1 \text{ bit/symbol}$$

Not very surprisingly, this is the **maximum** entropy we can get from one binary symbol (a bit). For instance, if $x_1 = 0$ is more probable than $x_2 = 1$ or vice versa, then $H(X) < 1$ bit. For a "perfect" channel, i.e. a channel that does not introduce any errors or the error probability: $p = 0$. This implies that

$$P(Y = 1 \mid X = 1) = 1$$

$$P(Y = 1 \mid X = 0) = 0$$

$$P(Y = 0 \mid X = 1) = 0$$

$$P(Y = 0 \mid X = 0) = 1$$

Inserting the conditional probabilities into equation (8.4) we obtain

$$I(X, Y) = H(Y) - H(Y \mid X) = H(Y) = 1 \text{ bit/symbol} \tag{8.5}$$

The conditional entropy, or in other words the uncertainty of the channel $H(Y \mid X) = 0$ and all the "input entropy" ("information"), is obtained on the output of the channel. By observing Y we know everything about X.

If a BSC has an error probability of $p = 0.1$, on the average 10% of the bits are erroneous, which corresponds to 10% of the bits being inverted. The transition probabilities can readily be calculated as

$$P(Y = 1 \mid X = 1) = 0.9$$

$$P(Y = 1 \mid X = 0) = 0.1$$

$$P(Y = 0 \mid X = 1) = 0.1$$

$$P(Y = 0 \mid X = 0) = 0.9$$

Again, using equation (8.4) the resulting average mutual information will be

$$I(X, Y) = H(Y) - H(Y \mid X) = 1 - 0.469 = 0.531 \text{ bit/symbol} \tag{8.6}$$

In this case, we only gain 0.531 bits of information about X by receiving the symbols Y. The uncertainty in the channel has "diluted" the information about the input symbols X. Finally, for the worst channel possible, the error probability is $p = 0.5$. On the average we get errors half of the time. This **is** the worst case, since if p is larger than 0.5, i.e. the channel inverts more than half the number of transmitted bits, we could simply invert Y and thus reduce the error probability. The transition probabilities are

$$P(Y = 1 \mid X = 1) = 0.5$$

$$P(Y = 1 \mid X = 0) = 0.5$$

$$P(Y = 0 \mid X = 1) = 0.5$$

$$P(Y = 0 \mid X = 0) = 0.5$$

Inserting into equation (8.4) we get

$$I(X, Y) = H(Y) - H(Y \mid X) = 1 - 1 = 0 \, \text{bit/symbol} \qquad (8.7)$$

Such a result can be interpreted as if no information about X progresses through the channel. This is due to the great uncertainty in the channel itself. In other words, whether we obtain our data sequence Y from a random generator or from the output of the channel it does not matter. We would obtain equal amounts of knowledge about the input data X in both cases.

The BSC is a simple channel; however, in a realistic case more complicated models often have to be used. These models may be non-symmetric and use M number of symbols rather than only 2 (0 and 1). An "analog" channel (continuous in amplitude) can be regarded as having an infinite number of symbols. Further, channel models can possess memory, introducing dependency between successive symbols.

8.1.2 The channel capacity

From the above discussion it is clear that every channel has a limited ability to transfer information (average mutual information), and that this limit depends on the transition probabilities, i.e. the error probability. The maximum average mutual information possible for a channel is called **the channel capacity**. For the discrete, memoryless channel, the channel capacity is defined as the following

$$C = \sup_{f_X(x)} I(X, Y) \qquad (8.8)$$

By changing the probability density distribution $f_X(x)$ of the symbols transmitted over the channel or by finding a proper **channel code**, the average mutual information can be maximized. This maximum value, the channel capacity, is a property of the channel and cannot be exceeded if we want to transmit information with arbitrarily low error rate. This is mainly what the channel coding theorem is about.

It is important to note that regardless of the complexity of the devices and algorithms, if we try to exceed the channel capacity by transmitting more information per time unit (per symbol) than the channel capacity allows, we will never be able to transmit the information error free. Hence, the channel capacity can be regarded as a universal "speed limit". If we try to exceed this information transfer limit, we will be "fined" in terms of errors, and the net amount of transmitted error-free information will never exceed the channel capacity.

Unfortunately, the channel coding theorem does not give any hint on how to design the optimum channel code needed to obtain the maximum information transfer capacity, i.e. the channel capacity. The theorem only implies that at least one such channel code exists. The search for the most effective channel codes has been pursued during the last 40 years, and some of these results will be presented later in this chapter.

Calculating the channel capacity for channels in "reality" is in many cases very hard or even impossible. Two simple, common, standard channels should, however, be mentioned: the BSC and the additive white Gaussian noise (AWGN).

Firstly, the BSC channel capacity can be represented as follows (Figure 8.1)

$$C = 1 + p\,\mathrm{lb}(p) + (1 - p)\,\mathrm{lb}(1 - p) \text{ bit/symbol} \tag{8.9}$$

Secondly, the common memoryless channel model for "analog" (continuous) signals is the **additive white Gaussian noise channel (AWGN)** shown in Figure 8.2, which is slightly more elaborate. The analog input signal is X and the output signal is Y. N is a white Gaussian noise signal, added to the input signal. The term "white" is used in the sense that the noise has equal spectral density at all frequencies, similarly to white light, which has an equal power at all wavelengths in the visible spectrum (contains equal amounts of all colors).

In the context of AWGN, the **signal-to-noise ratio (SNR)** is commonly used, which is simply the ratio between the signal power and the noise power (Ahlin and Zander, 1998)

$$\mathrm{SNR} = \frac{X}{N} \tag{8.10}$$

The SNR is often expressed in decibels (dB)

$$\mathrm{SNR_{dB}} = 10\,\mathrm{lg}(\mathrm{SNR}) \tag{8.11}$$

where $\mathrm{lg}(\ )$ is the logarithm base 10. Thus, the channel capacity of the AWGN can be shown to be (Cover and Thomas, 1991)

$$C = W\,\mathrm{lb}(1 + \mathrm{SNR}) = W\,\mathrm{lb}\!\left(1 + \frac{X}{N}\right)$$

$$= W\,\mathrm{lb}\!\left(1 + \frac{X}{WN_0}\right) \text{ bit/symbol} \tag{8.12}$$

where W is the bandwidth and N_0 is the spectral density of the noise in W/Hz. An interesting example is a common telephone subscriber loop. If we assume

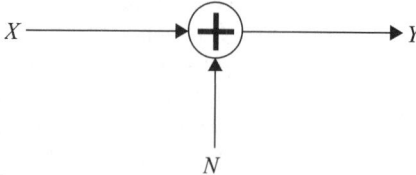

Figure 8.2 *The AWGN channel*

a bandwidth of $W = 4\,\text{kHz}$ and $\text{SNR} = 45\,\text{dB}$, the resulting channel capacity is therefore (approximately) $C = 60\,\text{kbit/symbol}$. Hence, the 57 600 baud dial up modems used today is approaching the theoretical upper limit. Some subscriber loops may of course have other bandwidth and SNR characteristics, which will affect the channel capacity.

Finally, to conclude this section, assume we have an information source with entropy H bit/symbol. In the previous chapter, the source coding theorem stated that to be able to reconstruct the information without errors, we must use an information rate R of

$$R \geq H \text{ bit/symbol}$$

This is, therefore a data compression limit. In this chapter, the channel coding theorem implies that to be able to transmit information error free at information rate R

$$C \geq R \text{ bit/symbol}$$

This is a communication speed limit.

8.2 Error-correcting codes

There are a number of different error-correcting codes. These codes can be divided into classes, depending on their properties. A rough classification is to divide the codes into two groups: **block codes** and **convolution codes**.

In this section, we will first discuss coding and properties of codes in general, followed by a presentation of some common block and convolution code algorithms.

8.2.1 Hamming distance and error correction

We shall begin with a brief illustration of a crude, yet simple **repetition code**.

Assume that we need to transmit information about a binary variable U. The probabilities of $u = 0$ and $u = 1$ are equal, which means $f_U(u = 0) = f_U(u = 1) = 0.5$. When transmitting this sequence of ones and zeros over a BSC with an error probability of $p = 0.1$, having a channel capacity $C = 0.531$ bit/symbol (equation (8.6)), we will get 10%-bit errors on the average.

By adding redundancy such that if we intend to transmit a "1", we repeat it twice. In other words, we transmit "111" and if we want to send a zero, we send the channel code word "000". If we get an error in 1 bit of a code word, we can still correct the error by making a majority vote when receiving the code. If there are more ones than zeros in a received code word, the information symbol sent was probably $u = 1$, and if there are more zeros than ones, $u = 0$ is the most likely alternative. This is simply called the **maximum likelihood (ML)** principle. The code word $\mathbf{X}$ is structured as

$$\mathbf{X} = [x_1 \quad x_2 \quad x_3] = [u \quad p_1 \quad p_2] \tag{8.13}$$

where u is the information bit, and p_1 and p_2 are denoted **parity bits** or **check bits**. If a code word is structured in such a way that the parity bits and information bits are in separate groups rather than a mixed order, the code word is said to have a **systematic form**. The coding process in this example will be very easy

$$x_1 = u$$

$$x_2 = p_1 = u \tag{8.14}$$

$$x_3 = p_2 = u$$

The decoding process will also be easy, a pure majority vote as outlined above. The length of the code word is often denoted by n and the number of information bits k, hence in our example above $n = 3$ and $k = 1$. Now, the **code rate** or code **speed** is obtained from

$$R = \frac{k}{n} \tag{8.15}$$

This can be seen as a "mix" ratio between the numbers of necessary information bits and the total number of bits in the code word (information bits + parity bits). In the first case above, transmitting U directly, without using any code, implies a rate of

$$R = \frac{1}{1} = 1 > C = 0.531$$

In this case, we have violated the channel coding theorem and consequently the length of the strings of symbols we send does not matter, as $n \to \infty$ we will still not be able to reduce the average error rate below 10%.

In the latter case, using the repetition code we have introduced redundancy and for this case

$$R = \frac{1}{3} = 0.333 < C = 0.531$$

It can be shown that for longer code words, i.e. as $n \to \infty$ the average error rate will tend to zero. Let us take a closer look at the underlying processes. Table 8.1 shows the eight possible cases for code word transmission and the associated transition probabilities expressed in terms of the error probability p

Table 8.1 *Transmission of simple 3-bit repetition code*

	X	Y	Probability	
(a)	$111 \to 111$		$(1-p)^3$	No errors
(b)	$111 \to 110, 101, 011$		$3(1-p)^2 p$	1-bit errors
(c)	$111 \to 001, 010, 100$		$3(1-p)p^2$	2-bit errors
(d)	$111 \to 000$		p^3	3-bit error
(e)	$000 \to 000$		$(1-p)^3$	No errors
(f)	$000 \to 001, 010, 100$		$3(1-p)^2 p$	1-bit errors
(g)	$000 \to 110, 101, 011$		$3(1-p)p^2$	2-bit errors
(h)	$000 \to 111$		p^3	3-bit error

of the BSC. From Table 8.1, in case (a) we intended to transmit $u = 1$ which was coded as 111. No errors occurred which has the probability of $(1 - p)^3 = (1 - 0.1)^3 = 0.729$ and 111 was received. This received code word was decoded using a majority vote, and found to be $\hat{u} = 1$. In case (b) a single bit error occurred and hence, 1 bit in the received code word is corrupt. This can happen in three ways. In case (c) 2-bit errors occur, which can also hit the code word in three different ways, and so on.

The interesting question which arises deals with the amount of cases in which this coding scheme will be able to correct the bit errors. In other words, how many cases can decode the right symbol in spite of the corrupt code word bits? From Table 8.1 it is clear to conclude that correct transmission will be achieved, i.e. $\hat{u} = u$ in cases: (a), (b), (e) and (f). In all other cases, the majority vote decoder will make errors and will not be able to correct the bit errors in the code word. Due to the majority decoding and the fact that we only transmit the code words 111 and 000, respectively, it is clear that errors will occur if there are more than $t = 1$ bit errors in the code word, hence we can define t as

$$t = \left\lfloor \frac{n}{2} \right\rfloor \tag{8.16}$$

where $\lfloor \ \rfloor$ is the "floor" operator, i.e. the first integer less than or equal to the argument. The total probability of erroneous transmission can now be expressed as

$$P_{\text{err}} = \sum_{i=t+1}^{n} \binom{n}{i} p^i (1-p)^{(n-i)} = \binom{3}{2} 0.1^2 (1 - 0.1) + \binom{3}{3} 0.1^3$$

$$= 3 \cdot 0.01 \cdot 0.9 + 0.001 = 0.028 \tag{8.17}$$

where

$$\binom{n}{i} = \frac{n!}{(n-i)! i!}$$

is the binomial coefficients, in other words, the number of ways i elements can be selected out of a set of n elements. As $n \to \infty$ in equation (8.17), $P_{\text{err}} \to 0$, in accordance to the channel coding theorem.

Intuitively from the above discussion, the greater the "difference" between the transmitted code words, the more bit errors can be tolerated, i.e. before a corrupt version of one code word seems to resemble another (erroneous) code word. For instance, assume that the code uses the two code words 101 and 001 instead of 111 and 000. In this case, one single bit error is enough for the decoder to make an erroneous decision.

The "difference" in coding context is denoted as the **Hamming distance** d. It is defined as the number of positions in which two code words differ. For example, the code words we used, 111 and 000 differ in three positions, hence the Hamming distance is $d = 3$. The two code words 101 and 001 have $d = 1$. It can be shown that a code having a minimum Hamming distance d between code words can **correct**

$$t = \left\lfloor \frac{d-1}{2} \right\rfloor \text{ bit errors in a code word} \tag{8.18}$$

Further, the same code can **detect** (but not correct)

$$\gamma = d - 1 \text{ bit errors in a code word} \tag{8.19}$$

If there are d-bit errors, we end up in another error-free code word and the error can never be detected. Techniques for detection of transmission errors are commonly used in **automatic repeat request** (ARQ) systems, where a detected error gives rise to a negative acknowledge signal, initiating a retransmission of the data. In some communication systems, the receiving equipment does not only output ones and zeros to the decoder. For instance, if a signal having poor strength is received, a third symbol, the **erasure** symbol, can be used. This is a "don't care" symbol saying the signal was too weak and cannot determine if the received signal was a 1 or a 0. In this case, the decoding algorithm can fill in the most probable missing bits

$$\rho = d - 1 \text{ erasures in a code word} \tag{8.20}$$

Finally when classifying a code, the triplet consists of code word length n, number of information bits k and minimum Hamming distance d and is commonly stated as

$$(n, k, d)$$

The number of parity bits is of course $n - k$. Using the minimum Hamming distance d and equation (8.18) we can figure out how many bit errors t (per code word) the code can handle. The example repetition code above is hence denoted (3, 1, 3).

It is worth noting that if there are more than t-bit errors in a code word, the coding will add even **more errors** and only make things worse. In such a case, a better code must be used. It might even be advantageous not to use any error-correcting code at all.

8.2.2 Linear block codes

For a general block code, we are free to select our code words arbitrarily, which for large blocks may result in extremely complicated encoders and decoders. If we demand the code to be linear, things become somewhat easier. The definition of a **linear code** is that the sum of the two code words $\mathbf{x}_i$ and $\mathbf{x}_j$ is also a code word. The addition of code words is performed component wise

$$\begin{aligned}
\mathbf{x}_i + \mathbf{x}_j &= [x_{i1} \ x_{i2} \ \cdots \ x_{in}] + [x_{j1} \ x_{j2} \ \cdots \ x_{jn}] \\
&= [x_{i1} + x_{j1} \ x_{i2} + x_{j2} \ \cdots \ x_{in} + x_{jn}] \\
&= [x_{m1} \ x_{m2} \ \cdots \ x_{mn}] = \mathbf{x}_m
\end{aligned} \tag{8.21}$$

Here the addition is performed according to the algebraic rules of the number system used. For the binary case, "$+$" means **binary** addition, i.e. addition modulo 2 or exclusive OR (XOR). Thus, $0 + 0 = 0, 1 + 0 = 0 + 1 = 1$ and $1 + 1 = 0$. For other number systems, similar rules shall apply. This algebraic topic is related to as the Galois fields ($GF(p)$) theory. In this chapter, we will however remain faithful to the binary number system. If an interest in this particular subject is not very high, then Galois is a trivial issue.

A linear code has the helpful property that the vicinity of all code words in the n-dimensional "code space" looks the same. Hence, when decoding, we can apply the same algorithm for all code words and there are many standard mathematical tools available for our help. When analyzing codes, it is common to assume that the zero code word (all zeros) has been sent. The zero code word is present in all binary linear block codes, since by adding a code word to itself, using equation (8.21)

$$\mathbf{x}_i + \mathbf{x}_i = [x_{i1} + x_{i1} \ x_{i2} + x_{i2} \ \cdots \ x_{in} + x_{in}] = [0 \ 0 \ \cdots \ 0] = \mathbf{0} \quad (8.22)$$

The coding process in a linear block code can be readily expressed as a matrix multiplication. The information word $\mathbf{u}$ (dim $1 \times k$) is multiplied by the **generator matrix G** (dim $k \times n$), thus obtaining the code word $\mathbf{x}$ (dim $1 \times n$)

$$\mathbf{x}_i = \mathbf{u}_i \mathbf{G} = \mathbf{u}_i [\mathbf{I} \ \vdots \ \mathbf{P}] \quad (8.23)$$

The generator matrix $\mathbf{G}$ defines the coding operation and can be partitioned into an identity matrix $\mathbf{I}$ (dim $k \times k$) and a matrix $\mathbf{P}$ (dim $k \times n - k$). In this way, the code will have a systematic form. The identity matrix simply "copies" the information bits u_i into the code word, while the $\mathbf{P}$ matrix performs the calculations needed to determine the parity bits p_j (compare to equation (8.14)).

During the transmission, the code word $\mathbf{x}$ will be subject to interference, causing bit errors. These bit errors are represented by the **error vector e** (dim $1 \times n$). The received code word $\mathbf{y}$ (dim $1 \times n$) can be expressed as

$$\mathbf{y} = \mathbf{x} + \mathbf{e} \quad (8.24)$$

(Remember we are still dealing with binary addition).

Finally, when decoding a linear block code, a syndrome-based decoding algorithm is commonly used. This algorithm calculates a **syndrome vector s** (dim $1 \times k$) by multiplying the received code word $\mathbf{y}$ by the **parity matrix H** (dim $k \times n$). The parity matrix can be partitioned into an identity matrix $\mathbf{I}$ (dim $k \times k$) and the matrix $\mathbf{Q}$ (dim $n - k \times k$)

$$\mathbf{s} = \mathbf{y}\mathbf{H}^{\mathrm{T}} = \mathbf{y}[\mathbf{Q} \ \vdots \ \mathbf{I}]^{\mathrm{T}} \quad (8.25)$$

For the binary case, if we set $\mathbf{Q} = \mathbf{P}^{\mathrm{T}}$, the syndrome vector will be equal to the zero vector (all-zero elements), i.e. $\mathbf{s} = \mathbf{0}$ if there are **no errors** in the received code word that this particular code is able to detect. This can be shown by

$$\mathbf{s} = \mathbf{y}\mathbf{H}^{\mathrm{T}} = (\mathbf{u}\mathbf{G} + \mathbf{e})\mathbf{H}^{\mathrm{T}} = \mathbf{u}\mathbf{G}\mathbf{H}^{\mathrm{T}} = \mathbf{u}[\mathbf{I} \ \vdots \ \mathbf{P}] \begin{bmatrix} \mathbf{Q}^{\mathrm{T}} \\ \cdots \\ \mathbf{I} \end{bmatrix}$$

$$= \mathbf{u}(\mathbf{Q}^{\mathrm{T}} + \mathbf{P}) = \mathbf{0} \quad (8.26)$$

where, of course, $\mathbf{e} = \mathbf{0}$ since there are no errors. If we have detectable errors, $\mathbf{s} \neq \mathbf{0}$ and the syndrome vector will be used as an index in a **decoding table**.

This decoding table shows the error vectors that are able to generate a specific syndrome vector. Commonly it is assumed that few bit errors in a code word is more likely than many errors. Hence, the suggested error vector $\hat{\mathbf{e}}$ (dim $1 \times n$) having the smallest number of ones (i.e. bit errors) is assumed to be equal to the true error vector $\mathbf{e}$. By adding the assumed error vector to the received code word including errors, a corrected code word $\hat{\mathbf{x}}$ (dim $1 \times n$) can be obtained

$$\hat{\mathbf{x}} = \mathbf{y} + \hat{\mathbf{e}} \tag{8.27}$$

From this corrected code word $\hat{\mathbf{x}}$, only simple bit manipulations are needed to extract the received information word $\hat{\mathbf{u}}$ (dim $1 \times k$).

Let us conclude this discussion on linear block codes with an example. Assume we have designed a linear block code LC(6, 3, 3), having $n = 6$, $k = 3$ and $d = 3$. The code words are in systematic form and look like

$$\mathbf{x} = [x_1 \ x_2 \ x_3 \ x_4 \ x_5 \ x_6] = [u_1 \ u_2 \ u_3 \ p_1 \ p_2 \ p_3] \tag{8.28}$$

where the parity bits are calculated using the following expressions

$$\begin{aligned} p_1 &= u_1 + u_2 \\ p_2 &= u_2 + u_3 \\ p_3 &= u_1 + u_2 + u_3 \end{aligned} \tag{8.29}$$

Using this information, it is straightforward to write down the generator matrix for this code

$$\mathbf{G} = \begin{bmatrix} \mathbf{I} & \vdots & \mathbf{P} \end{bmatrix} = \begin{bmatrix} 1 & 0 & 0 & 1 & 0 & 1 \\ 0 & 1 & 0 & 1 & 1 & 1 \\ 0 & 0 & 1 & 0 & 1 & 1 \end{bmatrix} \tag{8.30}$$

In a similar way, the parity matrix can also be formed

$$\mathbf{H} = \begin{bmatrix} \mathbf{P}^{\mathrm{T}} & \vdots & \mathbf{I} \end{bmatrix} = \begin{bmatrix} 1 & 1 & 0 & 1 & 0 & 0 \\ 0 & 1 & 1 & 0 & 1 & 0 \\ 1 & 1 & 1 & 0 & 0 & 1 \end{bmatrix} \tag{8.31}$$

Now, assume we want to transmit the information word 010, i.e. $\mathbf{u} = [0 \ 1 \ 0]$. The encoder will transmit the corresponding code word on the channel

$$\mathbf{x} = \mathbf{u}\mathbf{G} = [0 \ 1 \ 0] \begin{bmatrix} 1 & 0 & 0 & 1 & 0 & 1 \\ 0 & 1 & 0 & 1 & 1 & 1 \\ 0 & 0 & 1 & 0 & 1 & 1 \end{bmatrix} = [0 \ 1 \ 0 \ 1 \ 1 \ 1] \tag{8.32}$$

This time we are lucky, the channel is in a good mode, and no errors occur, i.e. $\mathbf{e} = \mathbf{0}$, thus

$$\mathbf{y} = \mathbf{x} + \mathbf{e} = \mathbf{x} = [0 \ 1 \ 0 \ 1 \ 1 \ 1] \tag{8.33}$$

In the decoder, the syndrome vector is first calculated as

$$\mathbf{s} = \mathbf{y}\mathbf{H}^{\mathrm{T}} = \begin{bmatrix} 0 & 1 & 0 & 1 & 1 & 1 \end{bmatrix} \begin{bmatrix} 1 & 0 & 1 \\ 1 & 1 & 1 \\ 0 & 1 & 1 \\ 1 & 0 & 0 \\ 0 & 1 & 0 \\ 0 & 0 & 1 \end{bmatrix} = \begin{bmatrix} 0 & 0 & 0 \end{bmatrix} = \mathbf{0} \tag{8.34}$$

Since the syndrome vector is equal to the zero vector, the decoder cannot detect any errors and no error correction is needed, $\hat{\mathbf{x}} = \mathbf{y}$. The transmitted information is simply extracted from the received code word by

$$\hat{\mathbf{u}} = \hat{\mathbf{x}}[\mathbf{I} \vdots \mathbf{0}]^{\mathrm{T}} = \begin{bmatrix} 0 & 1 & 0 & 1 & 1 & 1 \end{bmatrix} \begin{bmatrix} 1 & 0 & 0 \\ 0 & 1 & 0 \\ 0 & 0 & 1 \\ 0 & 0 & 0 \\ 0 & 0 & 0 \\ 0 & 0 & 0 \end{bmatrix} = \begin{bmatrix} 0 & 1 & 0 \end{bmatrix} \tag{8.35}$$

Now, assume we try to send the same information again over the channel; this time, however, we are not as lucky as before, and a bit error occurs in u_3. This is caused by the error vector

$$\mathbf{e} = \begin{bmatrix} 0 & 0 & 1 & 0 & 0 & 0 \end{bmatrix} \tag{8.36}$$

Hence, the received code word is

$$\mathbf{y} = \mathbf{x} + \mathbf{e} = \begin{bmatrix} 0 & 1 & 0 & 1 & 1 & 1 \end{bmatrix} + \begin{bmatrix} 0 & 0 & 1 & 0 & 0 & 0 \end{bmatrix} = \begin{bmatrix} 0 & 1 & 1 & 1 & 1 & 1 \end{bmatrix} \tag{8.37}$$

The syndrome calculation gives

$$\mathbf{s} = \mathbf{y}\mathbf{H}^{\mathrm{T}} = \begin{bmatrix} 0 & 1 & 1 & 1 & 1 & 1 \end{bmatrix} \begin{bmatrix} 1 & 0 & 1 \\ 1 & 1 & 1 \\ 0 & 1 & 1 \\ 1 & 0 & 0 \\ 0 & 1 & 0 \\ 0 & 0 & 1 \end{bmatrix} = \begin{bmatrix} 0 & 1 & 1 \end{bmatrix} \tag{8.38}$$

Using a decoding table (Table 8.2), we find that the eight proposed error vectors on line 4 may cause the syndrome vector obtained above.

Using the maximum likelihood argument, we argue that a single-bit error is more probable than a multi-bit error in our system. For this reason, we assume that the error vector in the first column to the left (containing only 1 bit) is the most probable error, hence

$$\hat{\mathbf{e}} = \begin{bmatrix} 0 & 0 & 1 & 0 & 0 & 0 \end{bmatrix} \tag{8.39}$$

The received code word is now successfully corrected (equation (8.27))

$$\hat{\mathbf{x}} = \mathbf{y} + \hat{\mathbf{e}} = \begin{bmatrix} 0 & 1 & 1 & 1 & 1 & 1 \end{bmatrix} + \begin{bmatrix} 0 & 0 & 1 & 0 & 0 & 0 \end{bmatrix}$$
$$= \begin{bmatrix} 0 & 1 & 0 & 1 & 1 & 1 \end{bmatrix} = \mathbf{x} \tag{8.40}$$

Table 8.2 *Decoding table for the code LC(6,3,3) in the example*

Syndrome	Possible error vectors							
000	000000	100101	010111	001011	110010	011100	101110	111001
101	100000	000101	110111	101011	010010	111100	001110	011001
111	010000	110101	000111	011011	100010	001100	111110	101001
011	001000	101101	011111	000011	111010	010100	100110	110001
100	000100	100001	010011	001111	110110	011000	101010	111101
010	000010	100111	010101	001001	110000	011110	101100	111011
001	000001	100100	010110	001010	110011	011101	101111	111000
110	101000	001101	111111	100011	011010	110100	000110	010001

Table 8.3 *Hamming distance between all possible error-free code words of the code LC(6,3,3) in the example*

	000000	100101	010111	001011	110010	011100	101110	111001
000000	0	3	4	3	3	3	4	4
100101	3	0	3	4	4	4	3	3
010111	4	3	0	3	3	3	4	4
001011	3	4	3	0	4	4	3	3
110010	3	4	3	4	0	4	3	3
011100	3	4	3	4	4	0	3	3
101110	4	3	4	3	3	3	0	4
111001	4	3	4	3	3	3	4	0

and the transmitted information can be extracted in the same way as in equation (8.35).

The example code above has a minimum Hamming distance between code words of $d = 3$. Using equation (8.18) we find that the code is only able to correct $t = 1$, i.e. single-bit errors. As pointed out earlier in this text, if there are more bit errors in a code word than t, the coding scheme will not be able to correct any errors, and it may in fact aggravate the situation. This can be illustrated by the following example. If we transmit $\mathbf{u} = [0\ 0\ 1]$, the code word will be $\mathbf{x} = [0\ 0\ 1\ 0\ 1\ 1]$. If a **double**-bit error situation occurs, $\mathbf{e} = [0\ 1\ 0\ 1\ 0\ 0]$ we will receive $\mathbf{y} = \mathbf{x} + \mathbf{e} = [0\ 1\ 1\ 1\ 1\ 1]$, i.e. the same code word as in the example above. This code word will be "corrected" and as above, we will obtain $\hat{\mathbf{u}} = [0\ 1\ 0] \neq \mathbf{u}$.

When calculating the error-correction performance t of a code using equation (8.18), we have been assuming that d is the **minimum** Hamming distance between code words. A deeper, but not too surprising analysis shows that the Hamming distance between all possible error-free code words in a given code vary. An example is the code $LC(6, 3, 3)$ used above, as shown in Table 8.3. In this particular example, it does not matter, because equation (8.18) gives the same result for both $d = 3$ and $d = 4$, but it is important to remember that the

minimum Hamming distance should be used in equation (8.18). An alternative way of finding the minimum Hamming distance of a code is to find the smallest number of column vectors in the parity matrix **H** (equation (8.31)) that need to be added to obtain the zero vector.

There are many ways to design the error correcting code. The main objective, however, is to find a code that can correct as many errors as possible, while requiring as few parity bits as possible. The code should also be easy to decode. One criteria, to be met when designing an effective code, is to choose the code words in such a way that the Hamming distances to the closest neighbors (i.e. the minimum Hamming distance) are the same for all code words. A code having this property is called a "perfect code". An example of a perfect code is the classical family of **Hamming Codes** (Haykin, 2001). This type of codes has been used since about 1948 and was initially used for error correction in long distance telephony systems. The parameters of these codes are

> Block length: $n = 2^m - 1$
> Number of information bits: $k = 2^m - m - 1$
> Hamming distance: $d = 3$

for $m = 2, 3, 4, \ldots$

From the above, it is obvious that all Hamming codes can correct single-bit errors only, disregarding the block length.

8.2.3 Cyclic codes, Bose, Chaudhuri, Hocquenghem codes

The class of **cyclic codes** is a sub-class of linear block codes. Instead of the linearity property where the sum of two code words yields another code word, a **cyclic shift** ("rotate") of the bit pattern is performed instead. This shift results in another code word. For example, if 111001 is a code word, then 110011 should also be a code word and so on.

Cyclic codes are often expressed using code **polynomials**. In these polynomials, a formal parameter x is used. **Note! This formal parameter x is something completely different from the code word bits x_i discussed earlier.** The formal parameter x is only a means of numbering and administrating the different bits in a cyclic code polynomial. The data bits of interest are the **coefficients** of these polynomials.

A code word $\mathbf{c} = [c_0 \; c_1 \; \cdots \; c_{n-1}]$ can hence be expressed as a code polynomial

$$c(x) = c_0 + c_1 x + c_2 x^2 + c_3 x^3 + \cdots + c_{n-1} x^{n-1} \tag{8.41}$$

An m step cyclic shift of an $n - 1$ bit code word is accomplished by a multiplication

$$x^m c(x) \quad \mod (x^n - 1) \tag{8.42}$$

which is performed modulo $x^n - 1$, i.e. after the polynomial multiplication by x^m a polynomial division by $(x^n - 1)$ should take place to keep the order of the resulting polynomial to $n - 1$ or below. In bit pattern words, this operation can be explained as the multiplication performs the "bit shifting", while the division makes the "bit rotating" (higher-order bits are rotated back as low-order bits).

Using this polynomial way of expressing vectors, a k-bit information vector can be written as a polynomial in the formal parameter x in a similar way to equation (8.41)

$$u(x) = u_0 + u_1 x + u_2 x^2 + u_3 x^3 + \cdots + u_{k-1} x^{k-1} \tag{8.43}$$

where $u_0, u_1, \ldots, u_{k-1}$ are the bits of the corresponding information vector (but the numbering is slightly different from the vector notation discussed earlier). Still remaining with binary numbers, modulo 2 addition applies. Instead of using a generator matrix to define the code, we now use a **generator polynomial** $g(x)$, and the coding process consists of a multiplication of polynomials

$$c(x) = u(x) g(x) \tag{8.44}$$

where $c(x)$ is the code word polynomial, corresponding to the code vector being transmitted over the channel. Unfortunately, this straightforward approach does not result in code words having systematic form. This can be accomplished by first shifting the information polynomial and then dividing this product by the generator polynomial

$$\frac{x^{n-k} u(x)}{g(x)} = q(x) + \frac{r(x)}{g(x)} \tag{8.45a}$$

where $q(x)$ is the quota and $r(x)$ the remainder. Rewriting equation (8.45a) we get

$$\begin{aligned} x^{n-k} u(x) &= q(x) g(x) + r(x) = c(x) + r(x) \\ &\Rightarrow c(x) = x^{n-k} u(x) - r(x) \end{aligned} \tag{8.45b}$$

The code word travels the channel as before and errors, represented by the **error polynomial** $e(x)$, may occur. The received data vector is represented by the polynomial $v(x)$

$$v(x) = c(x) + e(x) \tag{8.46}$$

When decoding the received word expressed as $v(x)$, calculation of the **syndrome polynomial** $s(x)$, used for error detection and correction, simply consists of division by the generator polynomial $g(x)$ or multiplication by the **parity polynomial** $h(x)$, where

$$h(x) = \frac{x^n - 1}{g(x)} \tag{8.47}$$

The syndrome polynomial is obtained as the remainder when performing the division

$$\frac{v(x)}{g(x)} = Q(x) + \frac{s(x)}{g(x)} \tag{8.48}$$

It is easy to show that the syndrome will depend on the error polynomial only, and if no errors are present, the syndrome polynomial will be 0. Using equations

(8.45), (8.46) and (8.48) we obtain

$$v(x) = c(x) + e(x) = q(x)g(x) + e(x) = Q(x)g(x) + s(x)$$
$$\Rightarrow e(x) = (Q(x) - q(x))g(x) + s(x) \tag{8.49}$$

This relation shows that the syndrome $s(x)$ is the remainder when dividing $e(x)$ by $g(x)$. Hence, we can calculate the syndromes for all possible error polynomials in advance. All unique syndromes can then successfully be used to identify the corresponding error polynomial $\hat{e}(x)$ and error correction of $v(x)$ can be performed as (in the binary case)

$$\hat{v}(x) = v(x) + \hat{e}(x) \tag{8.50}$$

where $\hat{v}(x)$ is the corrected message (hopefully) from which the information $\hat{u}(x)$ can be extracted. Once again, an example will illustrate the ideas and algorithms of binary, cyclic block codes. Assume we are using a cyclic code, $CC(7, 4, 3)$, i.e. a code having $n = 7$, $k = 4$ and $d = 3$. Using equation (8.18) we find that this code can correct single-bit errors, $t = 1$. Further, the code is defined by its code polynomial $g(x) = 1 + x + x^3$. In this example, we want to transmit the information 0010. This is expressed as an information polynomial

$$u(x) = x^2 \tag{8.51}$$

Multiplying this information polynomial by the generator polynomial, the code polynomial is obtained

$$c(x) = u(x)g(x) = x^2(1 + x + x^3) = x^2 + x^3 + x^5 \tag{8.52}$$

The corresponding bit sequence 0011010 is transmitted over the channel and this first time no errors occur, i.e. $e(x) = 0$ and

$$v(x) = c(x) + e(x) = c(x) = x^2 + x^3 + x^5 \tag{8.53}$$

At the decoder, we start by calculating the syndrome polynomial

$$\frac{v(x)}{g(x)} = \frac{x^5 + x^3 + x^2}{x^3 + x + 1} = x^2 + \frac{0}{x^3 + x + 1} = x^2 + \frac{s(x)}{x^3 + x + 1} \tag{8.54}$$

Since the remainder or the syndrome is zero, there are no errors and the received information is $\hat{u}(x) = x^2 = u(x)$. Now, we try this again, but this time a single-bit error occurs, having the corresponding error polynomial: $e(x) = x^5$, starting over from equation (8.53) we get (using binary addition)

$$v(x) = c(x) + e(x) = x^2 + x^3 + x^5 + x^5 = x^2 + x^3 \tag{8.55}$$

The syndrome calculation yields

$$\frac{v(x)}{g(x)} = \frac{x^3 + x^2}{x^3 + x + 1} = 1 + \frac{x^2 + x + 1}{x^3 + x + 1} = 1 + \frac{s(x)}{x^3 + x + 1} \tag{8.56}$$

Hence, this time the syndrome polynomial is $s(x) = x^2 + x + 1$. Using equation (8.48) and assuming the zero code polynomial $c(x) = 0$ (OK, since the code is linear, all code words have the same "vicinity"), we find it is easier to make a decoding table in advance.

$S(x)$	$\hat{e}(x)$
0	0
1	1
x	x
x^2	x^2
$x+1$	x^3
x^2+x	x^4
x^2+x+1	x^5
x^2+1	x^6

From this table, we can see that with the given syndrome $s(x) = x^2 + x + 1$, the corresponding single-bit error polynomial is $\hat{e}(x) = x^5$. Error correction yields

$$\hat{v}(x) = v(x) + \hat{e}(x) = x^2 + x^3 + x^5 \tag{8.57}$$

From this, the information $\hat{u}(x) = x^2$ can be obtained as before. In most cases, coding and decoding is not performed in the simple way outlined above, since there are smarter ways to implement the algorithms.

Shift registers and logic circuits have traditionally been used to implement cyclic code algorithms. Nowadays, these functions can readily be implemented as digital signal processor (DSP) software, using shift, rotate and Boolean computer instructions. Below, the basic ideas will be briefly presented. The underlying idea is that multiplication and division of polynomes can be viewed as **filtering operations**, and can be achieved using shift registers in structures resembling digital finite impulse response (FIR) and infinite impulse response (IIR) filters.

If we have a generator polynomial, $g(x) = g_0 + g_1 x + g_2 x^2 + \cdots + g_{n-k} x^{n-k}$, an information polynomial as in equation (8.43) and we want to perform the polynomial multiplication as in equation (8.44) to obtain a code vector as in equation (8.41) we get

$$c(x) = c_0 + c_1 x + c_2 x^2 + \cdots + c_{n-1} x^{n-1} = u(x) g(x)$$
$$= u_0 g_0 + (u_1 g_0 + u_0 g_1) x + (u_2 g_0 + u_1 g_1 + u_0 g_2) x^2 + \cdots \tag{8.58}$$

From the above expression it can be seen that the coefficients c_i can be obtained by a convolution of the coefficients of the information polynomial $u(x)$ and the coefficients of the generator polynomial $g(x)$

$$c_i = \sum_{j=0}^{i} g_j u_{i-j} \tag{8.59}$$

This operation is performed by an FIR filter (see Chapter 1), hence "filtering" the data bit sequence u_i using a binary FIR filter with tap weights g_j will perform the polynomial multiplication, otherwise known as "the coding operation" (equation (8.44)).

The syndrome calculation, division by $g(x)$ as in equation (8.48), can also be performed in a similar way. Consider the z-transform of the sequences u_i and g_j, respectively, denoted $U(z)$ and $G(z)$ (see Chapter 1). Since equation (8.59) is a convolution, this is equivalent to a multiplication in the z-transform domain, hence the z-transform of the code polynomial coefficients $C(z)$ can easily be obtained as

$$C(z) = U(z)\, G(z) \tag{8.60}$$

If we now require the generator polynomial to have $g_0 = 1$ and $g_{n-k} = 1$ it can be rewritten as

$$
\begin{aligned}
G(z) &= g_0 + g_1 z^{-1} + g_2 z^{-2} + \cdots + g_{n-k} z^{-n+k} \\
&= 1 + g_1 z^{-1} + g_2 z^{-2} + \cdots + z^{-n+k} = 1 + G'(z)
\end{aligned}
\tag{8.61}
$$

From Chapter 1 we recall that if an FIR filter having, for instance, the z-transform $G'(z)$, is put in a feedback loop, we have created an IIR filter with the transfer function

$$H(z) = \frac{1}{1 + G'(z)} = \frac{1}{G(z)} \tag{8.62}$$

This indicates that by feeding the data corresponding to $v(x)$ into a circuit consisting of a shift register in a feedback loop, we effectively achieve a polynomial division. The tap weights corresponding to $G'(z)$ are simply $g(x)$ but with $g_0 = 0$. A closer analysis shows that during the first n shifts, the quota $q(x)$ (equation (8.45a)) of the polynomial division is obtained at the output of the shift register. After these shifts, the remainder, i.e. the syndrome coefficients of $s(x)$, can be found in the delay line of the shift register. The syndrome can hence be read out in parallel and used as an index in a syndrome look-up table (LUT). Using information in this table, errors can be corrected. A special class of syndrome-based error-correcting decoders using the method above are known as **Meggit decoders**.

Figure 8.3 shows an encoder and channel (upper part) and decoder (lower part) for the cyclic code $g(x) = 1 + x + x^3$ in the previous example. The encoder and decoder consist of shift registers as outlined in the text above. It is an instructive exercise to mentally "step through" the coding and syndrome calculating processes. The syndrome is obtained after $n = 7$ steps.

There are a number of standardized cyclic codes, commonly denoted **cyclic redundancy check (CRC) codes**, some of them are

CRC-4:	$g(x) = 1 + x + x^4$
CRC-5:	$g(x) = 1 + x^2 + x^4 + x^5$
CRC-6:	$g(x) = 1 + x + x^6$
CRC-12:	$g(x) = 1 + x + x^2 + x^3 + x^{11} + x^{12}$
CRC-16:	$g(x) = 1 + x^2 + x^{15} + x^{16}$
CRC-CCITT:	$g(x) = 1 + x^5 + x^{12} + x^{16}$

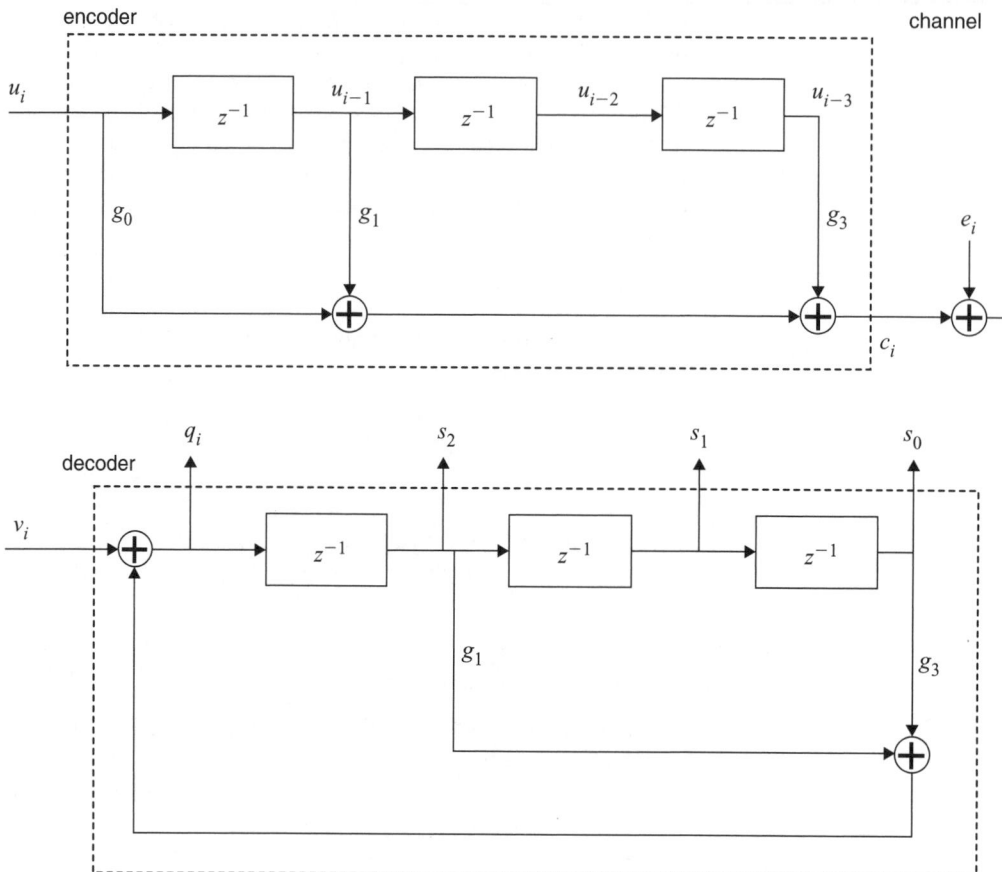

Figure 8.3 *Encoder and channel (upper) and decoder (lower) for a cyclic code having the generator polynomial* $g(x) = 1 + x + x^3$. *Encoder and syndrome calculating decoder built using shift register elements*

A special and important class of cyclic codes are the **BCH codes**, named after Bose, Chaudhuri and Hocquenghem. This class consists of a number of effective codes with moderate block length n. The parameters of these codes are

Block length:	$n = 2^m - 1$
Number of information bits:	$k \geq n - mt$
Minimum Hamming distance:	$d \geq 2t + 1$

for $m = 3, 4, 5, \ldots$

Reed–Solomon (RS) codes belong to a special type of BCH codes, working with groups of bits, in other words m-ary symbols rather than bits. The RS codes are very efficient and have the greatest possible minimum Hamming distance for a given number of check symbols and a given block length. The block length must be kept small in practice. Nevertheless, it is an important and popular type of code and is often used together with some other coding scheme, resulting in a **concatenated coding** system.

BCH and Reed–Solomon codes are, for instance, used in cellular radio systems, CD player systems and satellite communication systems, dating back to the 1970s.

8.2.4 Convolution codes

In the previous sections, we have discussed block codes, i.e. codes working with a structure of fixed length, consisting of information and parity bits. **Convolution codes** work in continuous "stream" fashion and "inserts" parity bits within the information bits according to certain rules. The convolution process is mainly a filtering process and is commonly implemented as some kind of shift register device (may be software).

A convolution encoder algorithm can be viewed as n binary FIR filters, having m steps long delay lines. The information bit sequence u_l to be coded is input to the filters. Each filter j has its own binary impulse response $g_i^{(j)}$ and hence, the output of each filter can be expressed as a convolution sum. The output of filter j is

$$c_l^{(j)} = \sum_{i=1}^{m} g_i^{(j)} u_{l-i+1} \quad l = 1, 2, \ldots, L \tag{8.63}$$

where $u_1, u_2, \ldots, u_L$ are the information bits and $g_i^{(j)}$ the jth **generator**, i.e. the impulse response of filter j. Note! The superscript is only an index, not an exponentiation operation. The code word bits $c_l^{(j)}$ will be arranged in the following way when transmitted over the channel

$$c_1^{(1)}, c_1^{(2)}, \ldots, c_1^{(n)}, c_2^{(1)}, c_2^{(2)}, \ldots, c_2^{(n)}, \ldots, c_{L+m}^{(1)}, c_{L+m}^{(2)}, \ldots, c_{L+m}^{(n)}$$

Thus, there will be $(L+m)n$ bits in the code word, and the rate of the code can be calculated by

$$R = \frac{L}{(L+m)n} \tag{8.64}$$

Further, the **constraint length** of the code, in other words the number of bits that may be affected by a given information bit is obtained from

$$n_s = (m+1)n \tag{8.65}$$

The practical algorithm (or hardware) operates as follows. Since the n delay lines will run in parallel, having the same input, only **one** delay line is needed. This line is shared by all the n filters. Initially, all the m elements in the delay line are set to zero. The first information symbol u_1 is fed to the delay line, and the outputs $c_1^{(j)}$ of all filters are calculated and sent to the output. Then, the delay line is shifted one step and the next input bit u_2 enters the circuit. The filter outputs $c_2^{(j)}$ are calculated and transferred to the output as before.

This scheme is repeated until the last information bit u_L is reached. After that point, only zeros are entered into the delay line for another m steps, thus

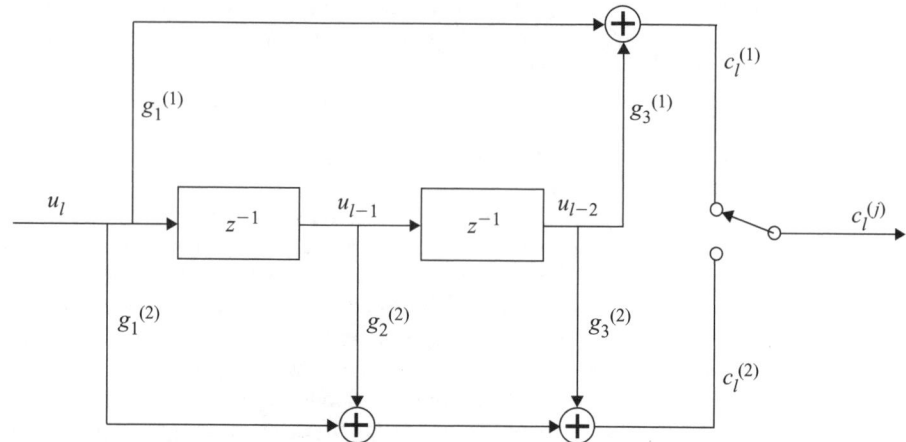

Figure 8.4 *Encoder for a simple convolution code. Parameters:* n $= 2$, k $= 1$, m $= 2$ *and generators,* g$^{(1)} = (1,0,1)$, g$^{(2)} = (1,1,1)$

clearing the entire delay line. During this period, the **tail** of the code is generated and transmitted through the channel.

Figure 8.4 shows an example of a convolution encoder for the code $(n, k, m) = (2, 1, 2)$, with generators $g^{(1)} = (1, 0, 1)$ and $g^{(2)} = (1, 1, 1)$. Assuming an input information bit sequence u_l is $(0, 1, 0, 1)$ which will result in the code word sequence $c_l^{(j)}$ (including tail) $(00, 11, 01, 00, 01, 11)$.

The example convolution code shown here is a simple one. Convolution-type codes have been used for space and satellite communications since the early 1960s. During the Voyager missions to Mars, Jupiter and Saturn during the 1970s, a couple of different convolution codes were used to ensure good data communication. Two of these codes that became NASA standards are

- The **Jet Propulsion Lab** convolution code $(2, 1, 6)$, with generators

$$g^{(1)} = (1, 1, 0, 1, 1, 0, 1)$$
$$g^{(2)} = (1, 0, 0, 1, 1, 1, 1)$$

- The **Linkabit** $(3, 1, 6)$ convolution code, with generators

$$g^{(1)} = (1, 1, 0, 1, 1, 0, 1)$$
$$g^{(2)} = (1, 0, 0, 1, 1, 1, 1)$$
$$g^{(3)} = (1, 0, 1, 0, 1, 1, 1)$$

8.2.5 Viterbi decoding

There are many ways to decode convolution-coded data. One of the most common methods is to use a **Viterbi decoder**, named after A. Viterbi. The idea of Viterbi decoding is to decode an entire sequence of information symbols and parity symbols at a time, rather than every single group (or block) of information and parity bits. Let us assume that the transmitted sequence of bits entering a channel is

$$\mathbf{c} = \left(\mathbf{c}_1, \mathbf{c}_2, \ldots, \mathbf{c}_{L+m}\right) = \left(c_1^{(1)}, c_1^{(2)}, \ldots, c_1^{(n)}, c_2^{(1)}, \ldots, c_{L+m}^{(n)}\right)$$

as above, and $\mathbf{c}_i = \left(c_i^{(1)}, c_i^{(2)}, \ldots, c_i^{(n)}\right)$ is the ith block. The received sequence, coming out of the channel is expressed in a similar way

$$\mathbf{v} = \left(\mathbf{v}_1, \mathbf{v}_2, \ldots, \mathbf{v}_{L+m}\right) = \left(v_1^{(1)}, v_1^{(2)}, \ldots, v_1^{(n)}, v_2^{(1)}, \ldots, v_{L+m}^{(n)}\right)$$

From the discussion on channels, we remember that for a communication channel, there is a set of transition probabilities. These probabilities are denoted $P\left(v_i^{(j)} \mid c_i^{(j)}\right)$ and are the conditional probabilities of receiving bit $v_i^{(j)}$ when transmitting $c_i^{(j)}$. If the bit error probability is independent for all bits, then the conditional probability of receiving block $\mathbf{v}_i$ when block $\mathbf{c}_i$ is transmitted is

$$P(\mathbf{v}_i \mid \mathbf{c}_i) = \prod_{j=1}^{n} P\left(v_i^{(j)} \mid c_i^{(j)}\right) \tag{8.66a}$$

In a similar way, the conditional probability of receiving the entire sequence $\mathbf{v}$, given that bit sequence $\mathbf{c}$ (including tail) is transmitted, is

$$P(\mathbf{v} \mid \mathbf{c}) = \prod_{i=1}^{L+m} P(\mathbf{v}_i \mid \mathbf{c}_i) \tag{8.67a}$$

Now, the task of a maximum likelihood (ML) decoder is to maximize (equation (8.67a)). Taking the logarithm of equations (8.66a) and (8.67a), respectively, we obtain

$$L(\mathbf{v}_i, \mathbf{c}_i) = \log(P(\mathbf{v}_i \mid \mathbf{c}_i)) = \sum_{j=1}^{n} \log\left(P\left(v_i^{(j)} \mid c_i^{(j)}\right)\right) \tag{8.66b}$$

$$L(\mathbf{v}, \mathbf{c}) = \log(P(\mathbf{v} \mid \mathbf{c})) = \sum_{i=1}^{L+m} \log(P(\mathbf{v}_i \mid \mathbf{c}_i)) \tag{8.67b}$$

Hence, given the received bit sequence $\mathbf{v}$, the ML decoder should find the bit sequence $\mathbf{c}$ that maximizes $L(\mathbf{v}, \mathbf{c})$ as in equation (8.67b). If we assume a binary symmetric channel (BSC) with error probability p, equation (8.66b) above can be written

$$L(\mathbf{v}_i \mid \mathbf{c}_i) = d_H(\mathbf{v}_i, \mathbf{c}_i) \log(p) + (n - d_H(\mathbf{v}_i, \mathbf{c}_i)) \log(1 - p)$$

$$= d_H(\mathbf{v}_i, \mathbf{c}_i) \log\left(\frac{p}{1-p}\right) + n \log(1 - p) \tag{8.68}$$

where $d_H(\mathbf{v}_i, \mathbf{c}_i)$ is the Hamming distance between the received binary sequence $\mathbf{v}_i$ and the transmitted sequence $\mathbf{c}_i$. The last term $n \log(1 - p)$ can be neglected, since it is a constant. Further, for error probability $0 < p < 1/2$, the constant factor $\log(p/(1-p)) < 0$. Hence, we define

$$L_i = -d_H(\mathbf{v}_i, \mathbf{c}_i) \tag{8.69}$$

and

$$S = \sum_{i=1}^{L+m} L_i = -\sum_{i=1}^{L+m} d_H(\mathbf{v}_i, \mathbf{c}_i) \tag{8.70}$$

From this we draw the conclusion, that finding the maximum likelihood sequence in the case of BSC is equivalent to finding the minimum sum of Hamming distances for the blocks. This fact is utilized in the Viterbi decoding algorithm. To be able to calculate the Hamming distances at the decoder, we of course need access to the received data $\mathbf{v}_i$ and the transmitted data $\mathbf{c}_i$. Unfortunately, $\mathbf{c}_i$ is not known at the receiving site (this is why we put up the channel in the first place). Hence, using our knowledge of the operation of the encoder, we create a local copy of expected data $\hat{\mathbf{c}}_i$ for our calculations. As long as $\hat{\mathbf{c}}_i = \mathbf{c}_i$ we are doing just fine, and the correct message is obtained.

To demonstrate the steps needed when designing a Viterbi decoder, we will use the previous convolution code example shown in Figure 8.4. First, we draw a **state transition diagram** of the encoder. The contents of the memory elements are regarded as the states of the device. Figure 8.5 shows the state transition diagram. The digits on the branches show in parenthesis the input symbol u_l and following that, the output code block $\mathbf{c}_l = \left(c_l^{(1)}, c_l^{(2)} \right)$.

Starting in state 00 (all memory elements cleared) and assuming an input information bit sequence $\mathbf{u} = (0, 1, 0, 1)$, it is straightforward to perform the encoding process using the state transition diagram below.

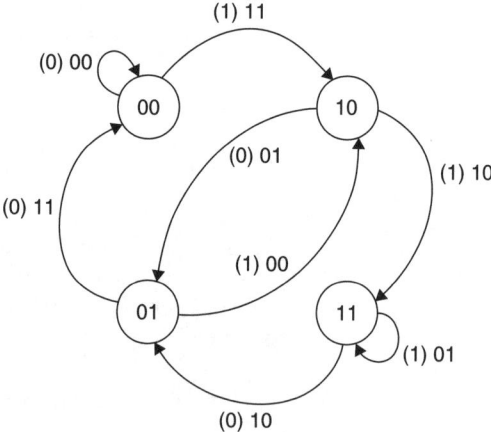

Figure 8.5 *State transition diagram for the convolution encoder shown in Figure 8.4. Digits in the nodes are the state numbers, i.e. the contents of the memory elements. The digits over the branches are the input information symbol* u_l *(in parenthesis) and the output code bits* $c_l^{(1)}$ *and* $c_l^{(2)}$

Input	State	Output code	
0	00	00	
1	00	11	
0	10	01	
1	01	00	
0	10	01	tail
0	01	11	tail
	00		

From the above table, the resulting code word sequence (including tail) is $\mathbf{c} = (00, 11, 01, 00, 01, 11)$. Now, this state transition diagram can be drawn as a **code trellis**, a diagram showing all possible transitions (y-axis) as a function of the step number l (x-axis) (see Figure 8.6). Since we know the initial state 00, the diagram will start to fan out from one state to the left to all the possible four states. We also know, that after the L-steps, there will be the m-steps of zeros, constituting the tail, and ensuring that we end up in state 00 again. This is why we have the fan in at the right end of the trellis.

If we assume that our transmitted code sequence $\mathbf{c}$ suffers 3-bit errors being exposed to an error vector $\mathbf{e} = (01, 10, 00, 10, 00)$, we receive $\mathbf{v} = (01, 01, 01, 10, 01, 11)$. Using the trellis of Figure 8.6 and the expression (8.70), we will illustrate the Viterbi decoding algorithm.

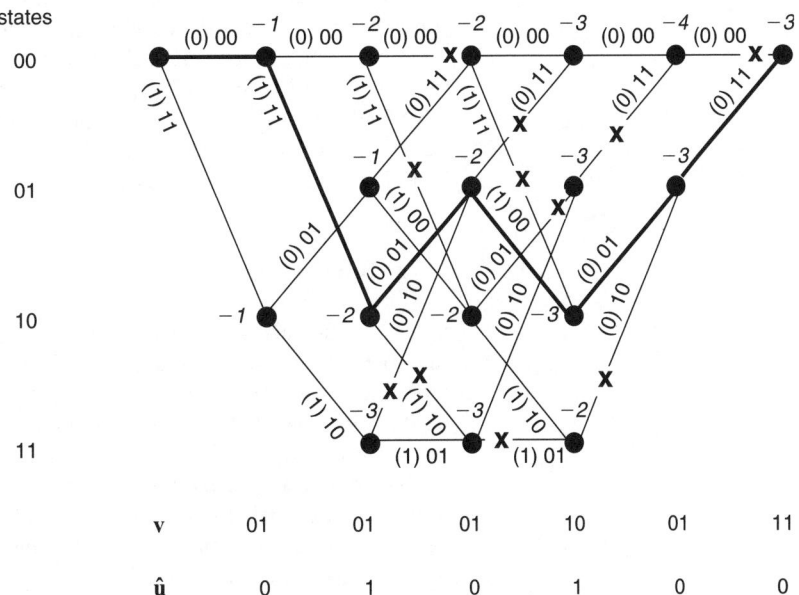

Figure 8.6 *Viterbi decoding algorithm for the example in the text, illustrated by a code Trellis. States on y-axis, steps on x-axis, numbers in italics are the negative accumulated Hamming distances, numbers in parenthesis are the estimated information symbols, numbers on branches are the expected code blocks*

Starting from the left, we know we are in the state 00. During the first step, the received block is $v_1 = (0, 1)$, which can be seen from the line under the trellis. Sitting in state 00 we know, from the state transition diagram in Figure 8.5, that at coding time, the encoder had two choices depending on the input information symbol. If $u_1 = 1$ the encoder could have moved to state 10 and generated the output code $c_1 = (1, 1)$, otherwise it would have stayed in state 00 and generated $c_1 = (0, 0)$. We have received $v_1 = (0, 1)$, so when calculating L_1 using equation (8.69) we find the negative Hamming distance to be -1 in both cases. These figures are found above the respective node points in italics.

We move to the next step. Here we received $v_2 = (0, 1)$. We calculate the **accumulated** Hamming distances for all the possible transitions. For instance, if we assume state 10 and transition to 11, this implies $u_2 = 1$ and $c_2 = (1, 0)$. The negative Hamming distance is -2, and since we already had -1 in state 10, it "costs" total -3 to reach state 11.

Starting in the third step we realize that there are two ways to reach the states. We simply proceed as before, but choose the "cheapest" way to a node in terms of accumulated Hamming distance. The more expensive way is **crossed** out. If both ways are equally expensive, either of them is crossed and represents a cut branch. For example, if we are looking for costs to reach state 01, there are two possibilities, either from state 10, cost -2 or from state 11, cost -5. We choose the former and cross-out the latter.

This process is repeated until we reach the end node to the right of the trellis. If we now backtrack through the trellis, there should be only one path back to the start node to the left. While backtracking, we can easily obtain the estimated information symbols $\hat{u}$, found at the bottom of Figure 8.6. In this particular case, we managed to correct all the errors, hence $\hat{u} = u$.

This was of course, a fairly simple example, but illustrated the basics of Viterbi decoding. A similar procedure is sometimes used in equalizers to counteract intersymbol interference (ISI) in, for instance, radio links (for instance cellular phones) caused by frequency selective fading (Proakis, 1989).

8.2.6 Interleaving

All the codes discussed above have the common property that only a limited number t of erroneous bits per block (or over the constraint length) can be corrected. If there are more bit errors, the coding process will most certainly make things worse, i.e. add even more errors.

Unfortunately, for some channels such as the radio channel, bit errors appear in **bursts** due to the fading phenomenon, and not randomly as assumed by the AWGN model. Hence, the probability of more than t consecutive errors to appear in a burst may be considerable. Commonly, the distance between bursts is long and the problem may be solved using, for instance, long block codes having large n, i.e. being able to correct a large number of bit errors. Unfortunately, the delay in the system increases with n and the complexity of most decoding algorithms increases approximately as n^3. For these reasons, long block codes are not desirable. A trick to make block codes of moderate length perform well even in burst error situations is to use **interleaving** (Ahlin and Zander, 1998).

Interleaving is a method of spreading the bit errors in a burst over a larger number of code blocks. In this way, the number of bit errors per block can be brought down to a level that can be handled by the fairly short code in use. Figure 8.7 shows the principle. We are using a block code of moderate length n, having k information bits. The incoming information symbols are encoded as before and the code words are stored row by row in the interleaving matrix. The matrix has l rows, which results in an **interleaving depth** l.

After storing l code words, the matrix is full. In Figure 8.7, $u_i^{(j)}$ the information bit i is in the code word j and in a similar way $p_m^{(j)}$ the parity bit m is in the code word j. The nl bits are output to the channel, but this time column by column, as

$$u_1^{(1)}, u_1^{(2)}, \ldots, u_1^{(l)}, u_2^{(1)}, u_2^{(2)}, \ldots, p_{n-k}^{(l)}$$

At the decoder site, the reverse action takes place. The bits arriving from the channel are stored column by column, and then read out to the decoder row by row. Decoding and error correction is performed in a standard manner. What

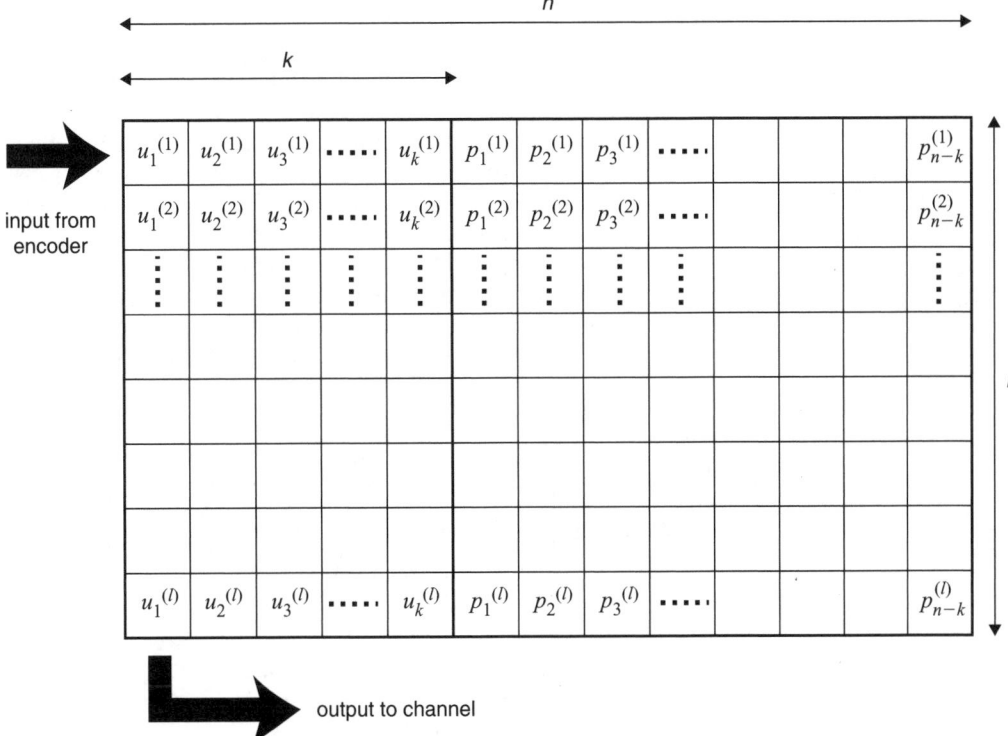

Figure 8.7 *Interleaving matrix, the encoder stores the code words row by row. The bits are then transmitted over the channel column by column. In the decoder, the reverse action takes place. Figure shows the systematic block code* (n, k)

we have achieved using this scheme is that a burst of N consecutive bit errors are spread out on l code words, where every code word contains

$$t_l = \frac{N}{l} \text{ bit errors per block} \tag{8.71}$$

If $t_l \leq t$ for the code in use, the errors will be corrected. If t is the number of bits per block the code can correct, hence we can correct a burst of length

$$N = tl \text{ bits} \tag{8.72}$$

This can be viewed as: given a block code (n, k) we have created a code (nl, kl).

There are smarter ways of achieving interleaving, for instance, **convolution interleaving**, which has a shorter delay.

8.2.7 Concatenated codes and turbo codes

A general problem with the traditional way of designing codes, like block codes and convolution codes relying on algebraic structures, is the block or constraint length required to approach the channel capacity. Long blocks imply transmission delays and complex decoders as pointed out above. Two ways to boost performance while avoiding too long blocks are the use of **concatenated codes** or **turbo codes**.

Concatenated coding rely on the old concept of "divide and conquer", and was first proposed by Forney in the 1970s. Instead of using one long and complex code, two smaller ones are used in "cascade" (see Figure 8.8). First, the information is processed by an "outer encoder", using, for instance, a Reed–Solomon code. The outer encoder is followed by interleaving and an "inner encoder" employing, e.g. a convolution code. At the receiving end, a corresponding arrangement can be found, an inner decoder, a de-interleaver and an outer decoder. The critical issue in such a system is to find a good combination of inner code and outer code. For example, if the inner code is not powerful enough but lets too many errors through, the outer code may be useless and

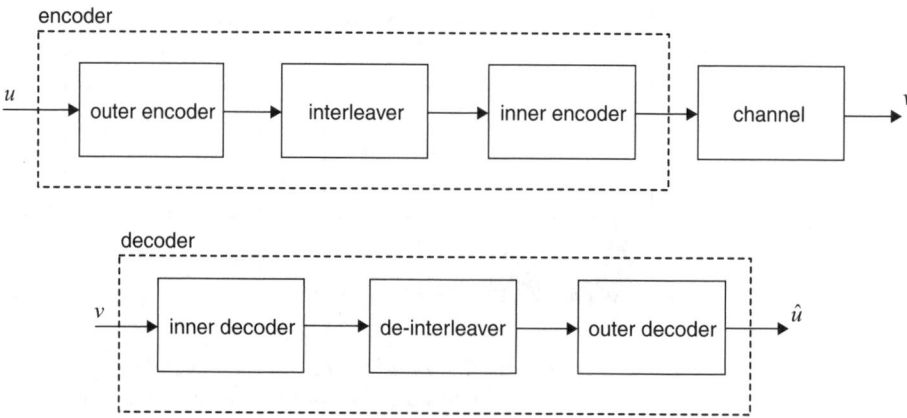

Figure 8.8 *Example of a system using concatenated coding*

the total performance will be poor. For example, concatenated codes and turbo codes are used in space communication applications and in cellular telephone system (e.g. GSM).

Turbo codes (Haykin, 2001) has a little bit of the same flavor as concatenated codes and can be viewed as a "parallel" way of doing concatenated coding. Another way of viewing turbo codes is as a mix between block codes and convolution codes. A simplified turbo system is shown in Figure 8.9.

The turbo encoder consists of two constituent encoders, an interleaver and a multiplexer. The incoming data u follows three paths; one direct to the multiplexer, one to the first encoder and one to the second encoder via an interleaver. Commonly, but not necessarily, the same error-correcting code is used by both encoders, typically a recursive systematic convolutional (RSC) code (Haykin, 2001) having short constraint length. The interleaver scrambles the data bits in a pseudo-random fashion. The idea is that errors that are likely to occur in the parity generated by one encoder should be very unlikely to appear in the other and vice versa. Data u and parity bits p_1 and p_2 from the encoders are multiplexed and transmitted over the communications channel.

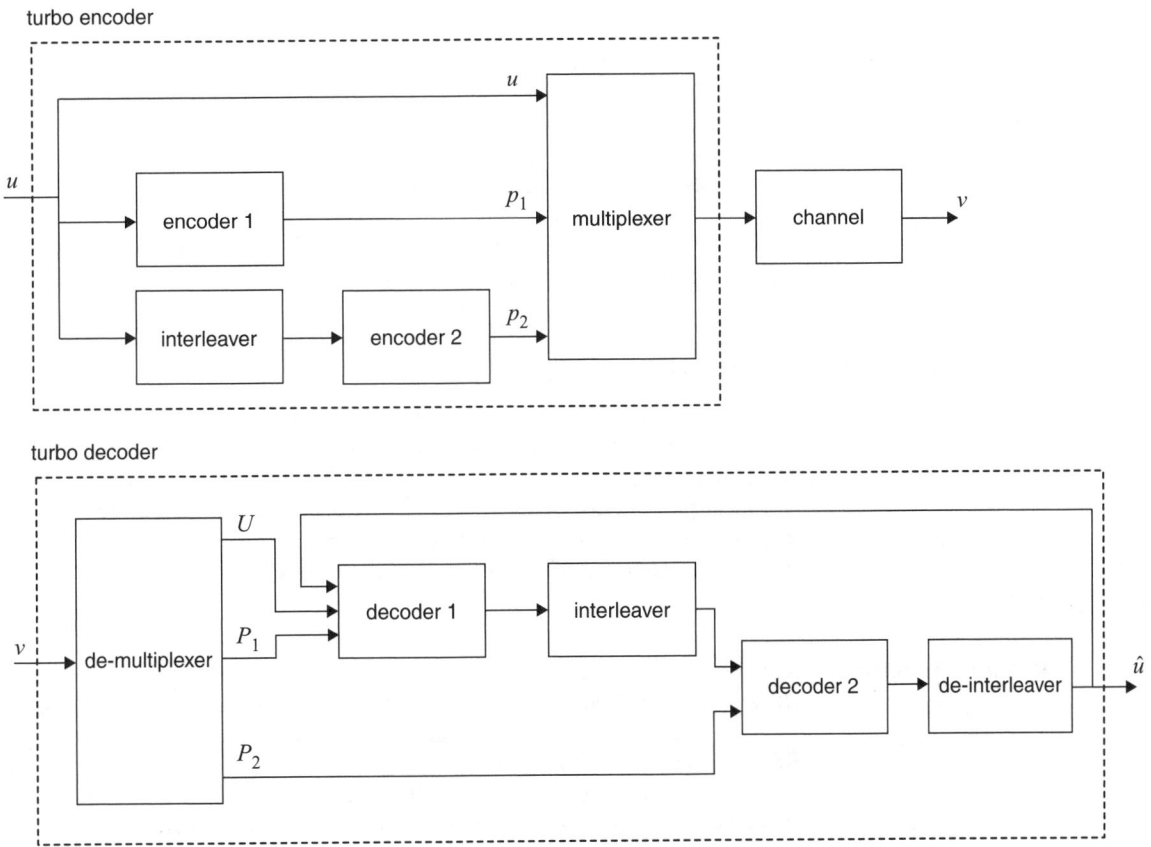

Figure 8.9 *A simplified turbo coding system, the upper part shows the turbo encoder and the lower part the turbo decoder*

At the receiving side, a demultiplexer separates the paths, noisy data U and noisy parity P_1 and P_2. The rest of the turbo decoder consists of two decoders, an interleaver and a de-interleaver. The decoders are of maximum a posteriori probability (MAP) type. Such a decoder generates the output symbol i.e. the most probable given the present input signal. Commonly the decoders are implemented using BCJR algorithm (Haykin, 2001), invented by Bahl, Cocke, Jelinke and Raviv. The BCJR algorithm is a more elaborate cousin to the Viterbi algorithm presented earlier. As can be seen from the figure, the turbo decoder has a feedback path and works iteratively. First, noisy data U and noisy parity symbols P_1 from the first encoder, is fed to the first decoder. The output of the first decoder is the best estimate of the transmitted data u taking the parity symbols P_1 into account. This estimate is passed through the interleaver, being of the same type as the interleaver in the turbo encoder. After the interleaving, the estimate of the first decoder is "in phase" with the received parity symbols P_2 originating from the second encoder. Using the parity symbols P_2, the transmitted data u is re-estimated by the second decoder. Finally, using a de-interleaver (doing the reverse action of the interleaver) a good estimate of the transmitted data u is obtained. This estimate is fed back to the first decoder again, and a re-estimation is taking place. This process is iterated until some given condition (e.g. a certain number of iterations) is met and a final estimate is produced.

This iterative process has actually given name to the turbo coding system, since the decoder circulates estimates in the same way as air in a turbo engine. In a turbo engine, the exhaust gas flow drives a turbine, driving a compressor to feed air into the intake manifold of the engine. Thus, the more revolutions per minute (RPM) you get, the more intake of air, which in turn gives more engine power, resulting in increasing RPM and so on.

Turbo coding is used in code division multiple access (CDMA) radio systems and performs well even in low SNR situations.

Summary   In this chapter we have presented:

- Basic communications theory, BSC, bit error probability and mutual information, AWGN and channel capacity
- Channel coding, Hamming distance, error detection and correction
- Linear block codes, Cyclic codes and BCH codes and Meggit decoders
- Convolution codes and the Viterbi decoder
- Interleaving, concatenated codes and turbo codes.

Review questions

R8-1 Draw the model for the BSC and identify the transitions probabilities.

R8-2 Explain the concept of channel capacity and its consequences. What is the expression for channel capacity in the AWGN case?

R8-3 Explain the following terms, Hamming distance, parity bits, systematic form, code rate, error correction and error detection.

R8-4 Why is decoding and analysis of linear block codes easier than for non-linear block codes?

R8-5 What special property does a cyclic code have compared to any linear block code?

R8-6 Explain the structure of a code word generated by a convolution encoder, then explain the idea with the tail.

R8-7 Explain the algorithm for the Viterbi decoder.

R8-8 When and why is interleaving used?

Solved problems

P8-1 Assume we have a BSC with bit error probability $p = 0.05$. If we introduce a repetition code with parameters $n = 5$, $k = 1$ calculate:

(a) The code rate
(b) The number of parity bits
(c) The maximum Hamming distance of the code
(d) The error-correction capacity
(e) The error-detection capacity
(f) The symbol error probability using error correction by majority vote decoding.

P8-2 If we transmit a byte 10110101, answer the following questions regarding the convolution code Jet Propulsion Lab (2, 1, 6):

(a) How many states are there in the encoder?
(b) What is the length of the tail?
(c) Calculate the constraint length.
(d) Calculate the code rate.
(e) Draw a block diagram of the convolution encoder.
(f) What is the code word on the output of the encoder?

P8-3 Write a MATLAB™ program to plot the channel capacity of a BSC according to equation (8.9).

P8-4 Assume that the channel capacity of a system can be expressed as in equation (8.12). For the noise power density $N_0 = 4 \cdot 10^{-21}$ W/Hz, write a MATLAB™ program that generates a three-dimensional plot showing how the channel capacity depends on the bandwidth and the signal power in the range 1 kHz–1 MHz and 10^{-18}–10^{-9} W, respectively. Use logarithmic scaling.

9 Digital signal processors

Background

The acronym **DSP** is used for two terms, **digital signal processing** and **digital signal processor**. Digital signal processing is to perform signal processing using digital techniques with the aid of digital hardware and/or some kind of computing device. (Signal processing can, of course, be analog as well.) A specially designed digital computer or processor dedicated to signal processing applications is called a digital signal processor.

In this chapter, we will focus on hardware issues associated with digital signal processor chips, and we will compare the characteristics of a DSP to a conventional, general-purpose microprocessor. (The reader is assumed to be familiar with the structure and operation of a standard microprocessor.) Furthermore, software issues and some common algorithms will be discussed.

Objectives

In this chapter we will discuss:

- System types, on-line, off-line, batch systems and real-time systems
- The multiply add accumulate operation, processing speed, architectures, microprocessors, DSP, field programmable gate arrays (FPGA) and application specific integrated circuits (ASIC)
- Fixed and floating-point format, numerical problems, truncation and rounding
- The DSP software development process
- Program and data structures, addressing modes, instruction repertoire and the state machine
- Implementation examples, finite impulse response and infinite impulse response filters.

9.1 System considerations

9.1.1 Applications and requirements

Signal processing systems can be divided into many different classes, depending on the demands. One way of classifying systems is to divide them into **off-line** or **batch systems** and **on-line** or **real-time systems**. In an off-line system, there is no particular demand on the data processing speed of the system, aside from the patience of the user. An example could be a data analysis system for long-term trends in thickness of the arctic ice cap. Data is collected and then stored on a data disc, for instance, and the data is then analyzed at a relatively slow pace. In a real-time system, on the other hand, the available processing time is highly limited and must be processed quickly, in synchronism with some external process. Typical examples are digital filtering of sampled analog signals, where the filtering algorithm must be completed within the sampling period t_s. Further, in many cases no significant delay between the input signal

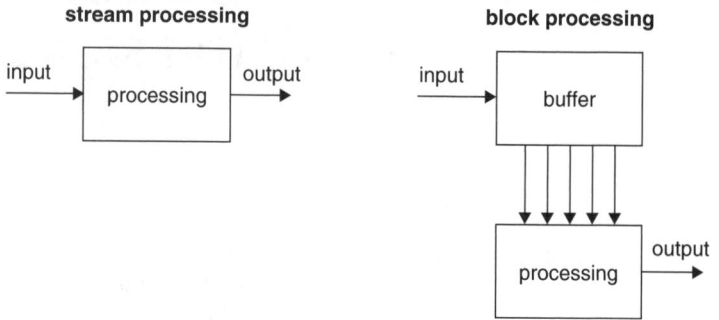

Figure 9.1 *Stream processing and block processing*

and the output signal will be allowed. This is especially true in digital control systems, where delays may cause instability of the entire control loop (which may include heavy machinery). Most applications discussed in this book belong to the class of real-time systems, hence processing speed is crucial.

Another way of classifying signal processing systems is to distinguish between **stream data** systems and **block data** systems (see Figure 9.1). In a stream data system, a continuous flow of input data is processed, resulting in a continuous flow of output data. The digital filtering system mentioned above is a typical stream data system. At every sampling instance, data is fed to the system and after the required processing time $t_p \leq t_s$ has completed, output data will be presented.

Examples of block data systems are spectrum analyzers based on fast Fourier transform (FFT) or channel decoders. In these cases, a block of data must first be inputted into the system before any computation can take place. After the processing is completed, a block of output data is obtained. A signal processing block data system often requires larger data memory than a stream data system. The most demanding applications can probably be found in the area of digital video processing. Such systems are real memory hoggers and/or require extremely high computational power. Quite often digital image processing systems are multiprocessor systems, and consist of a number of processors, dedicated subsystems and hardware. In this book, we will not consider the implementation problems associated with digital image processing systems.

Another way of classifying systems relates to the numerical resolution and the dynamic range, for instance, systems that use **fixed-point** or **floating-point** arithmetics. A floating-point system often makes life easier for the designer, since the need to analyze algorithms and input data in terms of numerical truncation and overflow problems does not exist as in a fixed-point design. Floating-point arithmetics may make things easier, but as pointed out in Chapter 2, it is worth noting that a 32-bit fixed-point system, for instance, can have a higher resolution than a 32-bit floating-point system. Another point is that most systems dealing with real world signals often have some analog and/or mechanical interfacing parts. These devices have a dynamic range and/or resolution being only fractions of what a floating-point signal processing system can achieve. Hence, in most practical cases, floating-point systems are "overkill". Today, fixed-point digital processors are still less expensive and execute faster than floating-point processors.

If we try to identify the most common arithmetic operation in digital signal processing (DSP) algorithms, we will find it to be the sequential calculation of a "scalar vector product" or a convolution sum

$$y(n) = \sum_{k=a}^{b} h(k)x(n-k) = \sum_{k=a}^{b-1} h(k)x(n-k) + h(b)x(n-b) \tag{9.1}$$

Hence, the typical operation is a repeated "multiply add accumulate (MAC)" sequence, often denoted **MAC**. This is found in, for instance, finite impulse response (FIR) and infinite impulse response (IIR) filter structures, where the input stream of samples $x(n)$ is convoluted with the impulse response coefficients $h(n)$ of the filter. The same situation can be found in a neural network node, where the coefficients are the weights w_{ij} or in an FFT algorithm. In the latter case, the coefficients are the "twiddle factors" W_N^{kn} (see Chapter 5). In correlators, modulators and adaptive filters, the input data sequence $x(n)$ is convoluted with another signal sequence instead of constant coefficients, otherwise the structure is the same. Further, vector and matrix multiplication, common in block coders and decoders, require the same kind of calculations.

To summarize the demands: in many digital signal processing applications we are dealing with real-time systems. Hence, computational speed, i.e. "number crunching" capacity is imperative. In particular, algorithms using the MAC-like operations should execute fast and numerical resolution and dynamic range need to be under control. Further, the power consumption of the hardware should preferably be low. Early DSP chips were impossible to use in battery-operated equipment, for instance mobile telephones. Besides draining the batteries in no time, the dissipated power called for large heat sinks to keep the temperature within reasonable limits. On top of this, the common "commercial" system requirements apply: the hardware must be reliable and easy to manufacture at low cost.

9.1.2 Hardware implementation

There are mainly four different ways of implementing the required hardware:

- conventional microprocessor
- DSP chip
- bitslice or wordslice approach
- dedicated hardware, field programmable gate array (FPGA), application specific integrated circuit (ASIC).

Comparing different hardware solutions in terms of processing speed on a general level is not a trivial issue. It is not only the clock speed or instruction cycle time of a processor that determines the total processing time needed for a certain signal processing function. The bus architecture, instruction repertoire, input/output (I/O) hardware, the real-time operating system and most of all, the software algorithm used will affect the processing time to a large extent. Hence, only considering **million instructions per second (MIPS)** or **million floating-point operations per second (MFLOPS)** can be very misleading. When trying to compare different hardware solutions in terms of speed, this should preferably be done using the actual application. If this is not possible, benchmark tests may be a solution. The ways these benchmark tests are designed and selected can of course always be subjects of discussion.

In this book, the aim is not to give exact figures of processing times, nor to promote any particular chip manufacturer. ("Exact" figures would be obsolete within a few years, anyhow.) The goal is simply to give some approximate, **typical** figures of processing times for some implementation models. In this kind of real-time system, processing time translates to the maximum sampling speed and hence the maximum bandwidth of the system. In this text, we have used a simple straightforward 10-tap FIR filter for benchmark discussion purposes

$$y(n) = \sum_{k=0}^{9} b_k x(n-k) \tag{9.2}$$

The first alternative is a **conventional microprocessor** system, for instance an IBM-PC Pentium-type system or some single-chip microcontroller board. By using such a system, development costs are minimum and numerous inexpensive system development tools are available. On the other hand, reliability, physical size, power consumption and cooling requirements may, however, present problems in certain applications. Another problem in such a system would be the operating system. General-purpose operating systems, for instance Windows™ (Microsoft) are not, due to their many unpredictable interrupt sources, well suited for signal processing tasks. A specialized real-time operating system should preferably be used. In some applications, no explicit operating system at all may be a good solution.

Implementing the FIR filter (equation (9.2)) using a standard general-purpose processor as above (no operating system overhead included) would result in a processing time of approximately $1 < t_p < 5\,\mu s$, which translates to a maximum sampling frequency i.e. $f_s = 1/t_s \leq 1/t_p$ of around 200 kHz–1 MHz. This in turn implies a 100–500 kHz bandwidth of the system (using the Nyquist criterion to its limit). The 10-tap FIR filter used for benchmarking is a very simple application. Hence, when using complicated algorithms, this kind of hardware approach is only useful for systems having quite low sampling frequencies. Typical applications could be low-frequency signal processing and systems used for temperature and/or humidity control, in other words, slow control applications.

The next alternative is a **DSP chip**. DSP chips are microprocessors optimized for signal processing algorithms. They have special instructions and built-in hardware to perform the MAC operation and have architecture based on multiple buses. DSPs of today are manufactured using complementary metal oxide semiconductor (CMOS) low voltage technology, yielding low power consumption, well below 1 W. Some chips also have specially designed interfaces for external analog-to-digital (A/D) and digital-to-analog (D/A) converters. Using DSP chips requires moderate hardware design efforts. The availability of development tools is quite good, even if these tools are commonly more expensive than in the case above. Using a DSP chip, the 10-tap FIR filter (equation (9.2)) would require a processing time of approximately $t_p \approx 0.5\,\mu s$, implying a maximum sampling frequency $f_s = 2$ MHz, or a maximum bandwidth of 1 MHz. More elaborate signal processing applications would probably use sampling frequencies of around 50 kHz, a typical sampling speed of many digital audio systems today. Hence, DSP chips are common in digital audio and telecommunication applications. They are also found in more advanced digital control systems in, for instance, aerospace and missile control equipment.

The third alternative is using **bitslice** or **wordslice** chips. In this case, we buy sub-parts of the processor, such as multipliers, sequencers, adders, shifters, address generators, etc., in chip form and design our own processor. In this way, we have full control over internal bus and memory architecture and we can define our own instruction repertoire. We, therefore, have to do all the microcoding ourselves. Building hardware this way requires great effort and is costly. A typical bitslice solution would execute our benchmark 10-tap FIR filter in about $t_p \approx 200$ ns. The resulting maximum sampling frequency is $f_s = 5$ MHz and the bandwidth 2.5 MHz. The speed improvement over DSP chips is not very exciting in this particular case, but the bitslice technique offers other advantages. For example, we are free to select the bus width of our choice and to define special instructions for special-purpose algorithms. This type of hardware is used in systems for special purposes, where power consumption, size and cost are not important factors.

The fourth alternative is to build our own system from gate level on silicon, using one or more **application specific integrated circuit (ASIC)** or **field programmable gate array (FPGA)**. In this case, we can design our own adders, multipliers, sequencers and so on. We are also free to use mainly any computational structure we want. However, quite often no conventional processor model is used. The processing algorithm is simply "hardwired" into the silicon. Hence, the resulting circuit cannot perform any other function. Building hardware in this way may be very costly and time consuming, depending on the development tools, skill of the designer and turn-around time of the silicon manufacturing and prototyping processes. Commonly, design tools are based on **very high-speed integrated circuit hardware description language (VHDL)** (Zwolinski, 2004) or the like. This simplifies the process of importing and reusing standard software defined hardware function blocks. Further, good simulation tools are available to aid the design and verification of the chip before it is actually implemented in silicon. This kind of software tools may cut design and verification times considerably, but many tools are expensive.

Using the ASIC approach, the silicon chip including prototypes must be produced by a chip manufacturer. This is a complicated process and may take weeks or months, which increases the development time. The FPGA on the other hand is a standard silicon chip that can be programmed in minutes by the designer, using quite simple equipment. The drawback of the FPGA is that it contains fewer circuit elements (gates) than an ASIC, which limits the complexity of the signal processing algorithm. On the other hand, more advanced FPGA chips are constantly released on the market. For instance, FPGAs containing not only matrices of programmable circuit elements, but also a number of DSP kernels are available today. Hence, the difference in complexity between FPGAs and ASICs is reduced. However, FPGAs are commonly not very well suited for large volume production, due to the programming time required.

There are mainly only two reasons for choosing the ASIC implementation method. Either we need the maximum processing speed, or we need the final product to be manufactured in very large numbers. In the latter case, the development cost per manufactured unit will be lower than if standard chips would have been used. An ASIC, specially designed to run the 10-tap benchmark FIR filter is likely to reach a processing speed (today's technology) in the vicinity of $t_p \approx 2$ ns, yielding a sampling rate of $f_s = 500$ MHz and a bandwidth of 250 MHz. Now we are approaching speeds required by radar and advanced video processing systems. Needless to say, when building such hardware in practice, many additional problems occur since we are dealing with fairly high-frequency signals.

If yet higher processing capacity is required, it is common to connect a number of processors, working in parallel in a larger system. This can be done in different ways, either in a **single instruction multiple data (SIMD)** or in a **multiple instruction multiple data (MIMD)** structure. In an SIMD structure, all the processors are executing the same instruction but on different data streams. Such systems are sometimes also called vector processors. In an MIMD system, the processors may be executing different instructions. Common for all processor structures is however the demand for communication and synchronization between the processors. As the number of processors grows, the communication demands grow even faster.

Large multiprocessor systems ("super computers") of this kind are of course very expensive and rare. They are commonly used for advanced digital image processing for solving hard optimization problems and for running large neural networks. One classical example of such a machine is the "**connection machine**" (CM-1) (Hillis, 1987). The CM-1 consists of 65 536 quite simple processors, connected by a packet-switched network. The machine is fed instructions from a conventional-type host computer and a specially designed computer language is used. The CM-1 has an I/O capacity of 500 Mbits/s and is capable of executing about 1000 MIPS. The machine is air cooled and dissipates about 12 kW. (One begins to think of the old electronic numerical integrator and computer (ENIAC), using electron tubes . . .).

An interesting thing is that this machine is only good at executing an appropriate type of algorithms, i.e. algorithms that can be divided into a large number of **parallel activities**. Consider our simple benchmark example, the 10-tap FIR filter. The algorithm can only be divided into 10 multiplications that can be performed simultaneously and four steps of addition (a tree of 5 groups + 2 groups + 1 group + 1 group) which has to be performed in a sequence. Hence, we will need one instruction cycle for executing the 10 multiplications (using 10 processors) and four cycles to perform the additions, thus a total of five cycles. Now, if the machine consists of 65 536 processors, each processor only has a capacity of $10^9/65\,536 = 0.015$ MIPS, which is not very impressive. If we disregard communication delays, etc., we can conclude that running the 10-tap FIR filter on this super computer results in 65 526 processors out of 65 536 which are idling. The processing time will be in the range of 300 μs, in other words, considerably slower than a standard (cheaper) personal computer (PC). Our benchmark problem is obviously too simple for this machine. This also illustrates the importance of "matching" the algorithm to the hardware architecture, and that MIPS alone may not be an appropriate performance measure.

9.2 Digital signal processors versus conventional microprocessors

9.2.1 Conventional microprocessors

9.2.1.1 Architecture

A conventional microprocessor commonly uses a **von Neumann** architecture, which means that there is only one common system bus used for transfer of both

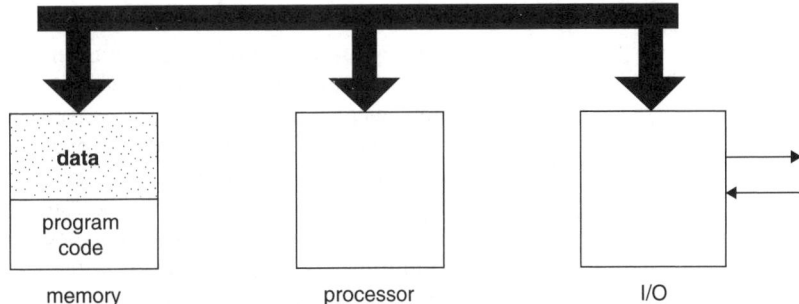

Figure 9.2 *von Neumann architecture, program code and data share memory*

instructions and data between the external memory chips and the processor (see Figure 9.2). The system bus consists of the three sub-buses: the data bus, the address bus and the control bus. In many cases, the same system bus is also used for I/O operations. In signal processing applications, this single bus is a bottleneck. Execution of the 10-tap FIR filter (equation (9.2)) will, for instance, require at least 60 bus cycles for instruction fetches and 40 bus cycles for data and coefficient transfers, a total of approximately 100 bus cycles. Hence, even if we are using a fast processor, the speed of the bus cycle will be a limiting factor.

One way to ease this problem is the introduction of **pipelining** techniques, which means that an **execution unit (EU)** and a **bus unit (BU)** on the processor chip work simultaneously. While one instruction is being executed in the EU the next instruction is fetched from memory by the BU and put into an instruction queue, feeding the instruction decoder. In this way, idle bus cycles are eliminated. If a jump instruction occurs in the program, a restart of the instruction queue has however to be performed, causing a delay.

Yet another improvement is to add a **cache memory** on the processor chip. A limited block (some thousand words) of the program code is read into the fast internal cache memory. In this way, instructions can be fetched from the internal cache memory at the same time as data is transferred over the external system bus. This approach may be very efficient in signal processing applications, since in many cases the entire program may fit in the cache, and no reloading is needed.

The execution unit in a conventional microprocessor may consist of an **arithmetic logic unit (ALU)**, a **multiplier**, a **shifter**, a **floating-point unit (FPU)** and some data and **flag** registers. The ALU commonly handles 2's complement arithmetics (see Chapter 2), and the FPU uses some standard Institute of Electrical and Electronics Engineers (IEEE) floating-point formats. The **binary fractions** format discussed later in this chapter is often used in signal processing applications but is not supported by general-purpose microprocessors.

Besides **program counter (PC)** and **stack pointer (SP)**, the **address unit (AU)** of a conventional microprocessor may contain a number of address and **segment** registers. There may also be an ALU for calculating addresses used in complicated addressing modes and/or handling virtual memory functions.

9.2.1.2 Instruction repertoire

The instruction repertoire of many general-purpose microprocessors supports quite exotic addressing modes which are seldom used in signal processing algorithms. On the other hand, instructions for handling such things like **delay lines** or **circular buffers** in an efficient manner are rare. The MAC operation often requires a number of computer instructions, and loop counters have to be implemented in software, using general-purpose data registers.

Further, instructions aimed for operating systems and multi-task handling may be found among "higher end" processors. These instructions often are of very limited interest in signal processing applications.

Most of the common processors today are of the **complex instruction set computer (CISC)** type, i.e. instructions may occupy more than one memory word and hence require more than 1 bus cycle to fetch. Further, these instructions often require more than 1 machine cycle to execute. In many cases, **reduced instruction set computers (RISC)**-type processors may perform better in signal processing applications. In an RISC processor, no instruction occupies more than one memory word; it can be fetched in 1 bus cycle and executes in 1 machine cycle. On the other hand, many RISC instructions may be needed to perform the same function as one CISC-type instruction, but in the RISC case, you can get the required complexity only when needed.

9.2.1.3 Interface

Getting analog signals into and out of a general-purpose microprocessor often requires a lot of external hardware. Some microcontrollers have built-in A/D and D/A converters, but in most cases, these converters only have 8- or 12-bit resolution, which is not sufficient in many applications. Sometimes these converters are also quite slow. Even if there are good built-in converters, there is always need for external sample-and-hold (S/H) circuits, and (analog) anti-aliasing and reconstruction filters.

Some microprocessors have built-in high-speed serial communication circuitry, **serial peripheral interface (SPI)** or $\mathbf{I^2C}^{\mathrm{TM}}$. In such cases we still need to have external converters, but the interface will be easier than using the traditional approach, i.e. to connect the converters in parallel to the system bus. Parallel communication will of course be faster, but the circuits needed will be more complicated and we will be stealing capacity from a common, single system bus.

The interrupt facilities found on many general-purpose processors are in many cases "overkill" for signal processing systems. In this kind of real-time application, timing is crucial and **synchronous programming** is preferred. The number of **asynchronous events**, e.g. **interrupts**, is kept to a minimum. Digital signal processing systems using more than a few interrupt sources are rare. One single interrupt source (be it timing or sample rate) or none is common.

9.2.2 Digital signal processors

9.2.2.1 Architecture

DSP chips often have a **Harvard**-type architecture (see Figure 9.3) or some modified version of Harvard architecture. This type of system architecture implies that there are at least two system buses, one for instruction transfers

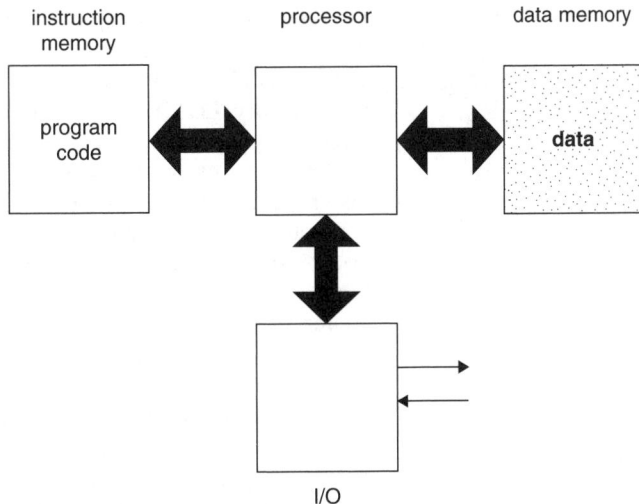

Figure 9.3 *Harvard architecture, separate buses and memories. I/O, data and instructions can be accessed simultaneously*

and one for data. Quite often, three system buses can be found on DSPs, one for instructions, one for data (including I/O) and one for transferring coefficients from a separate memory area or chip.

In this way, when running an FIR filter algorithm like in equation (9.2) instructions can be fetched at the same time as data from the delay line $x(n - k)$ is fetched and as filter coefficients b_k are fetched from coefficient memory. Hence, using a DSP for the 10-tap FIR filter, only 12 bus cycles will be needed including instruction and data transfers.

Many DSP chips also have internal memory areas that can be allocated as **data memory**, **coefficient memory** and/or **instruction memory**, or combinations of these. Pipelining is used in most DSP chips.

A **warning** should however be issued! Some DSP chips execute instructions in the pipeline in a parallel, "smart" fashion to increase speed. The result will in some cases be that instructions will **not** be executed in the same order as written in the program code. This may of course lead to strange behavior and cumbersome troubleshooting. One way to avoid this is to insert "dummy" instructions (for instance, no operation (NOP)) in the program code in the critical parts (consult the data sheet of the DSP chip to find out about pipeline latency). This will of course increase the execution time.

The execution unit consists of at least one (often two) arithmetic logic unit (ALU), a multiplier, a shifter, accumulators, and data and flag registers. The unit is designed with a high degree of **parallelism** in mind, hence all the ALUs, multipliers, etc., can be run simultaneously. Further, ALUs, the multiplier and accumulators are organized so that the MAC operation can be performed as efficiently as possible with the use of a minimum amount of internal data movements. Fixed-point DSPs handle 2's complement arithmetics and binary fractions format. Floating-point DSPs use floating-point formats that can be IEEE standard or some other non-standard format. In many cases, the ALUs can also handle both **wrap-around** and **saturation** arithmetics which will be discussed later in this chapter.

Many DSPs also have ready-made **look-up tables (LUT)** in memory (read only memory (ROM)). These tables may be A-law and/or μ-law for companding systems and/or sine/cosine tables for FFT or modulation purposes.

Unlike conventional processors having 16-, 32- or 64-bit bus widths, DSPs may have uncommon bus widths like 24, 48 or 56 bits, etc. The width of the instruction bus is chosen such that an RISC-like system can be achieved, i.e. every instruction only occupies one memory word and can hence be fetched in 1 bus cycle. The data buses are given a bus width that can handle a word of appropriate resolution, at the same time as extra high bits are present to keep overflow problems under control.

The address unit is complicated since it may be expected to run three address buses in parallel. There is of course a program counter and a stack pointer as in a conventional processor, but we are also likely to find a number of **index** and **pointer** registers used to generate data memory addresses. Quite often there is also one or two ALUs for calculating addresses when accessing delay lines (vectors in data memory) and coefficient tables. These pointer registers can often be **incremented** or **decremented** in a **modulo** fashion, which for instance simplifies building circular buffers. The AU may also be able to generate the specific **bit reverse** operations used when addressing butterflies in FFT algorithms.

Further, in some DSPs, the stack is implemented as a separate **last in first out (LIFO)** register file in silicon ("hardware stack"). Using this approach, pushing and popping on the stack will be faster, and no address bus will be used.

9.2.2.2 Instruction repertoire

The special **multiply add accumulate (MAC)** instruction is almost mandatory in the instruction repertoire of a DSP. This single instruction performs one step in the summation of equation (9.2), i.e. multiplies a delayed signal sample by the corresponding coefficient and adds the product to the accumulator holding the sum. Special instructions for rounding numbers are also common. On some chips, even special instructions for executing the Viterbi decoding algorithms are implemented.

There are also a number of instructions that can be **executed in parallel** to use the hardware parallelism to its full extent. Further, special prefixes or postfixes can be added to achieve **repetition** of an instruction. This is accomplished using a special **loop counter** implemented in hardware as a special loop register. Using this register in loops, instruction fetching can be completely unnecessary in some cases.

9.2.2.3 Interface

It is common to find built-in high-speed serial communication circuitry in DSP chips. These serial ports are designed to be directly connected to coder–decoders (CODECs) and/or A/D and D/A converter chips for instance. Of course, parallel I/O can also be achieved using one of the buses.

The interrupt facilities found on DSP chips are often quite simple, with a fairly small number of interrupt inputs and priority levels.

9.3 Programming digital signal processors

9.3.1 The development process

Implementing a signal processing function as DSP software often requires considerable design and verification efforts. If we assume that the system specification is established and the choice of suitable hardware is made, designing and implementing the actual DSP program remains.

As an example, the passband specifications for a filtering application may be determined, the filter type chosen and a transfer function $F(z)$ formulated using the z-transform. Now, starting from the transfer function, the remaining work can be described by the following checklist:

- Designing the actual **algorithm**, i.e. the method to calculate the difference equations corresponding to the transfer function
- Simulating and verifying the algorithm, using a general-purpose computer; this is often done on a non-real-time basis, using floating-point numbers and high-level programming language, for instance C++, C or Pascal
- Simulating and verifying the algorithm as above, but now using the same number format as will be used by the target DSP, e.g. fixed-point arithmetics
- "Translating" the algorithm computer code to the target program language applicable for the DSP, commonly C or assembly program code
- Simulating the target DSP program code, using a software simulator
- Verifying the function of the target DSP program code, using the "high-level" simulations as a reference
- Porting the DSP program code on the target hardware, using, e.g. an emulator
- Verifying and debugging the target system using the final hardware configuration, verifying real-time performance.

The steps above may have to be **iterated** a number of times to achieve a system having the desired performance. If unlucky, it may even turn out that the hardware configuration has to be changed to satisfy the system requirements.

When initially designing the algorithm, many alternative solutions may be evaluated to find the one that executes the best, using the selected hardware. Processor architecture, arithmetic performance and addressing possibilities, together with memory demands and I/O options will affect the algorithm design process.

To accomplish the "high-level" simulations using floating-point and for instance C++, C, Pascal or FORTRAN, it is also common to use some standard computational program packages like, MATLAB™ (MathWorks), Mathematica™ (Wolfram Research) or MathCad™ (MathSoft). General-purpose spreadsheet programs like EXCEL™ (Microsoft) has also proven to be handy in some situations. Input to the simulations may be data generated by a model or real measured and stored data.

During the "high-level" simulations, using the same number format as the target system, the aim is to pinpoint numerical, **truncation** and **overflow** problems. As an example, when working with integer arithmetics and limited word length, the two expressions $a(b - c)$ and $ab - ac$ being equivalent from a strict mathematical point of view, may give different results. Quite often, algorithms have to be repartitioned to overcome such problems. If the target system uses floating-point format, these simulations will in most cases be less cumbersome than in the case of fixed-point formats. Fixed-point DSPs are however common (today), since they are less expensive and run faster than floating-point chips. Further, the numerical performance is not only a matter of number format, in that other arithmetic aspects like using 2's complement or signed integer or binary fractions, etc. may also come into play. The results from these simulations will be used later in the development process as reference data.

"Translating" the algorithm into a program for the target DSP can be quite easy if the "high-level" simulations use the same language as the target system. One such example is using the program language C, which is available for a great number of computing platforms and for most DSP chips. It is however not uncommon that some macro language or pure assembly code is needed to program the target DSP. In the latter case, the program hence needs to be rewritten, commonly in assembly code. Even if high-level languages such as C are becoming more common for DSPs, assembly language is the only choice if peak performance is required.

The next step is to run the DSP program code on a host computer using a software simulator. The purpose of this step is to verify the basic operation of the software.

After having debugged the DSP program code, it has to be verified and the function has to be compared to the "high-level" simulations made earlier. "Every bit" must be the same. If there are any deviations, the reason for this must be found.

Finally, the DSP program code is moved to the intended hardware system using an emulator, programmable read only memory (PROM) simulator or erasable programmable read only memory (EPROM). The software is now executed using the dedicated hardware.

The last step is to verify the complete system including hardware and the DSP software. In this case, the entire function is tested in real time. It should be noted that some emulators do not execute at the same speed as a real processor. There may also be timing differences regarding, for instance, interrupt cycles. Hence, when verifying the system in real time, using the real processor is often preferred to an emulator.

9.3.2 Digital signal processing programming languages

As mentioned above, even if there are a number of **C cross-compilers** around today, it is still common to program DSPs using **assembly language**. Most human programmers are smarter than compilers. Hence, tedious assembly language programming has the potential of resulting in more compact and faster code. The development time will, however, be considerably longer compared to using C. For many DSPs, cross-compilers for C++ and JAVA are also available. It is not easy to see that C++ or JAVA is better choice for signal processing software than pure C language.

The DSP assembly instruction repertoire differs somewhat from one DSP to another. Commonly, the assembly instruction can, however, be divided into four typical groups.

(1) **Arithmetic and logical instructions**: In this group, we find instructions for adding, subtracting, multiplying and dividing numbers as well as the MAC instruction discussed earlier. There are also instructions for rounding,

shifting, rotating, comparing and obtaining absolute values, etc. Many instructions may come in different versions, depending on the number format used. There may be for instance multiplication of **signed** or **unsigned** numbers, and multiplication of **integers** or **binary fractions**. In this group we can also find the standard logical, bitwise AND, OR, NOT and XOR functions.

(2) **Bit manipulation instructions**: This group consists of instructions for setting, resetting and testing the state of single bits in registers, memory locations and I/O-ports. These instructions are handy for manipulating flags, polling external switches and controlling external indicators light emitting diodes (LEDs), etc.

(3) **Data transfer instructions**: Data transfer instructions are commonly MOVE, LOAD, STORE and so on, used to copy data to and from memory locations, registers and I/O-ports. In many cases, the source and destination may have different word lengths hence care must be exercised to make sure that significant data is transferred and that sign bits, etc. are properly set. Stack handling instructions also belong to this group.

(4) **Loop and program control instructions**: Typical instructions in this group are unconditional and conditional jump, branch and skip operations. The conditional instructions are in most cases linked to the state of the status bits in the flag register and occur in many different versions. Subroutine jumps and returns also belong to this group. Further, we can find instructions for manipulating the hardware loop counter and for software interrupts (traps) as well as STOP, RESET and no operation (NOP).

9.3.3 The program structure

If the functions available in a proper real-time operating system are not considered, the main structure of a typical DSP program is commonly a **timed loop** or an **idling loop** and one or more **interrupt service routines**. In both cases, the purpose is to get the processing synchronized to the sampling rate. The timed loop approach can be illustrated by the following pseudo-code:

```
reset:  initializing stuff
        start timer                              // sampling rate

t_loop: do
        {
                if(timer not ready)
                {
                    background processing
                }
                else
                {                                // timed sequence
                    restart timer
                    get input
                    process
                    send result to output
                }
        }forever
```

The execution time t_p of the timed program sequence above must of course not exceed the sampling period t_s. The approach using an **idling loop and interrupt routines** assumes that interrupts are generated at sampling rate, by for instance external circuitry like A/D converters, etc. This approach is shown below:

```
reset:  initializing stuff

idle:   do
        {
            background processing
        } forever

irq:                                    // timed sequence
            acknowledge interrupt
            get input
            process
            send result to output
            return from interrupt
```

In this case, the interrupt service routine must of course be executed within the sampling period t_s, i.e. between successive interrupt signals. The latter approach is a bit more flexible and can easily be expanded using more interrupt sources and more interrupt service routines. This can be the case in **multi-rate sampled systems**. One has to remember, however, that the more asynchronous events like interrupt and direct memory access (DMA) there are in a system, the harder it is to debug the system and to guarantee and verify real-time performance.

The most common DSP program has a constant inflow and outflow (stream system) of data and consists of "simple" sequences of operations. One of the most common operations is the MAC operation discussed earlier. There are typically very few data-dependent conditional jumps. Further, in most cases only **basic addressing modes** and **simple data structures** are used to keep up execution speed.

9.3.4 Arithmetic issues

The 2's complement (see Chapter 2) is the most common fixed-point representation. In the digital signal processing community, we often interpret the numbers as **fractions** (fractional) rather than integers. This means that we introduce a **binary point** (not decimal point) and stay to the "right" of the point instead of to the "left", as in the case of integers. Hence, the weights of the binary bits representing fractions will be $2^{-1}, 2^{-2}, \ldots$

Table 9.1 shows a comparison between fractional and integer interpretation of some binary 2's complement numbers.

Fractional and integer multiplication differs by a 1-bit left shift of the result. Hence, if for instance trying fractional multiplication using a standard microprocessor, or using a DSP that does not support fractions, the result must be adjusted by one step left shift, to obtain the correct result. This is because the standard multiplication rule assumes multiplication of integers.

Table 9.1 *Interpreting binary 2's complement as integer versus fraction*

2's complement	Integer	Fraction
0111	7	0.875
0110	6	0.750
0101	5	0.625
0100	4	0.500
0011	3	0.375
0010	2	0.250
0001	1	0.125
0000	0	0.000
1111	−1	−0.125
1110	−2	−0.250
1101	−3	−0.375
1100	−4	−0.500
1011	−5	−0.625
1010	−6	−0.750
1001	−7	−0.875
1000	−8	−1.00

For example, multiplying 0.5 times 0.5 using the standard binary multiplication rule yields.

```
Binary          Fraction        Integer

    0100            0.500            4
x 0100          x 0.500          x 4
    0000            0.250           16
    0000
    0100
  0000
0010000 =           0.125           16

1 bit left shift yields

0100000 =           0.250
```

Most digital signal processors can handle fractional as well as integer multiplication. Some DSPs have a flag to set depending on if integer or fractional rules apply, whereas other DSPs have separate instructions for the two kinds of multiplication. Yet, other DSPs have special devices and/or instructions to perform one step shift left.

A virtue of 2's complement is that when adding a sequence of numbers whose sum you know is within bounds, all overflows and carries "on the way" can be ignored. You will still come up with the correct result. This is very handy, since in most DSP algorithms coefficients can be scaled to guarantee that the output should be OK, thus overflows and carries can be ignored. For example

Fraction	Binary	Partial sums
.625	0101	0101
+.750	+0110	1011
+.375	+0011	1110
+.625	+0101	(1) 0011
-.875	+1001	1100
-.625	+1011	(1) 0111
=.875		=.875

We assume that the adder "wraps around". This will, of course, **not** work if the processor is using **saturating arithmetic**, i.e. it does not wrap around. Many DSPs can support both standard (wrap around) arithmetic and saturating arithmetic by setting flags, etc. If we have very large variations in, for instance, an input signal and over or underflow conditions cannot be eliminated, using saturating arithmetic we will be able to preserve the sign information and have a fair chance to recover. Saturating arithmetic basically works as a soft limiter (see Chapter 4) and can be useful in neural network algorithms.

Multiplication of two binary numbers, N-bit wide, yields a result being $2N$-bit wide. This is why accumulators in DSPs are at least $2N$-bit wide, to be able to harbor results and accumulated results. At some point though, one must truncate or round back down to N-bit wide numbers again.

Assume that we have an M-bits 2's complement fractional number $b_0, b_{-1}, b_{-2}, \ldots, b_{-M+1}$ having the weights $-2^0, 2^{-1}, 2^{-2}, \ldots, 2^{-M+1}$ where the leftmost bit is the sign bit. If we **truncate** this word down to a width of B-bits, we simply cut away all bits to the right of bit number $B-1$, i.e. the truncated word will be (assuming $M \geq B$) $b_0, b_{-1}, b_{-2}, \ldots, b_{-B+1}$.

In this case, the error caused by the truncation operation will be

$$0 \leq \varepsilon < 2^{-B+1} \tag{9.3}$$

The mean error will be

$$m_\varepsilon = 2^{-B} \tag{9.4}$$

The variance of the error, or in other words the "error noise power" will be

$$\sigma_\varepsilon^2 = \frac{2^{-2(B-1)}}{12} \tag{9.5}$$

The other possibility to reduce the word length is **rounding**. Preferably the method of "convergent rounding", also called round-to-nearest (even) number, should be used. This method works as follows:

(a) if $b_{-B}, b_{-B-1}, \ldots, b_{-M+1} < 1, 0, \ldots, 0$ then
   truncate as before

(b) if $b_{-B}, b_{-B-1}, \ldots, b_{-M+1} > 1, 0, \ldots, 0$ then
add $b_{-B+1} = 1$ and truncate as before

(c) if $b_{-B}, b_{-B-1}, \ldots, b_{-M+1} = 1, 0, \ldots, 0$ and $b_{-B+1} = 0$ then
truncate as before

(d) if $b_{-B}, b_{-B-1}, \ldots, b_{-M+1} = 1, 0, \ldots, 0$ and $b_{-B+1} = 1$ then
add $b_{-B+1} = 1$ and truncate as before

Most DSPs have some kind of rounding function, but not all support "convergent rounding". The error caused by the rounding operation will be

$$-2^{-B} \leq \varepsilon < 2^{-B} \tag{9.6}$$

The mean error in this case will be

$$m_\varepsilon = 0 \tag{9.7}$$

The variance of the error, or in other words the "error noise power" will be the same as for pure truncation, i.e. as equation (9.5) above. Rounding is preferred to truncation in most cases. Truncation is easier to implement and can be performed simply using standard bit masking techniques, if a truncation program instruction is not available.

9.3.5 Data structures and addressing modes

In most cases, quite simple data structures are used in DSP software. In a typical case, there is a data vector and a coefficient vector, or two or more data vectors. In some cases, matrices are used, but since most DSP chips do not support matrix operations directly, matrices are broken down to row or column vectors and processed as a set of vectors.

To access vector structures, most DSP chips have a number of pointer registers. These registers are used for **indirect addressing**, i.e. the content of the register is used as an address to point to a word in the data memory. Further, the register can be **auto-incremented**, which means that the contents of the register is incremented by 1, every time the register is used to point into the data memory. This can be done in two ways, using **pre-increment**, the register is first incremented and then used as a pointer, whereas if **post-increment** is used, the increment and pointing operations take place in the reverse order. There are of course also the corresponding operations for **auto-decrement**, i.e. decrementing the register by one.

Another addressing mode that is common is **indexed addressing**. In this case, the **effective address**, in other words the address pointed to in the data memory, is the sum of a **base register** and the **offset** which may be the contents of an **offset register**. In this way, the base register can be used to point to the starting element of a vector or table, and the offset register is used for the actual addressing. Now, the same piece of software can easily be used to process another table or vector and only the contents of the base register has to be replaced.

The addressing modes described above can be found in most conventional general-purpose microprocessors as well. In DSP chips, there are also some "specialized" addressing modes. One such mode is auto-increment or auto-decrement using an offset other than 1. A pointer register can, for instance, be auto-incremented by 5 or some other constant stored in a register.

Quite often, one or more of the data vectors in a DSP program are also used as a **delay line** in, for instance, FIR filters. This means that a vector in data memory of length M data words are used to store the present value of our sampled signal $x(n)$ and the $M - 1$ "old" sampled values $x(n-1), x(n-2), \ldots, x(n-M+1)$:

$0000 + M - 1$	$x(n - M + 1)$
$\vdots$	$\vdots$
$\vdots$	
$\vdots$	$\vdots$
0002	$x(n - 2)$
0001	$x(n - 1)$
0000	$x(n)$

where we have assumed that the delay-line vector starts at address 0000, in other words, the base address is 0000. Now, for every time the sampling clock ticks, a new sampled value will arrive and the entire delay line must be updated. This can be done in two ways. Either we are using a straightforward **static list** method, or a somewhat more complicated **dynamical list** method. The static list method is based on moving the contents of all the memory cells in the vector one step "upwards", except the oldest element at address $0000 + M + 1$ that will be dropped. The new element will then be stored at the bottom of the vector in address 0000. This method is easy to implement and understand. The drawback is, however, that for every new sample, we have to make $M - 1$ "extra" data moves, thus consuming time.

A smarter approach is the dynamical list method. In this case, a **circular buffer** and a **start-of-list pointer** are used. The logical start of the vector is pointed to by the start-of-list pointer and can be any address in the memory block, not only address 0000 as above. As soon as a new sample arrives, we only need to move the start-of-list pointer "downwards" one step. At this address, the oldest sample can be found, since we are now dealing with a circular list. The new sample is stored at this position, thus overwriting the oldest sample. In this way, no shuffling of data is needed like in the static list method. Only the start-of-list pointer needs to be decremented. To handle a circular buffer, pointers need to "wrap around" when reaching the top or bottom of the memory area allocated for the vector. This calls for **modular addressing**, supported by many DSPs. General-purpose processors do not support this type of addressing, hence extra program instructions are required to test the pointers and "wrap around" when needed. If the size of the buffer is 2^k, where k is a positive integer, modular addressing can be achieved by simply ANDing the address pointer with an appropriate bit mask. Many DSP chips have specialized addressing hardware, supporting almost any buffer length, e.g. 17 or 38, etc.

Dealing with, for instance, some FFT algorithms, using "butterfly"-type computing strategies, **bit reverse addressing**, supported by many DSPs is handy. The DSP (as with most other digital computers) performs the calculations sequentially. Equations (5.7) and (5.8) describe a four point FFT. It is very simple, but can serve as an example to demonstrate the ideas behind bit reverse addressing. Looking at equation (5.7), describing the computational "butterflies", we find that we need to access the input elements $x(0), x(1), \ldots$. The smartest way (in terms of calculation order) to access these elements is not in the "normal" sequence 0, 1, 2, 3 but in the bit reversed order 0, 2, 1, 3. If the transform has more than four input values, the gain in calculation time will of course be greater. So, bit reverse addressing is a way of calculating the pointer values so that an FFT algorithm accesses the elements of the input vector in the "smartest" order. The bit reverse addressing is obtained by "mirroring" the address word, so that the most significant bit becomes the least significant bit and vice versa. Table 9.2 shows a 3-bit normal and bit reverse addressing scheme.

9.3.6 The state machine

The **state machine** model is a common algorithm of designing stable software which is easy to modify and verify. The model is used in many different real-time application areas, and is certainly also usable in digital signal processing systems. In this section, a simple state machine implementation will be shown as an example.

Assume we would like to design a piece of software to receive Huffman-coded, variable length messages, as in the example in Figure 7.1. There are five possible messages having their own meaning as shown in Table 9.3.

Table 9.2 *"Normal" order addressing versus bit reverse-order addressing*

Normal order		Bit reverse order	
0	000	000	0
1	001	100	4
2	010	010	2
3	011	110	6
4	100	001	1
5	101	101	5
6	110	011	3
7	111	111	7

Table 9.3 *The five possible Huffman-coded messages in the state machine example*

Name	Code	Command
u_1	1	Stop
u_2	011	Forward
u_3	010	Reverse
u_4	001	Starboard
u_5	000	Port

Every time a complete message code is received, the corresponding command should be presented on a display. A real world system must also have a time-out function to take care of corrupt messages and lost bits. This function will, however, be omitted in this example for simplicity.

First, we identify the possible **events**. There are three possible events, namely: (1) no input symbol received, (2) a **zero** symbol input received and (3) a **one** symbol input received. The respective events are assigned with the numbers in parentheses. Second, we determine the different **states** of our system and the actions supposed to be taken in the respective state. The present state of the system is determined by a **state variable**. Assume we need the nine states presented below in our system:

State 1: wait for the first data bit of the coded message to arrive
State 2: print `stop` on the display
State 3: wait for the second data bit to arrive, given the first one was zero
State 4: wait for the third data bit to arrive, given the second one was one
State 5: wait for the third data bit to arrive, given the second one was zero
State 6: print `forward` on the display
State 7: print `reverse` on the display
State 8: print `starboard` on the display
State 9: print `port` on the display

Every state is typically implemented as a separate program segment, a function or a subroutine. The behavior of the entire program is then determined by the rules governing the execution order of the segments. These rules are commonly logical combinations of events. For our example, the rules are shown as a **transition diagram** (see Figure 9.4). A transition diagram is a very effective

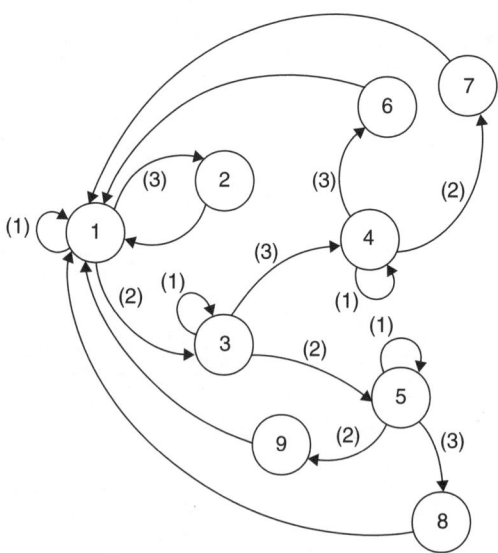

Figure 9.4 *Transition diagram for state machine example*

way of planning and documenting a computer program (and other activities as well). It is quite easy to make sure that you have really been thinking of every possible event, and that the software performs predictably in all situations.

The rings represent states and the arrows in between them are transitions taking place when the event, or combination of events, printed above the arrow occur. When the system is initialized, the state variable is set to 1 and the system starts in state 1. There are many different ways of implementing the system in software. In this example, three different approaches will be shown. They are all written using the programming language C/C++, but the method can, of course, be used with basically any computing language. If you are not fluent in C/C++, there are many good books on the market, for instance Heller (1997).

All of the program code examples shown below are built starting from a main (timed) loop, as in Section 9.3.3. The first version (example 1) is quite straight-forward and uses the `switch` statement to control the program flow. The name of the state variable is `statevar` and the event variable is named `eventvar`. Since we do not have any live input signals available, inputs are simulated using the keyboard by means of the `getchar()` function. This part of the code is, of course, different in a real world application. The state transitions, i.e. changes of the value of the state variable, take place directly in the code segment for the respective state. All variables are **global**. This programming style can be used for smaller applications and in systems where no major code changes are anticipated. Since no advanced pointer or addressing operations are involved, the code can easily be converted to rudimentary programming languages like assembly code. Note, one advantage of all state machine programming models is that good **debugging** help can easily be obtained by including, for instance, a `printf("%c", statevar)` statement in the main loop, printing the state variable for every iteration. The printout can then be compared to the transition diagram, to spot malfunctions:

```
void main(void)              {        // example 1
char statevar=1, eventvar;            // state and event variables

char c;                               // temporary input (dummy)

while(1==1)                  {        // main loop
    eventvar=1;                       // get input
    c=getchar();                      // from keyboard (dummy)
    if(c=='0') eventvar=2;            // determine event
    else if(c=='1') eventvar=3;
    switch(statevar)         {        // branch to code for current state
      case 1:                         // state 1
       if(eventvar==2) statevar=3;
       else if(eventvar==3) statevar=2;
       break;
      case 2:                         // state 2
       puts("STOP"); statevar=1;
       break;
      case 3:                         // state 3
       if(eventvar==2) statevar=5;
       else if(eventvar==3) statevar=4;
       break;
```

```
            case 4:                              // state 4
             if(eventvar==2) statevar=7;
             else if(eventvar==3) statevar=6;
             break;
            case 5:                              // state 5
             if(eventvar==2) statevar=9;
             else if(eventvar==3) statevar=8;
             break;
            case 6:                              // state 6
             puts("FORWARD"); statevar=1;
             break;
            case 7:                              // state 7
             puts("REVERSE"); statevar=1;
             break;
            case 8:                              // state 8
             puts("STARBOARD"); statevar=1;
             break;
            case 9:                              // state 9
             puts("PORT"); statevar=1;
             break;
            default:                             // error, restart
             statevar=1;

        }
      }
    }
```

In the next code example (example 2), some programming features which make the code easier to expand and maintain have been introduced. One such feature is assigning **names** to the states. In this example, quite meaningless names have been used like st1 for state number 1, etc. In a real world application more descriptive names could preferably be used, like await_first_bit for state number 1, and so on. Another feature is that the code has been broken down into separate **functions**, being called from the switch structure. Further, values are passed to functions using formal parameters, and the use of global variables is avoided. One advantage of this programming style is that the code for the different functions need not necessarily be stored in the same source code file, but can reside in different program modules, developed by different programmers. This is, of course, valuable when designing larger systems, and in applications where reuse of software modules is desirable:

```
enum statest1=1,st2,st3,st4,st5,st6,st7,st8,st9    // state names
     statevar=st1;                                 // state variable
enum eventnil=1,zero,one;                          // event names

event getevent(void) {                             // get input
  char c; c=getchar();                             // from keyboard (dummy)
  if(c=='0') return zero;                          // determine event
  if(c=='1') return one; return nil;
}

state code_1(event ev, state oldst) {              // state 1
  if(ev==zero) return st3;
  if(ev==one) return st2; return oldst;
}
```

```
state code_2(void) {                                  // state 2
  puts("STOP"); return st1;
}

state code_3(event ev, state oldst) {                 // state 3
  if(ev==zero)return st5;
  else if(ev==one) return st4; return oldst;
}

state code_4(event ev, state oldst) {                 // state 4
  if(ev==zero) return st7;
  else if(ev==one)return st6; return oldst;
}

state code_5(event ev, state oldst) {                 // state 5
  if(ev==zero) return st9;
  else if(ev==one) return st8; return oldst;
}

state code_6(void) {                                  // state 6
  puts("FORWARD"); return st1;
}

state code_7(void) {                                  // state 7
  puts("REVERSE"); return st1;
}

state code_8(void) {                                  // state 8
  puts("STARBOARD"); return st1;
}

state code_9(void) {                                  // state 9
  puts("PORT"); return st1;
}

void main(void)            {                          // example 2
  while(1==1)              {                          // main loop
   switch(statevar)        {                          // branch to code for current
                                                      //        state
      case st1: statevar=code_1(getevent(), statevar);
                break;
      case st2: statevar=code_2();  break;
      case st3: statevar=code_3(getevent(), statevar);
                break;
      case st4: statevar=code_4(getevent(), statevar);
                break;
      case st5: statevar=code_5(getevent(), statevar);
                break;
      case st6: statevar=code_6(); break;
      case st7: statevar=code_7(); break;
      case st8: statevar=code_8(); break;
      case st9: statevar=code_9(); break;
      default:  statevar=st1;                         // error, restart
   }
  }
}
```

Table 9.4 *Transition table for the state machine example, indices are event number and old state number, values in the table are new state numbers*

Event	Old state								
	1	2	3	4	5	6	7	8	9
1	1	1	3	4	5	1	1	1	1
2	3	1	5	7	9	1	1	1	1
3	2	1	4	6	8	1	1	1	1

The third approach, resulting in a quite complex program code, relies on **tables**. The transition diagram in Figure 9.4, governing the behavior of the program, can be expressed as a **transition table**, where the indices are the values of the event variable and the **old** state variable. The value found in the table is the **new** (next) value of the state variable (see Table 9.4).

Hence, by "calling" the transition table with the event number and the old state number, the new state number is obtained. The transition table in program code (example 3) below is denoted `trans[][]`, and the "call" to the table is performed in the main loop. Note that this procedure has a conceptual resemblance to the state–space approach used, for instance, in the signal model of the Kalman filter in Section 6.2.1. A great advantage with this programming style is that the entire program flow is governed by the contents of the transition table. This gives a very good overview of the program structure and behavior. Further, alternative transition tables can be stored in different header files. By replacing the header file and recompiling the source code, a program performing a new task can be obtained, without changing a single character in the source code file.

The program code for the different states is partitioned in functions, as in example 2. In this way, functions from different separately compiled program modules can be handled. This is, of course, valuable when designing larger systems. Pointers to the starting address of the different functions are stored in the **process table**, denoted `*proctab[]`. This table gives the connection between the state number and the starting address of the corresponding function, containing the necessary program code. The process table is accessed in line two of the main loop, and a vectorized function call is accomplished. Note that the main loop only contains two lines of code.

Yet another feature of the program code in example 3 is the **message table** `*mess[]`, containing the message strings to be printed on the display. This table could be stored in a separate header file. By replacing this table, display messages in foreign languages can be produced. There can, for instance, be different header files for messages in English, German, Spanish, and so on:

```
enum state{st1=1,st2,st3,st4,st5,st6,st7,st8,st9,MST} // state names
     statevar=st1;                                     // state variable
enum event{nil=1,zero,one,MEV};                        // events
enum command{ST,FOR,REV,STAR,POR};                     // commands

state trans[MEV-1][MST-1] = {                          // transition table
     {st1, st1, st3, st4, st5, st1, st1, st1, st1},
     {st3, st1, st5, st7, st9, st1, st1, st1, st1},
     {st2, st1, st4, st6, st8, st1, st1, st1, st1}};
```

```
char   *mess[] = {                                    // messages
       "STOP","FORWARD","REVERSE","STARBOARD","PORT"};

event getevent(void) {                                // get input
 char c; c=getchar();                                 // from keyboard (dummy)
 if(c=='0') return zero;                              // determine event
 if(c=='1') return one; return nil;
}
void code_1(void) {};                                 // state 1
void code_2(void) {puts(mess[ST]);}                   // state 2
void code_3(void) {};                                 // state 3
void code_4(void) {};                                 // state 4
void code_5(void) {};                                 // state 5
void code_6(void) {puts(mess[FOR]);}                  // state 6
void code_7(void) {puts(mess[REV]);}                  // state 7
void code_8(void) {puts(mess[STAR]);}                 // state 8
void code_9(void) {puts(mess[POR]);}                  // state 9

void (*proctab[MST-1]) () = {                         // process table
   code_1,code_2,code_3,code_4,code_5,code_6,code_7,code_8,code_9};

void main(void)              {                        // example 3
   while(1==1)               {                        // main loop
     statevar=trans[getevent()-1][statevar-1];        // get new state
     (*proctab[statevar-1]) ();                        // execute code for state
  }
}
```

As can be seen from the above program code examples, there are many ways to implement a state machine in software. Further, the programming language C/C++ is very flexible (but you can also produce marvelous errors).

9.4 Implementation examples

9.4.1 Finite impulse response-type filter

We will use a simple third-order low-pass FIR filter (see Chapter 1) as an example. The filter specification is 0 dB attenuation in the passband $0 < q < 0.2$ and an attenuation of at least 20 dB in the stopband $0.4 < q < 0.5$, where the frequencies are given in "fnosq", i.e. relative frequency (see Chapter 1, equation (1.8a)). Using a standard filter design program packet, the resulting filter has the transfer function

$$H(z) = b_0 + b_1 z^{-1} + b_2 z^{-2} + b_3 z^{-3} \tag{9.8}$$

where the filter coefficients are

$$b_0 = 1.000000$$
$$b_1 = 0.2763932$$
$$b_2 = 0.6381966$$
$$b_3 = b_1 = 0.2763932$$

The corresponding difference equation, i.e. the filter algorithm, can easily be obtained from the transfer function (9.8)

$$y(n) = b_0 x(n) + b_1 x(n-1) + b_2 x(n-2) + b_3 x(n-3) \tag{9.9}$$

The first thing to investigate is the risk of overflow during some steps in the calculation. Fortunately, this task is very simple for FIR filters. We can easily calculate the maximum gain of the filter. Since it is a straightforward low-pass filter, maximum gain will be found at frequency $q = 0$, or in other words "DC".

If we assume a constant DC input, i.e. $x(n) = x(n-1) = x(n-2) = 1$, the gain of the filter is simply the sum of the coefficients

$$G(0) = \sum_{i=0}^{3} b_i = 1.000000 + 0.2763932 + 0.6381966 + 0.2763932 = 2.2$$

(9.10)

If we, for instance, assume that the maximum word length of the input signal is 24 bits, and the coefficients are 24 bits, this yields 48 bits out from the multiplier. Further, if we assume that the accumulators are 56-bit wide, we have a margin of 8 bits, corresponding to a multiplication (i.e. gain) of 256 times. Since the maximum gain of the filter is roughly 2, we will never use more than 49 bits out of the 56 bits in the accumulators, hence no overflow problem will occur. If we would have suspected overflow problems, we could have **scaled** all the filter coefficients by a proper scaling constant to prevent overflow in the accumulator.

Now, our first algorithm will use a simple straightforward **static list** approach. The memory usage can be seen in Figure 9.5. There are three address spaces, the Y data memory address space, the X data memory address space and the P program memory address space. The corresponding letters are put in front of the hexadecimal addresses for clarity. We are using memory mapped I/O and the

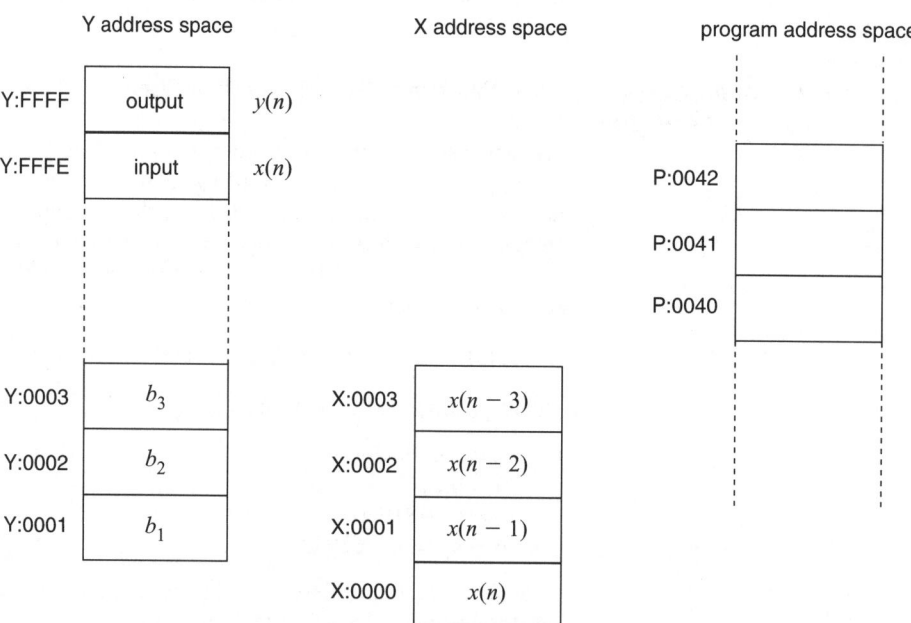

Figure 9.5 *Memory usage for the static list, third-order FIR filter example*

ports are mapped into the Y address space, where also the coefficient memory is located. The input port has address Y:FFFE and the output port Y:FFFF. The filter coefficients b_1, b_2 and b_3 are stored at locations Y:0001, Y:0002 and Y:0003, respectively. Since $b_0 = 1$, it has been omitted.

The delay line is mapped in the X data memory according to Figure 9.5, and the program code is of course stored in the program memory, mapped into P address space. The program starts at address P:0040. The algorithm can be described by the following pseudo-code:

```
init:   reset pointers
        load filter coefficients into Y memory
        clear delay line in X memory

loop:   do
        {
          get input data into delay line
          restart pointers
          clear accumulator A
          for 3 loops
          {
            get x value from delay line
            point to next x value
            get coefficient
            point to next coefficient
            multiply and accumulate in A
            move x element one step in delay line
          }
          add x(n) since this coefficient is 1
          round to 24 bits
          send to output
        } forever
```

As an example, assume we make an assembly program for the Motorola DSP 56001 digital signal processor chip ("DSP56000/56001 Digital Signal Processor User's Manual", 1989). The R0 register will be used as pointer to X memory. The R1 register will point to Y memory and the offset register N0 will hold the fixed one step offset used when moving elements in the delay-line one step. The first part of the algorithm is only initializations and will only be run once, after a reset. The actual filtering action takes place in the filter loop. This loop would normally be timed in some way, to be synchronized to the sampling rate. This detail has been omitted in this example, and we are running the loop at maximum speed. The actual assembly code for the program is shown below:

```
; initializing

; init pointers
P:0040  init:       MOVE #$3,R0         ; pointer for X data
P:0041              MOVE #$1,R1         ; pointer for Y data
P:0042              MOVE #$1,N0         ; offset, X pointer

; load filter coefficients into Y memory
P:0043              MOVE #.2763932,X1   ; load b₁
P:0045              MOVE X1,Y:(R1)+     ; to Y(1), increment pointer
```

```
P:0046                    MOVE #.6381966,X1      ; load b2
P:0048                    MOVE X1,Y:(R1)+        ; to Y(2), increment pointer
P:0049                    MOVE #.2763932,X1      ; load b3
P:004B                    MOVE X1,Y:(R1)         ; to Y(3)
          ; clear delay line vector in X memory
P:004C                    MOVE #$0,X1
P:004D                    MOVE X1,X:(R0)-        ; X(n-3)=0, decrement pointer
P:004E                    MOVE X1,X:(R0)-        ; X(n-2)=0, decrement pointer
P:004F                    MOVE X1,X:(R0)         ; X(n-1)=0
          ; the actual filter loop
          ; get input data
P:0050    floop:          MOVE Y:$FFFE,X1        ; from memory mapped port
P:0052                    MOVE X1,X:$0000        ; to X(n)
          ; restart pointers
P:0053                    MOVE #$3,R0            ; pointer to X mem, delay line
P:0054                    MOVE #$3,R1            ; pointer to Y mem, coefficients
          ; clear accumulator
P:0055                    CLR  A
          ; the convolution process
P:0056         DO         #$3,$005D             ; hw loop 3 times, exit to $005D
P:0058                    MOVE X:(R0)-,X0        ; get data, decrement pointer
P:0059                    MOVE Y:(R1)-,Y0        ; get coeff, decrement pointer
P:005A                    MAC X0,Y0,A            ; the MAC operation
          ; the data shuffling in the delay line
P:005B                    MOVE X:(R0),X0         ; get element
P:005C                    MOVE X0,X:(R0+N0)      ; move one step "upwards"
          END                                   ; end of loop
P:005D                    ADD X0,A               ; add x(n) since b0=1
          ; round result back to 24 bits and send to output
P:005E                    RND A                  ; convergent rounding
P:005F                    MOVE A1,Y:$FFFF        ; result in A1 to output
P:0060                    JMP $0050              ; get next sample
```

The first part of this software is the initialization of pointers and memory, which requires 16 words of program memory. The # character denotes an immediate constant and the expression MOVE X1,Y:(R1)+ means copy data from register X1 to Y memory at the address pointed to by R1, and post-increment of R1. This initialization part requires 36 machine cycles to execute. Using a 20 MHz clock, the cycle time is 100 ns, hence the initialization takes 3.6 µs. Maybe it is possible to find a smarter way of writing this piece of software, but it is not worth the effort, since this code is only executed once following a reset. It is better to concentrate on making the actual filter loop faster if possible, because the execution time of this loop is the upper limit sampling rate of the system.

In the filter loop, input data is first retrieved from the input port and sent to $x(n)$ in the delay line in X memory. The pointers for X memory (delay line) and Y memory (coefficients) are restarted to point to the oldest element $x(n-3)$ and the coefficient b_3, respectively. Since we will use accumulator A in the summation process, the accumulator is cleared.

The inner loop, performing the convolution in a sequential fashion, is implemented as a hardware loop, meaning that we use the internal hardware loop counter on the chip. This loop starts with DO and ends with END. After DO follow

Table 9.5 *The sequential calculation of the difference equation in which every line is an iteration of the inner loop*

X:0000	X:0001	X:0002	X:0003	A
$x(n)$	$x(n-1)$	$x(n-2)$	$x(n-3)$	0
$x(n)$	$x(n-1)$	$x(n-2)$	$x(n-2)$	$x(n-3)b_3$
$x(n)$	$x(n-1)$	$x(n-1)$	$x(n-2)$	$x(n-3)b_3 + x(n-2)b_2$
$x(n)$	$x(n)$	$x(n-1)$	$x(n-2)$	$x(n-3)b_3 + x(n-2)b_2$ $+x(n-1)b_1$
$x(n)$	$x(n)$	$x(n-1)$	$x(n-2)$	$x(n-3)b_3 + x(n-2)b_2$ $+x(n-1)b_1 + x(n)b_0$

the number of repetitions and the exit address, where execution continues after the loop is finished. Inside the inner loop, first an x value from the delay line is fetched as well as the corresponding filter coefficient. The heart of the algorithm is the MAC operation, i.e. multiplying the x value (in register X0) by the filter coefficient (in register Y0) and adding this product to the contents of accumulator A.

After the MAC operation follows the data shuffling needed to update the delay line. The present x value is moved "upwards" in the list, thus "aging" one sample time. After the inner loop has been processed three times, the calculation of the difference function (9.9) is almost finished. Only the term $x(n)b_0$ is missing. Since $b_0 = 1$, we simply add $x(n)$ to the accumulator A. The result is then rounded back from 56 to 24 bits using convergent rounding and then finally sent to the memory mapped output parallel port. The filter loop is now ready to fetch the next input sample. The sequential calculation of function (9.9) is shown in Table 9.5.

The filter loop code requires 18 words of program memory and executes in 96 machine cycles, corresponding to 9.6 μs at 20 MHz clock speed. This means that the maximum sampling rate using this software is 104 kHz. It is important to note that we have not used the possibilities of parallel data moves inherent in the chip. The DSP 56001 can do better, which we will be demonstrated in the next example.

We will now show a more sophisticated way of implementing the same filter (equation (9.9)) as above, using a **dynamic list** approach and a **modular programming** style. The dynamic list approach implies that we do not need to shuffle the delay-line data around. Instead, we implement the delay line as a circular buffer, using modulo addressing, and with the use of a start-of-list pointer. Thus, by moving the start-of-list pointer one step, the entire delay-line "ages" in a jiffy. The memory usage is shown in Figure 9.6. The main difference compared to the previous example is that we have added the coefficient b_0 to the Y memory vector to make the program more general. We cannot expect $b_0 = 1$ in the general case. Further, since we are now using a delay line in the form of a circular buffer, the position of the x values having different delays will vary; e.g. there is no fixed place for $x(n-2)$ in the buffer. It will change continuously according to the start-of-list pointer.

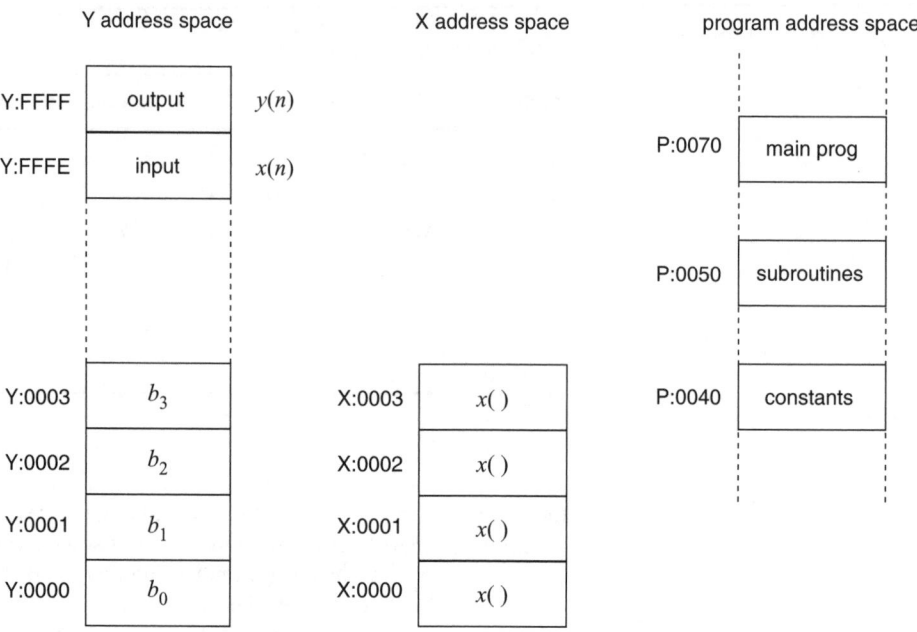

Figure 9.6 *Memory usage for the dynamic list, third-order FIR filter example*

Another improvement is a more structured use of program memory. Starting from address P:0040 there is a block of constants, followed by an area of subroutine code from P:0050 and at address P:0070, the main program code can be found. Hence, this address is also the entry point of the system. The constant area contains the filter coefficients. In this example, we are not using immediate-type program instructions to load coefficients. In many cases, it is advantageous not to "embed" constants in the code. It is better to collect all constants in an easily found block in program memory. This is especially true if we need to change the constants frequently and/or if the constants are used in many different places of the program. The algorithm can be described by the following pseudo-code:

```
init:      reset pointers
           load coefficients into Y memory from program memory
           clear delay line in X memory
           return

filter:    get input data into delay line
           clear accumulator A
           for 4 loops
           {
               get x value from delay line
               point to next x value
               get coefficient
               point to next coefficient
               multiply and accumulate in A
               move x element one step in delay line
           }
```

```
                       round to 24 bits
                       send to output
                       return

            main:      init
                       do
                       {
                         filter
                       } forever
```

In the assembly code, register R4 will be used to point to coefficients in Y memory and R0 will point to *x* values in the delay-line vector in X memory. Since we will now be working with circular buffers, modulo addressing will be needed. For this reason, the modulo registers M4 and M0 will be loaded with 4. Further, since we always go through the entire delay line from oldest to newest *x* value once every sample period, we do not need an explicit start-of-list pointer. We automatically know where the list starts.

Further, in this example, we will use the parallel execution feature which makes the entire "inner" loop only **one** instruction long. For this reason, we will use the REP (repeat) function instead of the DO END hardware loop. The assembly program is shown below:

```
          ; coefficients
P:0040              DC    .9999999          ; b₀
P:0041              DC    .2763932          ; b₁
P:0042              DC    .6381966          ; b₂
P:0043              DC    .2763932          ; b₃


          ; initializing
          ; init pointers and modulo registers
P:0050    init:     MOVE #$0043,R1          ; temporary pointer
P:0051              MOVE #$0003,R4          ; pointer for Y memory
P:0052              MOVE #$0000,R0          ; pointer for X memory
P:0053              MOVE #$0003,M0          ; modulo 4, circ addressing X
P:0054              MOVE M0,M4              ; modulo 4, circ addressing Y
          ; load coefficients to Y and clear delay line in X
P:0055              MOVE #$0,X0             ; set X0=0
P:0056        DO         #$4,$005A          ; hw loop 4 times, exit to $005A
P:0058              MOVE P:(R1)-,Y0         ; get coeff from program memory
P:0059              MOVE X0,X:(R0)+ Y0,Y:(R4)-
                                            ; clear X, increment pointer
                                            ; store coeff, decrement pointer
              END
P:005A              RTS                     ; return from subroutine
          ; the filter

P:0060    filter:   MOVEP   Y:FFFE,X:(R0)-   ; get input, store in delay line
          ; clear accumulator get x value to X0 and coeff to Y0
P:0061              CLR    A X:(R0)-,X0 Y:(R4)-,Y0
P:0062        REP #$3                        ; repeat next instr 3 times
P:0063              MAC X0,Y0,A X:(R0)-,X0 Y:(R4)-,Y0
                                            ; MAC operation
                                            ; get next x value
                                            ; get next coeff
P:0046              MACR X0,Y0,A            ; MAC operation and rounding
```

```
                    ; send to output
P:0047                       MOVEP    A1,Y:$FFFF              ; to output
P:0048                       RTS                             ; return from subroutine

                    ; main program
P:0070      main:           JSR $0050                       ; run init
P:0071                      JSR $0060                       ; run filter
P:0072                      JMP $0071                       ; forever
```

This software makes use of the possibilities of **parallel execution** in which up to three instructions can be executed simultaneously under certain circumstances. These instructions are written on the same line. For instance, MOVE X0,X: (R0)+ Y0,Y:(R4) - means copy contents of X0 to X memory address pointed to by R0 and post-increment R0 and copy contents of Y0 to Y memory address pointed to by R4 and post-decrement R4.

Another example is CLR A X:(R0)-,X0 Y:(R4)-,Y0 which means clear accumulator A and copy contents of X memory address pointed to by R0 to X0 and post-decrement R0 and copy contents of Y memory address pointed to by R4 to Y0 and post-decrement R4.

Yet another (the best one of them all) MAC X0,Y0,A X:(R0)-,X0 Y:(R4)-Y0 which means multiply contents of X0 (x value) by contents of Y0 (coefficient) and add to contents of accumulator A store in A and copy contents of X memory address pointed to by R0 to X0 and post-decrement R0 and copy contents of Y memory address pointed to by R4 to Y0 and post-decrement R4.

Making a table of the same type as Table 9.5 this example is left as an exercise to the reader. The program occupies a total of 25 words, or program memory of which four words are constants, 11 words are used by the init routine and seven by the filter routine and three by the main program. The init function executes in 62 machine cycles, i.e. 6.2 μs and the filter in 64 cycles, or in other words 6.4 μs. This means that the maximum sampling rate is 156 kHz.

9.4.2 Infinite impulse response-type filter

In this example a second-order low-pass IIR filter (see Chapter 1) will be used. The filter specification is 0 dB attenuation in the passband $0 < q < 0.1$ and an attenuation of at least 30 dB in the stopband $0.4 < q < 0.5$, where the frequencies are given in "fnosq", i.e. relative frequency (see Chapter 1, equation (1.8a)). Using a standard filter design program packet, the resulting filter has the transfer function

$$H(z) = \frac{b_0 + b_1 z^{-1} + b_2 z^{-2}}{1 - a_1 z^{-1} - a_2 z^{-2}} \tag{9.11}$$

where the filter coefficients are

$b_0 = 1.000000$
$b_1 = 1.79941$
$b_2 = 1.000000$
$a_1 = -0.299624$
$a_2 = 0.195021$

This is a "combined" IIR and FIR filter (see Chapter 1, Figure 1.10) and the corresponding difference equation is

$$y(n) = b_0 x(n) + b_1 x(n-1) + b_2 x(n-2) + a_1 y(n-1) + a_2 y(n-2)$$

$$(9.12)$$

A standard method is to divide this expression into two equations representing the FIR and IIR portions to simplify the implementation. Starting out from equation (9.11) we can separate the IIR and FIR parts

$$H(z) = \frac{Y(z)}{X(z)} = \frac{E(z)}{X(z)} \frac{Y(z)}{E(z)} = F(z)G(z) = \frac{b_0 + b_1 z^{-1} + b_2 z^{-2}}{1 - a_1 z^{-1} - a_2 z^{-2}}$$

$$= \frac{1}{1 - a_1 z^{-1} - a_2 z^{-2}} (b_0 + b_1 z^{-1} + b_2 z^{-2})$$

$$(9.13)$$

The IIR part $F(z)$ has the difference equation

$$e(n) = x(n) + a_1 e(n-1) + a_2 e(n-2) \qquad (9.14)$$

and the FIR part $G(z)$

$$y(n) = b_0 e(n) + b_1 e(n-1) + b_2 e(n-2) \qquad (9.15)$$

Hence, we do not put $x(n), x(n-1), \ldots$ in the delay line, as it is smarter to put the intermediate signal $e(n)$ into the delay line, i.e. $e(n), e(n-1), \ldots$, where $e(n)$ is the output from the IIR part of the filter (see also Figure 1.10, Chapter 1).

Since we are now dealing with an IIR filter, having poles in the transfer function, we must make sure the filter is stable, in other words, that the poles are within the unit circle in the complex z plane. There are many standard methods to check stability, but even if the filter is stable, it may have an oscillatory impulse response and some **resonant peak** with a very high gain. Two problems arise: firstly, the high gain may amplify weak round-off and truncation noise in the filter to considerable output levels ("phantom" output) and secondly, overflow problems are likely to appear. In many cases it is harder to determine good scaling factors for IIR filters than for FIR filters, where the magnitude of the signals in the filter are easier to calculate. For these reasons, IIR structures often have low orders, typically two or three. If a higher-order IIR filter is needed, a number of **cascaded** second-order filters are used. The same program code can be used (subroutine) for all the cascaded filters with only separate delay lines and coefficient vectors needed. Note, considering the round-off noise, reordering the cascaded IIR filters may change the noise level.

Simulating the algorithm is probably the best way to pinpoint noise and overflow problems in IIR-type filters. Running such a simulation of equations (9.14) and (9.15) the maximum expected gain is found to be about 3.5 times. The filter is well damped, and no significant "ringing" can be seen. Hence, no problems are expected inside the filter itself. There are, however, external subsystems to consider as well. If our input signals originate from an A/D converter and the output is connected to a D/A converter, this "extra" gain may cause problems. To avoid overflow in the D/A converter, we decide to reduce the maximum gain of the filter by a factor 2. This is done by scaling the filter

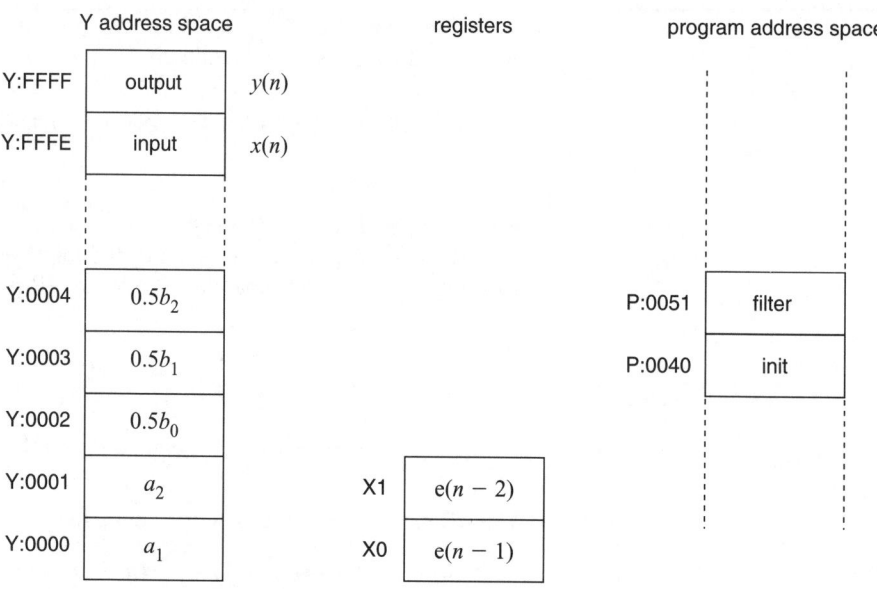

Figure 9.7 *Memory usage for the second-order IIR filter example*

coefficients b_0, b_1 and b_2 of the FIR part with a scaling factor 0.5. The scaled filter coefficients are

$$0.5b_0 = 0.500000$$
$$0.5b_1 = 0.89970$$
$$0.5b_2 = 0.500000$$
$$a_1 = -0.299624$$
$$a_2 = 0.195021$$

In this last example, a non-standard "smart"-type algorithm using a circular buffer approach for the coefficients and two registers X0 and X1 for the delay line will be described. This algorithm is not very "neat" from a programming point of view, but it is **quick** and **compact**. The memory usage can be seen in Figure 9.7. The two accumulators A and B are used to store intermediate results. The assembly code for the algorithm is shown below:

```
               ; initializing
               ; init pointer for coefficients
P:0040  init:       MOVE #$0,R4              ; pointer for Y memory
P:0041              MOVE #$4,M4              ; modulo 5 addressing in Y mem
               ; load coefficients into Y memory
P:0042              MOVE #-.299624,X0        ; load a₁
P:0044              MOVE X0,Y:(R4)+
P:0045              MOVE #.195021,X0         ; load a₂
P:0047              MOVE X0,Y:(R4)+
P:0048              MOVE #.500000,X0         ; load 0.5b₀
P:004A              MOVE X0,Y:(R4)+
```

```
P:004B                  MOVE #.29970,X1        ; load 0.5b₁
P:004D                  MOVE X1,Y:(R4)+
P:004E                  MOVE X0,Y:(R4)+        ; load 0.5b₂ = 0.5b₀
           ; clear delay line
P:004F                  MOVE #$0,X0            ; e(n-1) = 0
P:0050                  MOVE X0,X1             ; e(n-2) = 0
           ; the filter
P:0051     filter:      CLR A                  ; clear accumulator A
P:0052                  CLR B Y:(R4)+,Y0       ; clear accumulator B
                                               ; get a₁ to register Y0
P:0053                  MOVEP Y:$FFFE,A1       ; get input to accumulator A
P:0054                  MAC X0,Y0,A Y:(R4)+,Y0 ; x(n)+a₁e(n-1) to A
                                               ; get a₂ to register Y0
P:0055                  MACR X1,Y0,A Y:(R4)+,Y0 ; x(n)+a₁e(n-1)+ a₂e(n-2) to A
                                               ; round to 24 bits
                                               ; get 0.5b₀ to register Y0
P:0056                  MOVE A1,Y1             ; e(n) to register Y1
P:0057                  MAC Y0,Y1,B Y:(R4)+,Y0 ; 0.5b₀e(n) to B
                                               ; get 0.5b₁ to register Y0
P:0058                  MAC X0,Y0,B Y:(R4)+,Y0 ; 0.5b₀e(n)+0.5b₁e(n-1) to B
                                               ; get 0.5b₂ to register Y0
P:0059                  MACR X1,Y0,B X0,X1     ; 0.5b₀e(n)+0.5b₁e(n-1)+
                                               ; +0.5b₂e(n-2)to B
                                               ; e(n-1) to X1 (time passes by)
P:0060                  MOVE A1,X0             ; e(n) to X0
P:0061                  MOVEP B1,Y:$FFFF       ; output
P:0062                  JMP $0051              ; next sample
```

The program occupies 29 words of program memory, of which 17 words are for the initialization procedure, and 12 for the filter. Initialization executes in $3.6\ \mu s$ and the filtering function in $6.2\ \mu s$ assuming 20 MHz clock speed. Hence 160 kHz sampling rate is maximum.

9.5 Future systems and chips

The above, simplified program code segments are only intended as examples, and the classical but elegant DSP chip used is one out of many. It is not very risky to predict that coming DSP chips will be faster, more complex and cheaper. In some respects, there is also a merging of conventional microprocessor chip technology, DSPs and FPGA structures taking place. FPGAs with a number of DSP kernels on-chip are on the market today, and there are more to come. Applications implemented using FPGAs will probably become more common. There have also been general-purpose microprocessor chips around for a while, having MAC instructions and other typical DSP features. New, improved simulators, compilers and other development tools are constantly being launched on the market, making life easier for the designer.

DSP chips are used in many embedded systems today in large volume consumer products like cellular mobile telephones. As the processing speed of the DSPs increases, new applications will develop continuously. One interesting area is **radio technology**. An average cellular mobile telephone today contains a number of DSP chips. Classical radio electronics circuitry occupies only a small fraction of the total printed circuit board area. As DSP chips get faster, more of the radio electronics circuits will disappear and typical radio functions like filtering, mixing, oscillating, modulation and demodulation will be implemented as DSP software rather than hardware. This is true for radios in general, not only cellular mobile telephones. For instance, radio systems based on **wideband code division multiple access (WCDMA)** and **ultra wideband (UWB)** will depend heavily on DSP technology. The technology of **software defined radio (SDR)** is not yet mature, but will grow in importance as the DSP chips get faster.

DSPs will improve but still the **programming** of DSPs will be more complex and demanding. Not only are good programming skills required, but also considerable knowledge in signal theory, numeric methods, algorithm design and mathematics.

Summary

In this chapter the following issues have been treated:

- System types, processing speed, hardware architectures and hardware types
- Fixed and floating-point format, numerical problems, truncation and rounding
- The DSP software development process
- Program and data structures, the state machine
- Implementation examples.

Review questions

R9-1 What are the properties and requirements of a stream process and a batch process considering execution speed and memory demand?

R9-2 Explain the MAC operation. Why is it so important?

R9-3 What are the pros and cons of general-purpose microprocessors, DSP chips, FPGA circuits and ASICs from a digital signal processing point of view?

R9-4 Explain the von Neumann and Harvard computer architectures. State some pros and cons?.

R9-5 Explain the terms "wrap-around" and "saturating" arithmetics.

R9-6 What is bit reverse addressing? When is it used?

R9-7 Explain the data structures "static list" and "dynamic list".

R9-8 What is the difference between "truncation" and "rounding"?

Solved problems

P9-1 Assume that you are given the task of writing the software for a digital averaging filter having the transfer function $H(z) = \sum_{k=0}^{N-1} b_k z^{-k}$, where the filter coefficients are given by $b_k = 1/N$. The DSP available has an input bus width of 16 bits and the accumulator has 24 bits. What is the maximum length N of the filter to avoid overflow, without applying scaling?

P9-2 The sampling rate is 200 kHz for the input of the filter in P9-1 above. Determine the maximum allowed execution time of the main loop. How much data memory (in bytes) is required for the filter?

P9-3 Assume, for simplicity, that we are using a processor with 4-bit data width. Further, 2's complement fractional representation is used, as in Table 9.1. Our idea is to build a simple digital filter expressed by the difference equation $y(n) = c(ax(n) + bx(n-1))$, where $a = 0.225$,

$b = -0.125$ and $c = 4$. The first implementation of the filter is made directly from the difference equation as above. What is the output of the filter if $x(n) = 1$ for all n? We are not satisfied with the performance of the filter so we change the order of the calculations to $y(n) = cax(n) + cbx(n-1)$. What is the output of the filter using this implementation? What would the ideal output of the filter be? What is going on?

P9-4 Using the state machine approach, draw a state transition diagram and determine the corresponding state transition table for a system able to decode the Morse code for the letters "a", "e", "n" and "t". The Morse codes are "a" = dot–dash, "e" = dot, "n" = dash–dot, "t" = dash. Each transmitted letter is followed by silence for 1 s. The length of a dot is 0.1 s and a dash is 0.3 s.

Appendix 1 Solutions to problems

Chapter 1 **P1-1**

From Section 1.2.3, we know that a word length of $n = 9$ bits corresponds to $9 \cdot 6 = 54$ dB dynamic range, and that $n = 10$ bits gives $10 \cdot 6 = 60$ dB. Hence, we choose $n = 10$ bits word length per sample.

Further, from Section 1.2.2, we know that the sampling frequency has to be $f_s > 2f_{max}$, and since the maximum frequency of the analog input signal is 10 kHz, we get $f_s > 20$ kHz, i.e. 20 000 samples per second. Since each sample is 10 bits, we get a bit rate $R > 10 \cdot 20\,000 = 200$ kbits/s.

P1-2

The transfer function will be

$$H(z) = \frac{Y(z)}{X(z)} = \frac{z - a}{(z - b)(z - c)} = \frac{z - a}{z^2 - z(b + c) + bc}$$

$$= \frac{z^{-1} - az^{-2}}{1 - z^{-1}(b + c) + z^{-2}bc} \tag{A1.1.1}$$

The frequency function is obtained by setting $z = e^{j\Omega}$, where $\Omega = 2\pi f/f_s$, and by inserting into equation (A1.1.1) and using Euler's formula $e^{j\phi} = \cos(\phi) + j\sin(\phi)$, we get

$$H(\Omega) = \frac{e^{-j\Omega} - ae^{-j2\Omega}}{1 - e^{-j\Omega}(b + c) + e^{-j2\Omega}bc}$$

$$= \frac{\cos(\Omega) - j\sin(\Omega) - a\cos(2\Omega) + ja\sin(2\Omega)}{1 - (b + c)\cos(\Omega) + j(b + c)\sin(\Omega) + bc\cos(2\Omega) - jbc\sin(2\Omega)}$$

$$= \frac{(\cos(\Omega) - a\cos(2\Omega)) - j(\sin(\Omega) + a\sin(2\Omega))}{(1 - (b + c)\cos(\Omega) + bc\cos(2\Omega)) + j((b + c)\sin(\Omega) - bc\sin(2\Omega))} \tag{A1.1.2}$$

The gain function is $A(\Omega) = |H(\Omega)|$, and inserting equation (A1.1.2) gives

$$A(\Omega) = \frac{\sqrt{(\cos(\Omega) - a\cos(2\Omega))^2 + (\sin(\Omega) + a\sin(2\Omega))^2}}{\sqrt{(1 - (b + c)\cos(\Omega) + bc\cos(2\Omega))^2 + ((b + c)\sin(\Omega) - bc\sin(2\Omega))^2}}$$

$$= \frac{\sqrt{1 + 2a - 2a(\cos(\Omega)\cos(2\Omega) - \sin(\Omega)\sin(2\Omega))}}{\sqrt{1 + (b+c)^2 + b^2c^2 - 2(b+c)\cos(\Omega) + 2bc\cos(2\Omega) + 2(b+c)(\cos(\Omega)\cos(2\Omega) - \sin(\Omega)\sin(2\Omega))}}$$

$$= \frac{\sqrt{1 + 2a - 2a\cos(3\Omega)}}{\sqrt{1 + (b+c)^2 + b^2c^2 - 2(b+c)\cos(\Omega) + 2bc\cos(2\Omega) + 2(b+c)\cos(3\Omega)}}$$

$$= \sqrt{\frac{1 + 2a(1 - \cos(3\Omega))}{1 + (b+c)^2 + b^2c^2 - 2(b+c)(\cos(\Omega) - \cos(3\Omega)) + 2bc\cos(2\Omega)}}$$

where we have used the identities

$$\cos^2(\phi) + \sin^2(\phi) = 1 \quad \text{and}$$

$$\cos(\alpha)\cos(\beta) - \sin(\alpha)\sin(\beta) = \cos(\alpha + \beta)$$

The phase shift function is $\phi(\Omega) = \angle H(\Omega)$, and inserting equation (A1.1.2) yields

$$\phi(\Omega) = \arctan\left(\frac{-(\sin(\Omega) + a\sin(2\Omega))}{\cos(\Omega) - a\cos(2\Omega)}\right)$$

$$- \arctan\left(\frac{(b+c)\sin(\Omega) - bc\sin(2\Omega)}{1 - (b+c)\cos(\Omega) + bc\cos(2\Omega)}\right)$$

Finally, the difference equation can be obtained by taking the inverse z-transform of the transfer function (A1.1.1)

$$H(z) = \frac{Y(z)}{X(z)} = \frac{z^{-1} - az^{-2}}{1 - z^{-1}(b+c) + z^{-2}bc}$$

$$\Rightarrow Y(z) = X(z)z^{-1} - aX(z)z^{-2} + (b+c)Y(z)z^{-1} - bcY(z)z^{-2}$$

$$\Rightarrow y(n) = x(n-1) - ax(n-2) + (b+c)y(n-1) - bcy(n-2)$$

P1-3

The transfer function of the FIR filter is equation (1.43)

$$H(z) = b_0 + b_1z^{-1} + b_2z^{-2} + \cdots + b_Mz^{-M} \tag{A1.1.3}$$

Let M be even for the total number of taps to be odd. Rewrite equation (A1.1.3) as

$$H(z) = z^{-(M/2)}\left(b_0z^{M/2} + b_1z^{(M/2)-1} + \cdots \right.$$

$$\left. + b_{M/2} + \cdots + b_{M-1}z^{-(M/2)+1} + b_Mz^{-(M/2)}\right)$$

Setting $z = e^{j\Omega}$ and using the symmetry $b_n = b_{M-n}$ we get

$$H(z) = e^{-j(M/2)\Omega} \left(b_0 e^{j(M/2)\Omega} + b_1 e^{j((M/2)-1)\Omega} + \cdots + b_{M/2} + \cdots \right.$$

$$\left. + b_1 e^{-j((M/2)+1)\Omega} + b_0 e^{-j(M/2)\Omega} \right)$$

$$= e^{-j(M/2)\Omega} \left(b_0 \left(e^{j(M/2)\Omega} + e^{-j(M/2)\Omega} \right) \right.$$

$$\left. + b_1 \left(e^{j((M/2)-1)\Omega} + e^{-j((M/2)+1)\Omega} \right) + \cdots + b_{M/2} \right)$$

Using the identity $e^{j\alpha} + e^{-j\alpha} = 2\cos(\alpha)$, we get

$$H(z) = e^{-j(M/2)\Omega} \left(2b_0 \cos\left(\frac{M}{2}\Omega\right) + 2b_1 \cos\left(\left(\frac{M}{2} - 1\right)\Omega\right) + \cdots + b_{M/2} \right)$$

$$= e^{-j(M/2)\Omega} F(\Omega)$$

Since the function $F(\Omega)$ is real, it does not contribute to the phase shift and the phase shift function of the filter will be linear (see also equation (1.58))

$$\phi(\Omega) = -\frac{M}{2}\Omega$$

P1-4

Starting with equation (1.73) and substituting equation (1.75), $s \rightarrow f_s(1 - z^{-1})$, gives

$$G(s) = \frac{b_0}{a_0 + a_1 s + a_2 s^2}$$

$$\Rightarrow H(z) = \frac{b_0}{a_0 + a_1 f_s(1 - z^{-1}) + a_2 f_s^2(1 - z^{-1})^2}$$

$$= \frac{b_0}{a_0 + a_1 f_s(1 - z^{-1}) + a_2 f_s^2(1 - 2z^{-1} - z^{-2})}$$

$$= \frac{b_0}{a_0 + a_1 f_s + a_2 f_s^2 - z^{-1}(a_1 f_s + 2a_2 f_s^2) - z^{-2} a_2 f_s^2}$$

$$= \frac{b_0}{a_0 + a_1 f_s + a_2 f_s^2}$$

$$\cdot \frac{1}{1 - z^{-1}\dfrac{a_1 f_s + 2a_2 f_s^2}{a_0 + a_1 f_s + a_2 f_s^2} - z^{-2}\dfrac{a_2 f_s^2}{a_0 + a_1 f_s + a_2 f_s^2}}$$

Finally, inserting the numerical values we get

$$H(z) = 0.511 \frac{1}{1 - 0.017z^{-1} + 0.128z^{-2}}$$

P1-5

There are many ways to write the program; one simple approach is presented below:

```
% problem 1-5

Fs=40e3;                    % sampling frequency

W=[8e3, 12e3]/(Fs/2);       % passband spec

[b,a]=butter(4,W,'z');      % design Butterworth
                            % note the order is 2*4=8
[d,c]=cheby1(4,1,W,'z');    % design Chebyshev

figure(1)                   % make Bode plots
freqz(b,a,60,Fs)            % Butterworth in 60 points
hold on                     % both in the same diagram
freqz(d,c,60,Fs)            % Chebyshev

figure(2)                   % make pole/zero plot
pzmap(b,a)                  % for Butterworth

figure(3)                   % make pole/zero plot
pzmap(d,c)                  % for Chebyshev
```

The Bode plots of the two filters are shown in Figure A1.1, the solid line is the Butterworth filter, while the dotted one is the Chebyshev filter. In Figure A1.2 the pole–zero plot for the Butterworth filter is shown. Finally, Figure A1.3 shows the pole–zero plot for the Chebyshev filter.

Chapter 2   P2-1

The μ-law equation (2.2) is given by $y = (\log(1 + \mu x))/(\log(1 + \mu))$, setting $\mu = 0$ means that we need to solve a limit problem. Approximate the denominator and the numerator using Maclaurin expansions (1.22)

$$\text{numerator: } \log(1 + \mu x) \approx 0 + \frac{1}{1 + \mu x} x$$

$$\text{denominator: } \log(1 + \mu) \approx 0 + \frac{1}{1 + \mu}$$

(higher-order terms can be ignored, since they will tend to zero even faster)
    Inserting the approximations yields (for small μ)

$$y = \frac{\log(1 + \mu x)}{\log(1 + \mu)} \approx \frac{x(1 + \mu)}{1 + \mu x}$$

Setting $\mu = 0$ we obtain $y \approx (x(1 + \mu))/(1 + \mu x) = x$, i.e. a linear function.

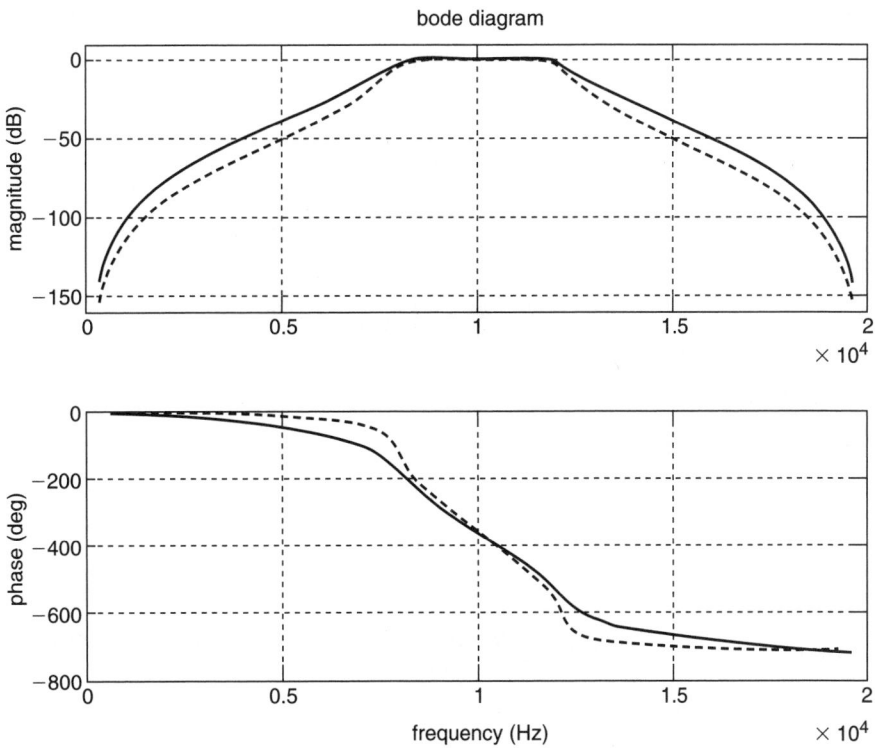

Figure A1.1 *Bode plot of the Butterworth filter (solid line) and the Chebyshev filter (dotted line)*

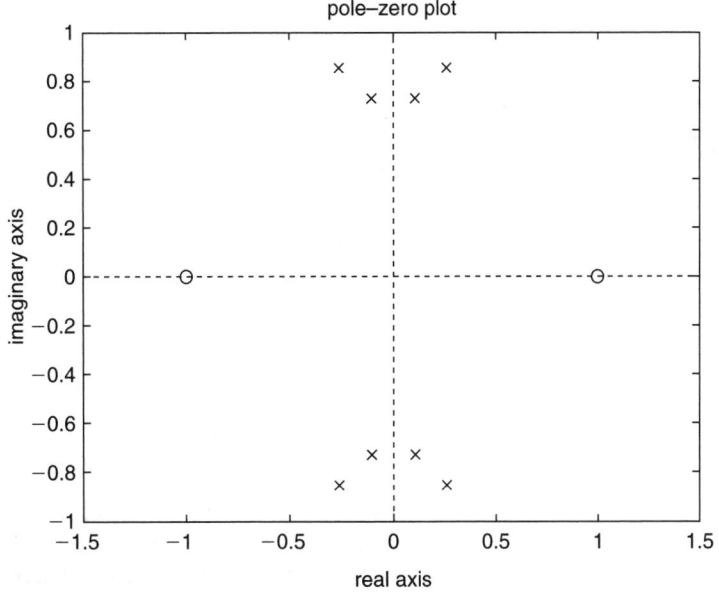

Figure A1.2 *Pole–zero plot of the Butterworth filter*

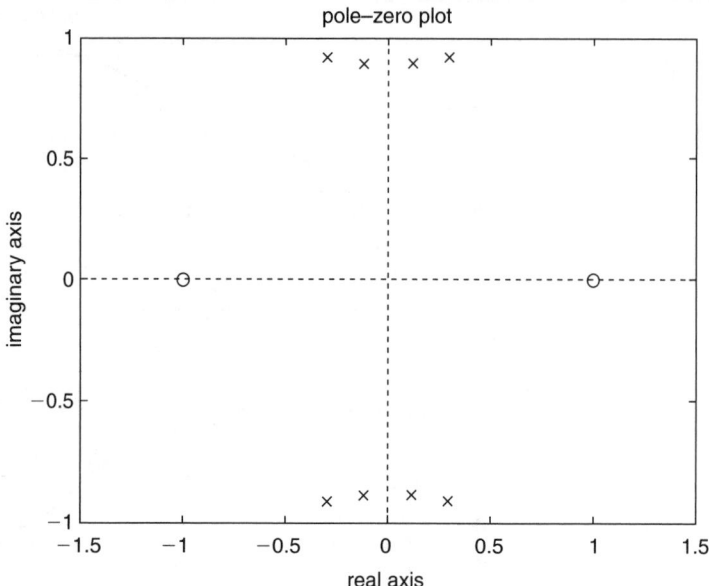

Figure A1.3 *Pole–zero plot of the Chebyshev filter*

P2-2

A 14-bit ADC having 5 V FS has a LSB step of $FS/2^n = 5/2^{14}$ V. An offset of 3 LSB is hence

$$3\frac{5}{2^{14}} = 0.18\,\text{mV}.$$

P2-3

There are many ways to write the program; one simple approach is presented below:

```
% problem 2-3
% comb filter, k=9

% transfer function
b=[1, 0, 0, 0, 0, 0, 0, 0, 0, 1];    % numerator
a=[1, 1, 0, 0, 0, 0, 0, 0, 0, 0];    % denominator

SYS=TF(b,a,-1);                       % convert model
pzmap(SYS)                            % plot poles and zeros
```

Figure A1.4 shows the pole–zero plot. At $z = -1$ there is a pole located right on the unit circle. The pole is cancelled by a zero; however, the pole or zero may drift away slightly due to numerical truncation problems. A pole outside the unit circle means an unstable system, and pole (not cancelled) on the unit circle represents an oscillator (resonator).

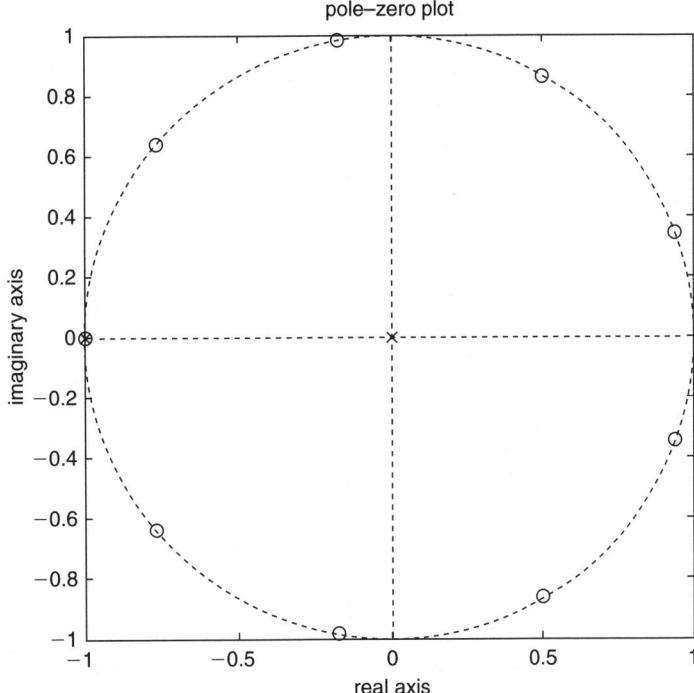

Figure A1.4 *Pole–zero plot of the comb filter with* k $= 9$

P2-4

There are many ways to write the program; one simple approach is presented below:

```
% problem 2-4
% comb filter, k=9

% transfer function
b=[1, 0, 0, 0, 0, 0, 0, 0, 0, 1];      % numerator
a=[1, 1, 0, 0, 0, 0, 0, 0, 0, 0];      % denominator
freqz(b,a)                             % plot Bode diagram
```

Figure A1.5 shows the Bode plot of the comb filter.

Chapter 3   **P3-1**

For the cancellation to work perfectly we require $d_k = n_{0k}$. Since the system is linear we can assume $s_k = 0$. The MSE can be found from equation (3.9)

$$\xi = \mathrm{E}\big[d_k^2\big] + \mathbf{W}^\mathrm{T}\mathbf{R}\mathbf{W} - 2\mathbf{P}^\mathrm{T}\mathbf{W}$$
$$= \mathrm{E}\big[d_k^2\big] + \mathbf{W}^\mathrm{T}\mathrm{E}\big[\mathbf{X}_k\mathbf{X}_k^\mathrm{T}\big]\mathbf{W} - 2\mathrm{E}\big[d_k\mathbf{X}_k\big]\mathbf{W}$$

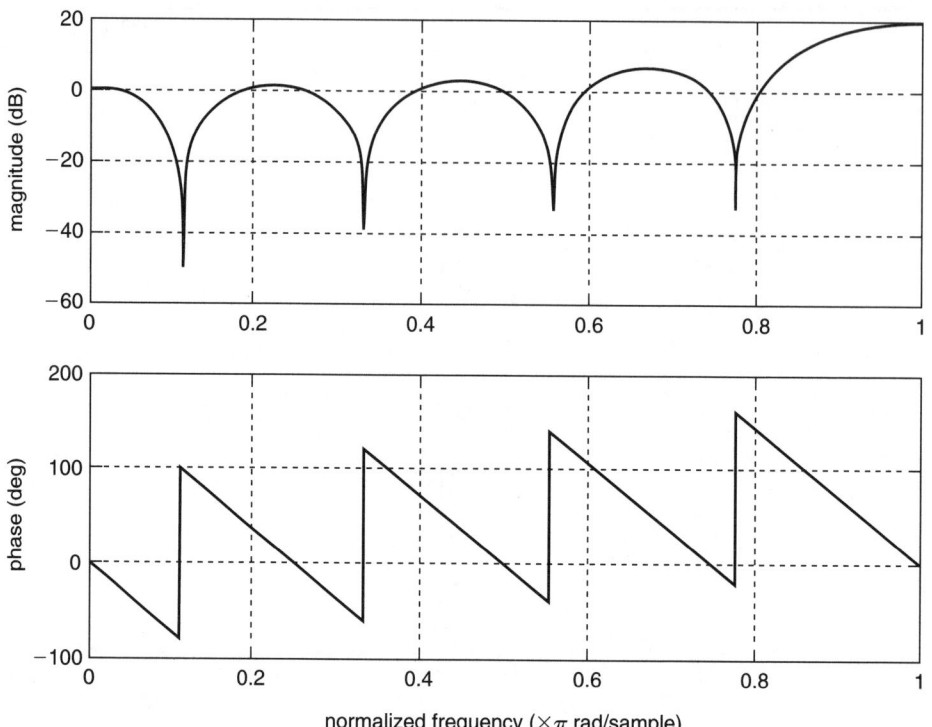

normalized frequency ($\times \pi$ rad/sample)

Figure A1.5 *Bode plot of the comb filter with* k $= 9$

where the mean square of the desired signal is

$$E[d_k^2] = E\left[\sin^2\left(\pi\frac{k}{7}\right)\right] = \frac{1}{2}E\left[1 - \cos\left(2\pi\frac{k}{7}\right)\right]$$

$$= \frac{1}{2} - \frac{1}{2}\frac{1}{N}\sum_{k=1}^{N}\cos\left(2\pi\frac{k}{7}\right) = \frac{1}{2}$$

and the input correlation matrix is obtained by

$$E[\mathbf{X}_k\mathbf{X}_k^{\mathrm{T}}] = \begin{bmatrix} E[x_k x_k] & E[x_k x_{k-1}] \\ E[x_{k-1} x_k] & E[x_{k-1} x_{k-1}] \end{bmatrix} = \mathbf{R}$$

$$E[x_k x_k] = E\left[\cos^2\left(\pi\frac{k}{7}\right)\right] = \frac{1}{2}E\left[1 + \cos\left(2\pi\frac{k}{7}\right)\right]$$

$$= \frac{1}{2} + \frac{1}{2}\frac{1}{N}\sum_{k=1}^{N}\cos\left(2\pi\frac{k}{7}\right) = \frac{1}{2}$$

$$E[x_k x_{k-1}] = E[x_{k-1} x_k] = E\left[\cos\left(\pi\frac{k}{7}\right)\cos\left(\pi\frac{k-1}{7}\right)\right]$$

$$= \frac{1}{2}E\left[\cos\left(\pi\frac{1}{7}\right) + \cos\left(\pi\frac{2k+1}{7}\right)\right]$$

$$= \frac{1}{2} \cos\left(\pi\frac{1}{7}\right) + \frac{1}{2}\frac{1}{N}\sum_{k=1}^{N}\cos\left(\pi\frac{2k+1}{7}\right) = \frac{1}{2}\cos\left(\pi\frac{1}{7}\right)$$

$$\mathrm{E}[x_{k-1}x_{k-1}] = \mathrm{E}\left[\cos^2\left(\pi\frac{k-1}{7}\right)\right] = \frac{1}{2}\mathrm{E}\left[1 + \cos\left(2\pi\frac{k-1}{7}\right)\right]$$

$$= \frac{1}{2} + \frac{1}{2}\frac{1}{N}\sum_{k=1}^{N}\cos\left(2\pi\frac{k-1}{7}\right) = \frac{1}{2}$$

$$\mathbf{R} = \frac{1}{2}\begin{bmatrix} 1 & \cos\left(\frac{\pi}{7}\right) \\ \cos\left(\frac{\pi}{7}\right) & 1 \end{bmatrix}$$

The cross-correlation vector is

$$\mathrm{E}[d_k\mathbf{X}_k] = \begin{bmatrix} \mathrm{E}[d_kx_k] \\ \mathrm{E}[d_kx_{k-1}] \end{bmatrix} = \mathbf{P}$$

$$\mathrm{E}[d_kx_k] = \mathrm{E}\left[\sin\left(\pi\frac{k}{7}\right)\cos\left(\pi\frac{k}{7}\right)\right] = \frac{1}{2}\mathrm{E}\left[\sin\left(\pi\frac{2k}{7}\right)\right]$$

$$= \frac{1}{2}\frac{1}{N}\sum_{k=1}^{N}\sin\left(\pi\frac{2k}{7}\right) = 0$$

$$\mathrm{E}[d_kx_{k-1}] = \mathrm{E}\left[\sin\left(\pi\frac{k}{7}\right)\cos\left(\pi\frac{k-1}{7}\right)\right]$$

$$= \frac{1}{2}\mathrm{E}\left[\sin\left(\pi\frac{2k-1}{7}\right) + \sin\left(\pi\frac{1}{7}\right)\right]$$

$$= \frac{1}{2}\frac{1}{N}\sum_{k=1}^{N}\sin\left(\pi\frac{2k-1}{7}\right) + \frac{1}{2}\sin\left(\pi\frac{1}{7}\right) = \frac{1}{2}\sin\left(\frac{\pi}{7}\right)$$

$$\mathbf{P} = \frac{1}{2}\begin{bmatrix} 0 \\ \sin\left(\frac{\pi}{7}\right) \end{bmatrix}$$

which finally gives the MSE function

$$\xi = \mathrm{E}[d_k^2] + \mathbf{W}^{\mathrm{T}}\mathbf{R}\mathbf{W} - 2\mathbf{P}^{\mathrm{T}}\mathbf{W}$$

$$= \frac{1}{2} + \frac{1}{2}[w_0 \quad w_1]\begin{bmatrix} 1 & \cos\left(\frac{\pi}{7}\right) \\ \cos\left(\frac{\pi}{7}\right) & 1 \end{bmatrix}\begin{bmatrix} w_0 \\ w_1 \end{bmatrix}$$

$$- 2\frac{1}{2}[0 \quad \sin\left(\frac{\pi}{7}\right)]\begin{bmatrix} w_0 \\ w_1 \end{bmatrix}$$

$$= \frac{1}{2} + \frac{1}{2}\left(w_0\left(w_0 + w_1\cos\left(\frac{\pi}{7}\right)\right) + w_1\left(w_0\cos\left(\frac{\pi}{7}\right) + w_1\right)\right)$$

$$- w_1\sin\left(\frac{\pi}{7}\right) = \frac{1}{2} + \frac{1}{2}(w_0^2 + w_1^2) + w_0w_1\cos\left(\frac{\pi}{7}\right) - w_1\sin\left(\frac{\pi}{7}\right)$$

P3-2

The LMS algorithm

$$\mathbf{W}_{k+1} = \mathbf{W}_k + 2\mu\varepsilon_k\mathbf{X}_k = \mathbf{W}_k + 2\mu(d_k - \mathbf{X}_k^{\mathsf{T}}\mathbf{W}_k)\mathbf{X}_k$$

$$= \begin{bmatrix} w_{0k} \\ w_{1k} \end{bmatrix} + 2\mu\sin\left(\frac{\pi}{7}\right)\begin{bmatrix} x_k \\ x_{k-1} \end{bmatrix}$$

$$- 2\mu\begin{bmatrix} x_k & x_{k-1} \end{bmatrix}\begin{bmatrix} w_{0k} \\ w_{1k} \end{bmatrix}\begin{bmatrix} x_k \\ x_{k-1} \end{bmatrix}$$

$$= \begin{bmatrix} w_{0k} + 2\mu\left(x_k\sin\left(\frac{\pi}{7}\right) - x_k^2 w_{0k} - x_k x_{k-1} w_{1k}\right) \\ w_{1k} + 2\mu\left(x_{k-1}\sin\left(\frac{\pi}{7}\right) - x_{k-1}x_k w_{0k} - x_{k-1}^2 w_{1k}\right) \end{bmatrix}$$

Hence

$$w_{0k+1} = w_{0k} + 2\mu\left(\cos\left(\pi\frac{k}{7}\right)\sin\left(\frac{\pi}{7}\right) - \cos^2\left(\pi\frac{k}{7}\right)w_{0k}\right.$$

$$\left. - \cos\left(\pi\frac{k}{7}\right)\cos\left(\pi\frac{k-1}{7}\right)w_{1k}\right)$$

$$w_{1k+1} = w_{1k} + 2\mu\left(\cos\left(\pi\frac{k-1}{7}\right)\sin\left(\frac{\pi}{7}\right)\right.$$

$$\left. - \cos\left(\pi\frac{k-1}{7}\right)\cos\left(\pi\frac{k}{7}\right)w_{0k} - \cos^2\left(\pi\frac{k-1}{7}\right)w_{1k}\right)$$

P3-3

A suggested MATLAB™ program, running the filter, plotting the MSE and updating the weights using LMS, is shown below:

```
% adap.m
% parameters and variables

N=1000;                                    % number of steps
k=1:N;                                     % initialize time index

mu=0.05;                                   % training constant
dk=sin(pi*k/7);                            % desired signal
X=[cos(pi*k/7); cos(pi*(k-1)/7)];          % input to combiner
```

```
W=zeros(2,N);                        % weight vectors
ksi=zeros(1,N);                      % MSE
y=zeros(1,N);                        % filter output

R=[1 cos(pi/7); cos(pi/7)1];         % auto-correlation matrix
P=[0; sin(pi/7)];                    % cross-correlation vector
Edk=0.5;                             % mean square of desired signal

% main loop
for k=1:N
  % run filter
  y(k)=X(1:2,k)'*W(1:2,k);

  % find MSE
  ksi(k)=Edk+W(1:2,k)'*R*W(1:2,k)-2*P'*W(1:2,k);

  % run LMS
  W(1:2,(k+1))=W(1:2,k)+2*mu*(dk(k)-y(k))*X(1:2,k);
end

% plot result
figure(1)
subplot(2,1,1)                       % output from filter
plot(y)
Title('y(k), output from filter')

subplot(2,1,2)                       % output after cancellation
plot(y-dk)
Title('eps(k), output after cancellation')

figure(2)
subplot(2,1,1)                       % weight w0
plot(W(1,1:N))
Title('weight w0')
subplot(2,1,2)                       % weight w1
plot(W(2,1:N))
Title('weight w1')

figure(3)
plot(ksi)                            % MSE
Title('MSE, learning curve')
```

Figure A1.6 shows the output $y(n)$ from the filter (upper plot) and the output after cancellation $y(n) - d_k(n)$ (lower plot). Figure A1.7 shows convergence of the two weights w_0 (upper plot) and w_1 (lower plot). Figure A1.8 shows the "learning curve", i.e. the minimum square error (MSE).

P3-4

The Wiener solution is given by

$$\mathbf{W}^* = \mathbf{R}^{-1}\mathbf{P}$$

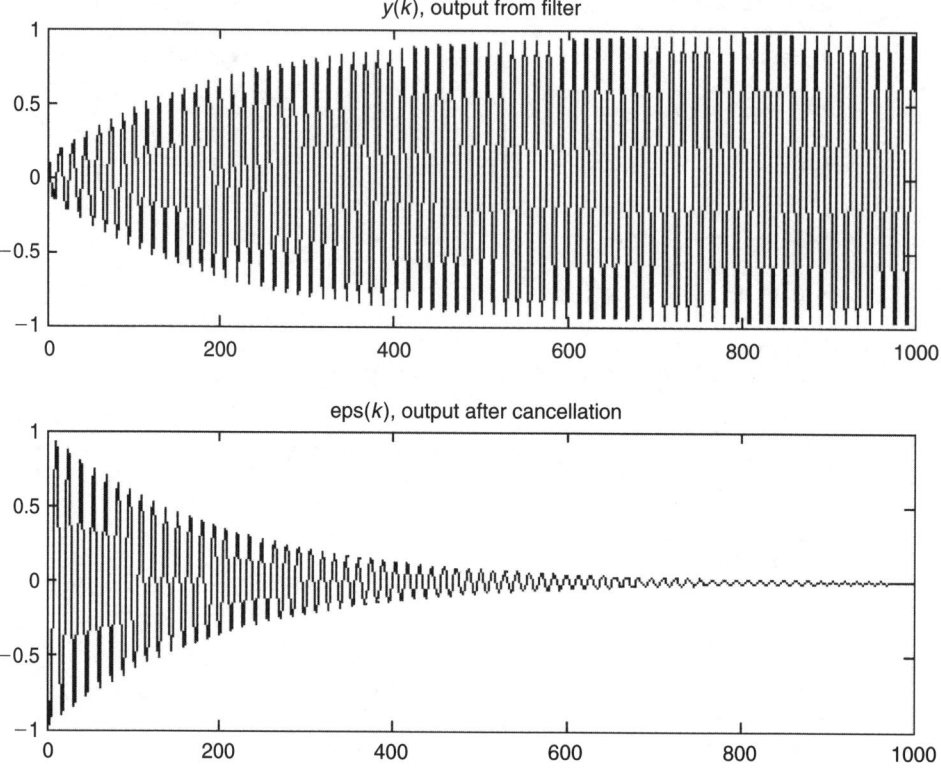

Figure A1.6 *Interference cancellation system, output from the filter (upper plot) and the output after cancellation (lower plot)*

This expression can be evaluated easily using the MATLAB™ command:

```
R\P
ans =
  -2.0765
   2.3048
>>
```

which corresponds well to the final values of the weights obtained in P3-3.

Chapter 4 P4-1

The program may look like (for $N = 3$) below:

```
% pr41.m

                                          % test vector
X=[0 0 0 0 1 2 3 4 4 4 3 -8 1 0 0 1 1 0 0 0 1 1 1 0 0 0 0];
N=3;                                      % filter length
Y=zeros(size(X));                         % output vector
```

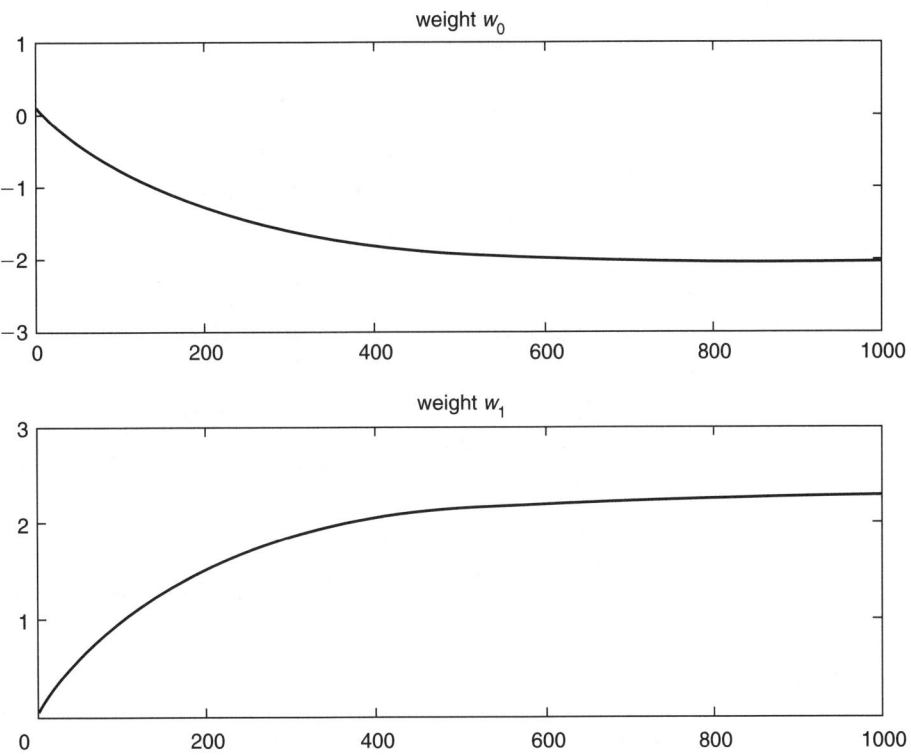

Figure A1.7 *Convergence of the filter weights,* w_0 *(upper plot) and* w_1 *(lower plot)*

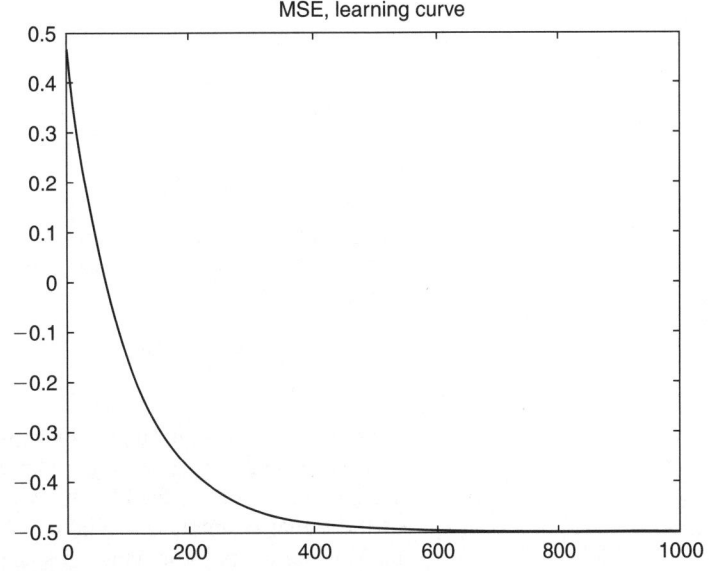

Figure A1.8 *MSE as a function of iteration, the "learning curve"*

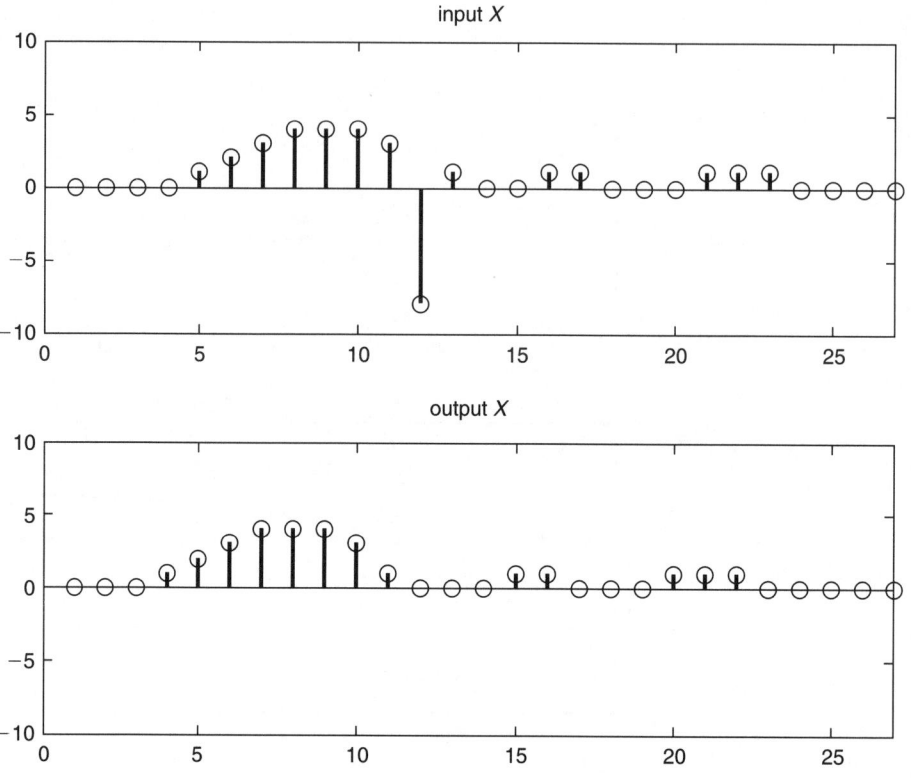

Figure A1.9 *The input (upper plot) and output (lower plot) of the median filter with filter length* N = 3

```
for i=1:(length(X)-N)
    Y(i)=median (X(i:(i+(N-1))));          % filter
end

subplot(2,1,1)
stem(X)                                     % plot input
axis([0 length(X)-10 10])

subplot(2,1,2)
stem(Y)                                     % plot output
axis([0 length(X)-10 10])
```

Figure A1.9 shows the input (upper plot) and output (lower plot) of the median filter for $N = 3$. The short one-sample transient pulse is removed, since one sample out of three in the filtering window is regarded as an impulse. The rest of the signal is untouched. Figure A1.10 is input and output for $N = 5$ and the two-sample pulse is removed since with this filter length it is an impulse. Finally, in Figure A1.11 the filter length is $N = 7$ and the three-sample pulse is also removed.

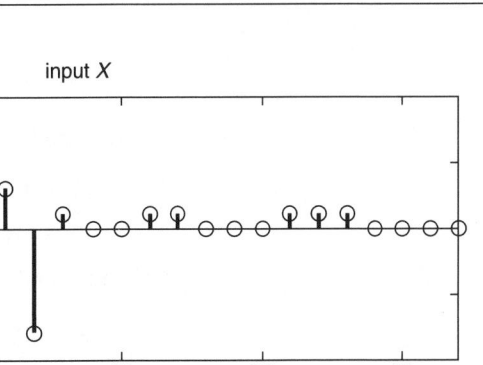

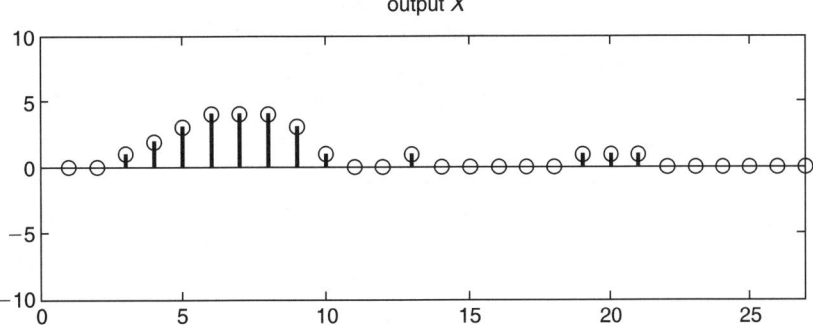

Figure A1.10 *Median filter, length* N = 5

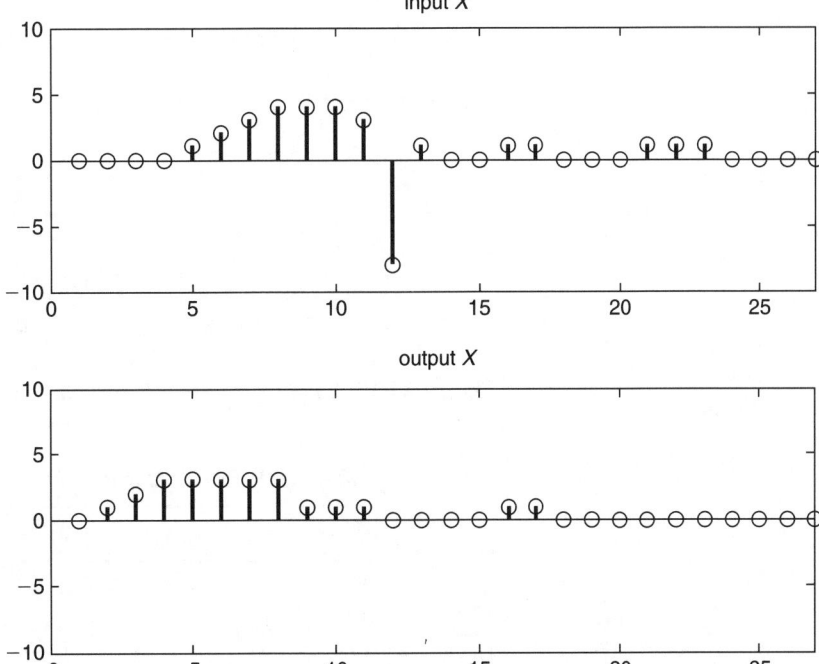

Figure A1.11 *Median filter, length* N = 7

P4-2

Equation (4.17) gives the required preprocessing of the input signals

$$y_1 = x_1 + x_2$$
$$y_2 = (x_1 + x_2)^2$$

and the node parameters are given by

$$w_{13} = 2 \quad w_{23} = -1 \quad \varphi_3 = -0.5$$

where we assume that the node number is 3. Figure A1.12 shows the preprocessing and the feedforward node, which is assumed to use a hard limiter activation function. The extended truth table is shown in Table A1.1.

P4-3

The sigmoid function (4.11) is

$$f(x) = \frac{1}{1 + e^{-x/T}}$$

If $x < 0 \Rightarrow -x/T \to \infty$ when $T \to 0 \Rightarrow e^{\infty} \to \infty \Rightarrow f(x) \to 0$ but, on the other hand, if $x > 0 \Rightarrow -x/T \to -\infty$ when $T \to 0 \Rightarrow e^{-\infty} \to 0 \Rightarrow f(x) \to 1$ hence, this compares to $a = 1$ and $b = 0$ in the hard limiter equation (4.9).

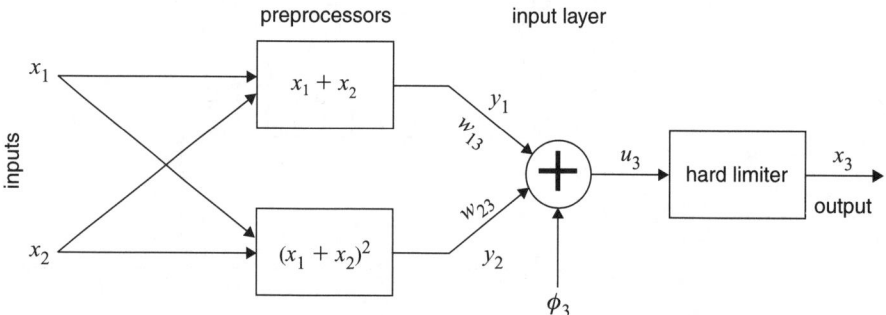

Figure A1.12 *Single feedforward node and preprocessors to solve the XOR problem*

Table A1.1 *Extended truth table for XOR function*

x_1	x_2	y_1	y_2	u_3	x_3
0	0	0	0	−0.5	0
0	1	1	1	0.5	1
1	0	1	1	0.5	1
1	1	2	4	−0.5	0

Chapter 5   P5-1

See Figure A1.13.

P5-2

Starting out from equations (5.80a) and (5.80b), $x(t) = a(t)\cos(\phi(t))$ and $y(t) = a(t)\sin(\phi(t))$. From equation (5.85a)

$$\sqrt{x^2(t) + y^2(t)} = \sqrt{a^2(t)\cos^2(\varphi(t)) + a^2(t)\sin^2(\varphi(t))}$$

$$= a(t)\sqrt{\cos^2(\varphi(t)) + \sin^2(\varphi(t))} = a(t)$$

where we have used the identity $\cos^2(\alpha) + \sin^2(\alpha) \equiv 1$, further, having equation (5.85b) in mind

$$\frac{y(t)}{x(t)} = \frac{a(t)\sin(\varphi(t))}{a(t)\cos(\varphi(t))} = \tan(\varphi(t)) \Rightarrow \varphi(t) = \arctan\left(\frac{y(t)}{x(t)}\right)$$

P5-3

Starting from equation (5.72)

$$R(s_0, s_1') = \int_0^T s_0(t)s_1'(t)\,\mathrm{d}t = a\int_0^T \cos(2\pi f_0 t)\cos(2\pi f_1 t)\,\mathrm{d}t$$

$$= \frac{a}{2}\int_0^T (\cos(2\pi(f_1 - f_0)t) + \cos(2\pi(f_1 + f_0)t))\,\mathrm{d}t$$

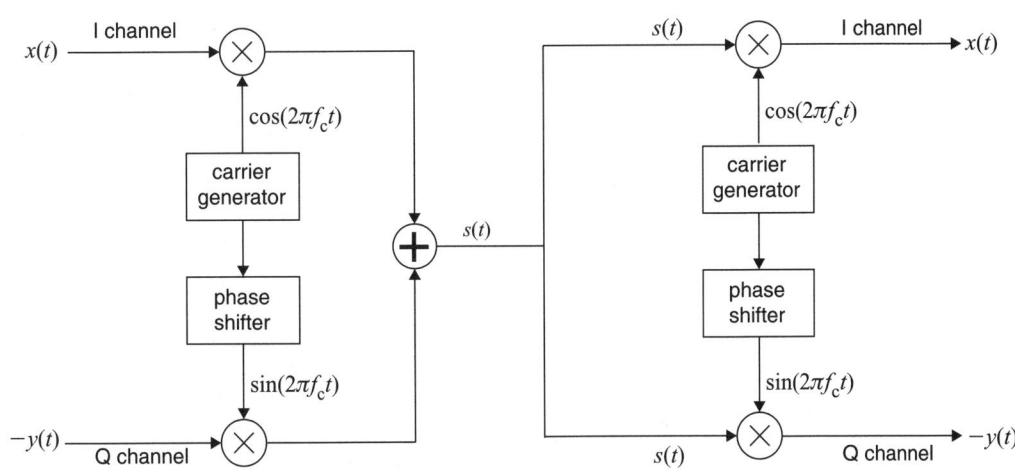

Figure A1.13 *I/Q-modulator and demodulator*

$$= \frac{a}{2} \left[\frac{\sin{(2\pi(f_1 - f_0)t)}}{2\pi(f_1 - f_0)} + \frac{\sin{(2\pi(f_1 + f_0)t)}}{2\pi(f_1 + f_0)} \right]_0^T \tag{A1.5.1}$$

$$= \frac{a}{4\pi} \left(\frac{\sin{(2\pi(f_1 - f_0)T)}}{(f_1 - f_0)} + \frac{\sin{(2\pi(f_1 + f_0)T)}}{(f_1 + f_0)} \right)$$

$$= \frac{a}{4\pi}(F_A(f_1, f_0) + F_B(f_1, f_0)) = 0$$

Setting $\Delta f = |f_1 - f_0|$ in equation (A1.5.1) and realizing that since the denominator $f_1 + f_0 \gg f_1 - f_0$, the second term $F_B(f_1, f_0) \ll F_A(f_1, f_0)$; hence, for high frequencies (compared to the "modulation frequency" $1/T$) $F_B(f_1, f_0)$ can be neglected and equation (A1.5.1) can be approximated as

$$R(s_0, s_1') = \frac{a}{4\pi}F_A(f_1, f_0) = \frac{a}{4\pi}\frac{\sin{(2\pi\Delta f T)}}{\Delta f} = 0$$

$$\Rightarrow 2\pi\Delta f T = k\pi \Rightarrow \Delta f T = k \Rightarrow \Delta f = \frac{k}{2T}$$

where $k = 1, 2, 3, \dots$ and equation (5.73) is derived. (Commonly, $k = 1$ is chosen.)

P5-4

A suggested MATLAB™ program is shown below:

```
% pr54.m

N=1024;                       % 1024 samples
n=1:N;

omi=0.01;                     % intelligence frequency
omc=2;                        % carrier frequency

s=(1+cos(omi*n)).*cos(omc*n); % AM modulation

w=(hamming(N)');              % Hamming window vector

figure(1)                     % plot FFT with rectangular window
X=fft(s,N);                   % FFT
M=abs(X);                     % get magnitude
plot(M(300:350))              % plot interesting part of spectrum

figure(2)                     % plot FFT with Hamming window
x=w*s;                        % windowing operation
X=fft(x,N);                   % FFT
M=abs(X);                     % get magnitude
plot(M(300:350))              % plot interesting part of spectrum
```

Figure A1.14 shows the spectrum using a rectangular windowing function (i.e. no windowing, just a sample vector of limited length). All three signal

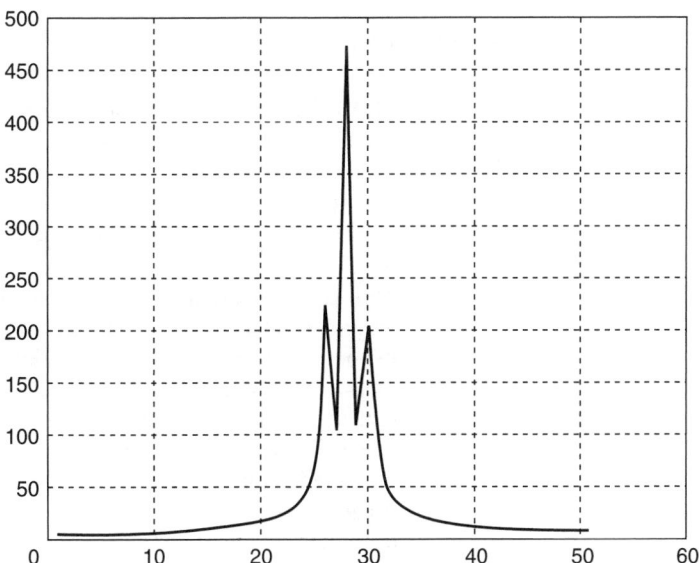

Figure A1.14 *FFT using rectangular windowing function, all frequency components visible*

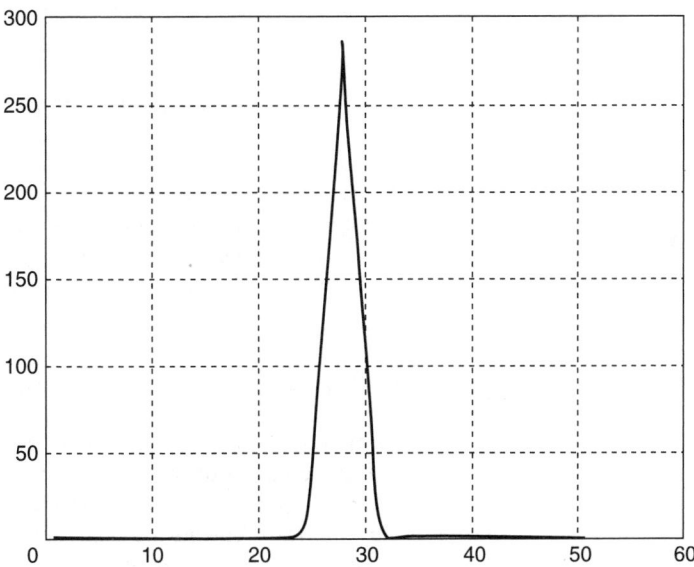

Figure A1.15 *FFT using Hamming windowing function, frequency resolution too coarse*

components are visible. Figure A1.15 shows the spectrum obtained using Hamming windowing of the sampled sequence. Due to the wider lobes of the Hamming spectrum, the weak frequency components are "drowned" in the main lobe from the strong component. We cannot see that there are actually three frequencies involved.

Chapter 6   P6-1

Measurement-update equation (6.48) updates the estimate of the state vector variable using the latest incoming measured data

$$\hat{\mathbf{x}}(n\,|\,n) = \hat{\mathbf{x}}(n\,|\,n-1) + \mathbf{C}(n\,|\,n-1)\,\mathbf{H}(n)(\mathbf{H}^{\mathrm{T}}(n)\,\mathbf{C}(n\,|\,n-1)\,\mathbf{H}(n)$$
$$+\,\mathbf{R}(n))^{-1}(\mathbf{z}(n) - \mathbf{H}^{\mathrm{T}}(n)\,\hat{\mathbf{x}}(n\,|\,n-1)) \tag{6.48}$$

This equation calculates the estimate of the state vector $\hat{\mathbf{x}}$ at time n based on all available measurements at time n. The equation works recursively, i.e. the new estimate is obtained by taking the previous one $\hat{\mathbf{x}}(n\,|\,n-1)$ based on measurements up to time $n-1$, i.e. all measurements except the last one just received. The actual measured data is $\mathbf{z}(n)$ from which the expected measured value is subtracted. The observation matrix $\mathbf{H}(n)$ governs the coupling between measured values and the state vector variable. The result from the subtraction is called the innovation: $\mathbf{z}(n) - \mathbf{H}^{\mathrm{T}}(n)\,\hat{\mathbf{x}}(n\,|\,n-1)$. The innovation is multiplied by a "gain constant" $\mathbf{C}(n\,|\,n-1)\,\mathbf{H}(n)(\mathbf{H}^{\mathrm{T}}(n)\,\mathbf{C}(n\,|\,n-1)\,\mathbf{H}(n) + \mathbf{R}(n))^{-1}$ and used to update the estimate. The "gain constant" consists of the observation matrix, the measurement noise vector $\mathbf{R}(n)$ (model of the noise inherent in the measurement process) and the error covariance matrix $\mathbf{C}(n\,|\,n-1)$. The error covariance matrix is a quality measure of the estimate.

The error covariance matrix is also updated recursively, using equation (6.49).

$$\mathbf{C}(n\,|\,n) = \mathbf{C}(n\,|\,n-1) - \mathbf{C}(n\,|\,n-1)\,\mathbf{H}(n)(\mathbf{H}^{\mathrm{T}}(n)\,\mathbf{C}(n\,|\,n-1)\,\mathbf{H}(n)$$
$$+\,\mathbf{R}(n))^{-1}\mathbf{H}^{\mathrm{T}}(n)\,\mathbf{C}(n\,|\,n-1) \tag{6.49}$$

Note! The measured value itself is not used in this equation. Hence, if the measurement noise is predetermined, the error covariance matrix can be calculated for all instants of time in advance, and stored as a table to speed up the execution of time critical software.

There are two time-update equations, such as equations (6.50) and (6.51).

$$\hat{\mathbf{x}}(n+1\,|\,n) = \mathbf{F}(n)\,\hat{\mathbf{x}}(n\,|\,n) \tag{6.50}$$

$$\mathbf{C}(n+1\,|\,n) = \mathbf{F}(n)\,\mathbf{C}(n\,|\,n)\,\mathbf{F}^{\mathrm{T}}(n) + \mathbf{G}(n)\,\mathbf{Q}(n)\,\mathbf{G}^{\mathrm{T}}(n) \tag{6.51}$$

In equation (6.50) the estimate is updated recursively one step at a time, using the model of the process (of which we are estimating the state vector variable). The model is represented by the state transition matrix $\mathbf{F}(n)$. So far, no measurements are taken into account in that we are only updating by "dead reckoning" using the process model. Equation (6.51) updates the error covariance matrix using the model and taking the process noise into account. $\mathbf{Q}(n)$ is the model of the process noise, while the matrix $\mathbf{G}(n)$ makes the coupling between the process noise and the state vector variable.

Case 1: excellent measurement signal quality, $\mathbf{R}(n) = 0$

Equation (6.48) turns into

$$\hat{\mathbf{x}}(n\,|\,n) = \hat{\mathbf{x}}(n\,|\,n-1) + \mathbf{C}(n\,|\,n-1)\,\mathbf{H}(n)(\mathbf{H}^{\mathrm{T}}(n)\,\mathbf{C}(n\,|\,n-1)$$
$$\mathbf{H}(n))^{-1}(\mathbf{z}(n) - \mathbf{H}^{\mathrm{T}}(n)\,\hat{\mathbf{x}}(n\,|\,n-1))$$

$$= \hat{\mathbf{x}}(n \,|\, n-1) + \mathbf{C}(n \,|\, n-1)\,\mathbf{H}(n)(\mathbf{H}(n))^{-1}(\mathbf{C}(n \,|\, n-1))^{-1}$$
$$(\mathbf{H}^{\mathrm{T}}(n))^{-1}(\mathbf{z}(n) - \mathbf{H}^{\mathrm{T}}(n)\,\hat{\mathbf{x}}(n \,|\, n-1))$$
$$= \hat{\mathbf{x}}(n \,|\, n-1) + (\mathbf{H}^{\mathrm{T}}(n))^{-1}(\mathbf{z}(n) - \mathbf{H}^{\mathrm{T}}(n)\,\hat{\mathbf{x}}(n \,|\, n-1))$$
$$= (\mathbf{H}^{\mathrm{T}}(n))^{-1}\mathbf{z}(n)$$

From this, we see that the estimate is calculated directly from the measurement signal.

Equation (6.49) turns into

$$\mathbf{C}(n \,|\, n) = \mathbf{C}(n \,|\, n-1) - \mathbf{C}(n \,|\, n-1)\,\mathbf{H}(n)(\mathbf{H}^{\mathrm{T}}(n)\,\mathbf{C}(n \,|\, n-1)$$
$$\mathbf{H}(n))^{-1}\mathbf{H}^{\mathrm{T}}(n)\mathbf{C}(n \,|\, n-1)$$
$$= \mathbf{C}(n \,|\, n-1) - \mathbf{C}(n \,|\, n-1)\,\mathbf{H}(n)(\mathbf{H}(n))^{-1}(\mathbf{C}(n \,|\, n-1))^{-1}$$
$$(\mathbf{H}^{\mathrm{T}}(n))^{-1}\mathbf{H}^{\mathrm{T}}(n)\mathbf{C}(n \,|\, n-1)$$
$$= \mathbf{C}(n \,|\, n-1) - \mathbf{C}(n \,|\, n-1) = 0$$

which is not surprising, since our measurements are noise free and perfect, with no estimation error.

Equation (6.50) is not affected, since it does not deal with $\mathbf{R}(n)$. This is also true for equation (6.51). The internal process (signal) model of the filter is updated.

So, in short, the output of the filter will rely solely on the input signal and the internal model in the filter is updated.

Case 2: extremely bad measurement signal quality, $\mathbf{R}(n) \to \infty$

Equation (6.48) turns into

$$\hat{\mathbf{x}}(n \,|\, n) = \hat{\mathbf{x}}(n \,|\, n-1)$$

That is, as we do not pay any attention at all to the incoming measurements (since the quality is so poor), the estimate is not updated.

Equation (6.49) turns into

$$\mathbf{C}(n \,|\, n) = \mathbf{C}(n \,|\, n-1)$$

which means that the error covariance is not changed, i.e. the quality of the estimate is not increased, since no useful measured data is available.

Equation (6.50) updates the estimate according to the internal signal model, and equation (6.51) updates the error covariance matrix in a way predicted by the process model.

So, in short, the output of the filter will rely solely on the internal signal model and the incoming measured data will be ignored.

P6-2

Inserting equation (6.48) into equation (6.50) we get

$$\hat{\mathbf{x}}(n+1 \,|\, n) = \mathbf{F}(n)\,\hat{\mathbf{x}}(n \,|\, n-1) + \mathbf{F}(n)\,\mathbf{C}(n \,|\, n-1)\,\mathbf{H}(n)(\mathbf{H}^{\mathrm{T}}(n)$$
$$\mathbf{C}(n \,|\, n-1)\,\mathbf{H}(n) + \mathbf{R}(n))^{-1}(\mathbf{z}(n) - \mathbf{H}^{\mathrm{T}}(n)\,\hat{\mathbf{x}}(n \,|\, n-1))$$

$$(\mathrm{A.1.6.1})$$

By defining $\mathbf{K}(n) = \mathbf{F}(n)\,\mathbf{C}(n\,|\,n-1)\,\mathbf{H}(n)(\mathbf{H}^{\mathrm{T}}(n)\,\mathbf{C}(n\,|\,n-1)\,\mathbf{H}(n) + \mathbf{R}(n))^{-1}$, equation (A1.6.1) can be rewritten as

$$\hat{\mathbf{x}}(n+1\,|\,n) = \mathbf{F}(n)\,\hat{\mathbf{x}}(n\,|\,n-1) + \mathbf{K}(n)(\mathbf{z}(n) - \mathbf{H}^{\mathrm{T}}(n)\,\hat{\mathbf{x}}(n\,|\,n-1))$$

$$(A.1.6.2)$$

Equations (6.52) and (6.53) follow.

In a similar way, equation (6.49) is inserted into equation (6.51) to obtain the Riccatti equation (6.54)

$$\begin{aligned}
\mathbf{C}(n+1\,|\,n) = {} & \mathbf{F}(n)(\mathbf{C}(n\,|\,n-1) - \mathbf{C}(n\,|\,n-1)\mathbf{H}(n)(\mathbf{H}^{\mathrm{T}}(n)\,\mathbf{C}(n\,|\,n-1) \\
& \mathbf{H}(n) + \mathbf{R}(n))^{-1}\mathbf{H}^{\mathrm{T}}(n)\,\mathbf{C}(n\,|\,n-1))\mathbf{F}^{\mathrm{T}}(n) \\
& + \mathbf{G}(n)\,\mathbf{Q}(n)\,\mathbf{G}^{\mathrm{T}}(n)
\end{aligned}$$

Case 1: excellent measurement signal quality, $\mathbf{R}(n) = 0$

The Kalman gain equation (6.53) will turn into

$$\mathbf{K}(n) = \mathbf{F}(n)\,\mathbf{C}(n\,|\,n-1)\,\mathbf{H}(n)(\mathbf{H}^{\mathrm{T}}(n)\,\mathbf{C}(n\,|\,n-1)\,\mathbf{H}(n))^{-1}$$

and the output will be equation (6.52)

$$\begin{aligned}
\hat{\mathbf{x}}(n+1\,|\,n) = {} & \mathbf{F}(n)\,\hat{\mathbf{x}}(n\,|\,n-1) \\
= {} & \mathbf{F}(n)\,\hat{\mathbf{x}}(n\,|\,n-1) + \mathbf{F}(n)\,\mathbf{C}(n\,|\,n-1)\,\mathbf{H}(n)(\mathbf{H}^{\mathrm{T}}(n)\,\mathbf{C}(n\,|\,n-1) \\
& \mathbf{H}(n))^{-1}(\mathbf{z}(n) - \mathbf{H}^{\mathrm{T}}(n)\,\hat{\mathbf{x}}(n\,|\,n-1)) \\
= {} & \mathbf{F}(n)\,\hat{\mathbf{x}}(n\,|\,n-1) + \mathbf{F}(n)\,\mathbf{C}(n\,|\,n-1)\,\mathbf{H}(n)(\mathbf{H}(n))^{-1} \\
& (\mathbf{C}(n\,|\,n-1))^{-1}(\mathbf{H}^{\mathrm{T}}(n))^{-1}(\mathbf{z}(n) - \mathbf{H}^{\mathrm{T}}(n)\,\hat{\mathbf{x}}(n\,|\,n-1)) \\
= {} & \mathbf{F}(n)\,\hat{\mathbf{x}}(n\,|\,n-1) + \mathbf{F}(n)(\mathbf{H}^{\mathrm{T}}(n))^{-1}(\mathbf{z}(n) - \mathbf{H}^{\mathrm{T}}(n)\,\hat{\mathbf{x}}(n\,|\,n-1)) \\
= {} & \mathbf{F}(n)(\mathbf{H}^{\mathrm{T}}(n))^{-1}\mathbf{z}(n)
\end{aligned}$$

The Riccatti equation (6.54) will be

$$\begin{aligned}
& \mathbf{C}(n+1\,|\,n) \\
& = \mathbf{F}(n)(\mathbf{C}(n\,|\,n-1) - \mathbf{C}(n\,|\,n-1)\,\mathbf{H}(n)(\mathbf{H}^{\mathrm{T}}(n)\,\mathbf{C}(n\,|\,n-1) \\
& \quad \mathbf{H}(n))^{-1}\mathbf{H}^{\mathrm{T}}(n)\,\mathbf{C}(n\,|\,n-1))\mathbf{F}^{\mathrm{T}}(n) + \mathbf{G}(n)\,\mathbf{Q}(n)\,\mathbf{G}^{\mathrm{T}}(n) \\
& = \mathbf{F}(n)(\mathbf{C}(n\,|\,n-1) - \mathbf{C}(n\,|\,n-1)\,\mathbf{H}(n)(\mathbf{H}(n))^{-1}(\mathbf{C}(n\,|\,n-1))^{-1} \\
& \quad (\mathbf{H}^{\mathrm{T}}(n))^{-1}\mathbf{H}^{\mathrm{T}}(n)\,\mathbf{C}(n\,|\,n-1))\mathbf{F}^{\mathrm{T}}(n) + \mathbf{G}(n)\,\mathbf{Q}(n)\,\mathbf{G}^{\mathrm{T}}(n) \\
& = \mathbf{G}(n)\,\mathbf{Q}(n)\,\mathbf{G}^{\mathrm{T}}(n)
\end{aligned}$$

In this case, as in P6-1, the output of the filter will be based completely on the measured input signal. The error covariance will be equal to the impact of the process noise on the "reality", making it deviate from the expected "perfect" model.

Case 2: extremely bad measurement signal quality, $\mathbf{R}(n) \to \infty$

The Kalman gain equation (6.53) will turn into

$$\mathbf{K}(n) = 0$$

i.e. the innovation is completely disconnected from the filter (see Figure 6.3). The output equation (6.52) will be

$$\hat{\mathbf{x}}(n+1 \mid n) = \mathbf{F}(n)\,\hat{\mathbf{x}}(n \mid n-1)$$

i.e. the output signal is only based on the internal process model (compare to P6-1). Finally, the Riccatti equation (6.54)

$$\mathbf{C}(n+1 \mid n) = \mathbf{F}(n)(\mathbf{C}(n \mid n-1)\,\mathbf{F}^{\mathrm{T}}(n) + \mathbf{G}(n)\,\mathbf{Q}(n)\,\mathbf{G}^{\mathrm{T}}(n)$$

The error covariance will be updated according to what could be expected taking the process model into account, and added to this is the uncertainty of the process due to the process noise.

In this case, as in P6-1, the output of the filter will be the based completely on the internal model.

P6-3

The program may look something like the below:

```
% pr63.m

% signal model
F=[1 1; 0 1];               % transition matrix
G=[0; 1];                   % drive vector
H=[1; 0];                   % observation vector
w=0.15;                     % amplitude process noise
v=10;                       % amplitude measurement noise
x=[0; 1];                   % state vector, reality

% simulation params
N=100;                      % sim length
x1=zeros(N);                % plot vector
x2=zeros(N);                % plot vector
z=zeros(N);                 % measurement vector

% reality
for k=1:N
  x=F*x+G*w*randn;          % update state vector
      z(k)=H'*x+v*randn;    % measurement
  x1(k)=x(1);
  x2(k)=x(2);
end

%plot
clf reset;                  % clear graph
subplot(3,1,1);             % plot velocity
plot(x2);
axis([1 N 0 10]);
title('velocity')
subplot(3,1,2);             % plot position
plot(x1);
axis([1 N 0 200]);
title('position')
```

```
subplot(3,1,3);                  % plot measurement
plot(z);
axis([1 N 0 200]);
title('measured')
```

The plots should look the same as in Figure 6.4. The output of the program is the vector **z**, containing all the measured data.

P6-4

The program may look something like the below. **Note!** The program used in P6-3 above must be run first in order to obtain the measurement data in vector **z**:

```
% pr64.m

% signal model
F=[1 1; 0 1];                    % transition matrix
G=[0; 1];                        % drive vector
H=[1; 0];                        % observation vector

% filter
Q=0.2;                           % variance, process noise
R=4;                             % variance, measurement noise
xhat=[0; 0];                     % estimated state vector
C=[1 1; 1 1];                    % covariance matrix
Kalm=[1; 1];                     % Kalman gain matrix

% simulation params
N=100;                           % sim length
x1=zeros(N);                     % plot vector
x2=zeros(N);                     % plot vector


% filter
for k=1:N
                                 % Riccati equation
  C=F*(C-C*H*inv(H'*C*H+R)*H'*C)*F+G*Q*G';
                                 % Kalman gain
  Kalm=F*C*H*inv(H'*C*H+R);
                                 % estimation
  xhat=F*xhat+Kalm*(z(k)-H'*xhat);

  x1(k)=xhat(1);
  x2(k)=xhat(2);

end

figure;
subplot(3,1,1);                  % plot estimated velocity
plot(x2);
axis([1 N 0 10]);
title('estimated velocity')
subplot(3,1,2);                  % plot estimated position
```

```
plot(x1);
axis([1 N 0 200]);
title('estimated position')
```

The plots should look the same as in Figure 6.5.

Chapter 7 P7-1

Decompressing the string `0t0h0e0_1h3m3_5e0o0r6_8s3s` (start with an empty dictionary) yields

```
    Dictionary                  Decompressed string

0       nothing                 the_theme_theorem_theses
1       t
2       h
3       e
4       _
5       th
6       em
7       e_
8       the
9       o
A       r
B       em_
C       thes
D       es
E
F
```

P7-2

There are alternative solutions; one is presented in Figure A1.16. The Huffman code thus obtained is

```
1       1
2       011
3       010
4       00011
5       001
6       00010
7       00001
8       00000
```

P7-3

The entropy is calculated by $H = -\sum_{i=1}^{8} p_i \, \mathrm{lb}(p_i) = 2.1658$ bits/symbol. The average symbol length is obtained by $L = \sum_{i=1}^{8} p_i l_i = 2.1840$ bits/symbol. The coding efficiency is $\eta = H/L = 2.1658/2.1840 = 0.992$ and the redundancy, $R = 1 - (H/L) = 1 - \eta = 1 - (2.1658/2.1840) = 0.008$ bits/symbol.

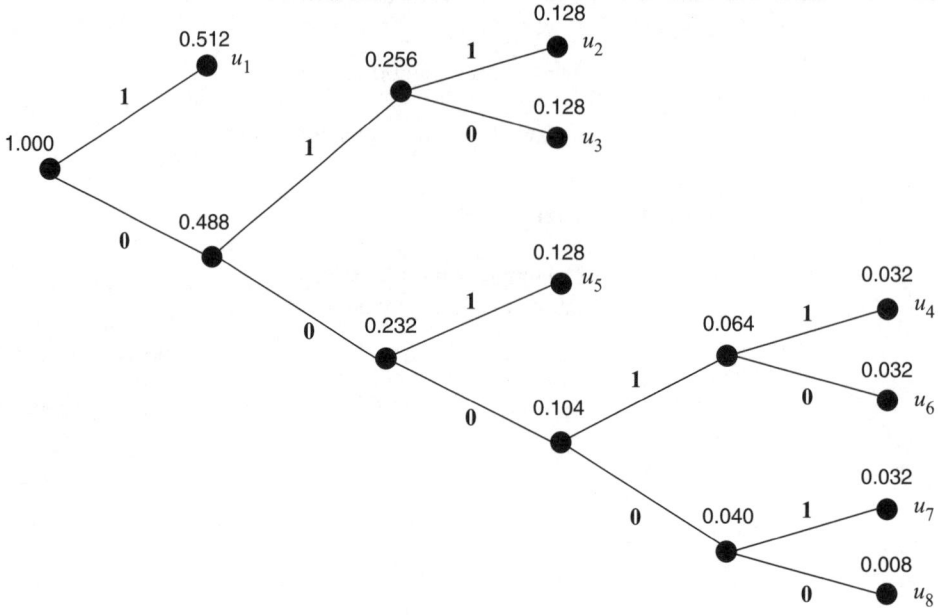

Figure A1.16 *Binary tree for Huffman code (there are other possible solutions)*

P7-4

The program may look something like the below:

```
% pr74.m

% simulate delta modulator

n=1:500;                          % sample number

omega=pi/100;                     % angular frequency
A=1;                              % amplitude

% source
x=A*cos(omega*n);                 % signal

% encoder
delta=.08;
s=zeros(1,length(x)+1);           % output
xhat=zeros(1,length(x)+1);        % predicted x

xhat(1)=0;                        % initialize
s(1)=0;
for i=2:length(x)
  xhat(i)=xhat(i-1)+s(i-1);       % prediction
  d=x(i)-xhat(i);                 % quantization
  if d>=0
    s(i)=delta;
  else
```

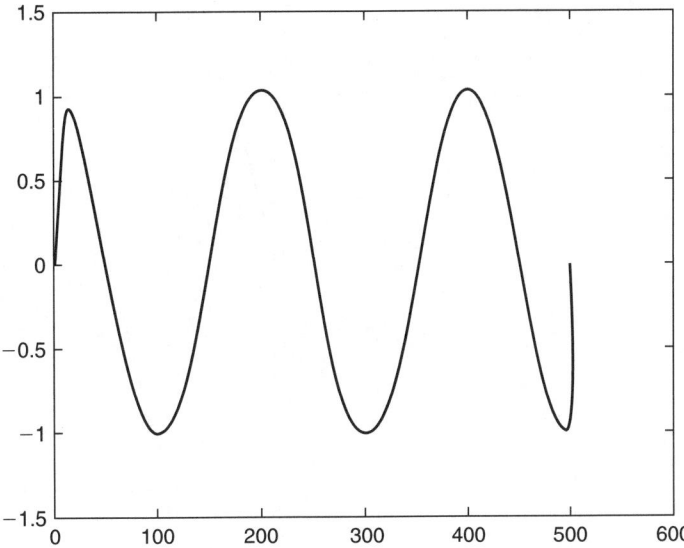

Figure A1.17 *Output of properly working delta modulator*

```
        s(i)=-delta;
    end
end

% decoder
y=zeros(1,length(x)+1);          % decoded output y
y(1)=0;                          % initialize
for i=2:length(x)
  y(i)=y(i-1)+s(i-1);
end
plot(y)                          % plot output
```

Running the system with initial settings as in the code example above, everything works fine. The output is shown in Figure A1.17. If the amplitude is increased to $A = 5$, **slope overload** occurs (see Figure A1.18). The problem is that the stepwise approximation produced by the delta modulator cannot keep up with the slope of the input signal. The same problem occurs if the frequency of the input signal increases. The remedy is to use a larger step size. In Figure A1.19, **granularity error** is shown. In this case, the amplitude of the signal is small, $A = 0.1$, and the too-large step size causes excessive ripple.

Chapter 8 P8-1

(a) The code rate $R = k/n = 1/5 = 0.2$
(b) The number of parity bits $n - k = 5 - 1 = 4$
(c) The maximum Hamming distance, since $k = 1$ there are only two code words 00000 and 11111, hence only one Hamming distance, $d = 5$
(d) Error-**correction** capacity $t = \lfloor (d - 1)/2 \rfloor = \lfloor 2 \rfloor = 2$ bits/code word
(e) Error-**detection** capacity $y = d - 1 = 5 - 1 = 4$ bits/code word

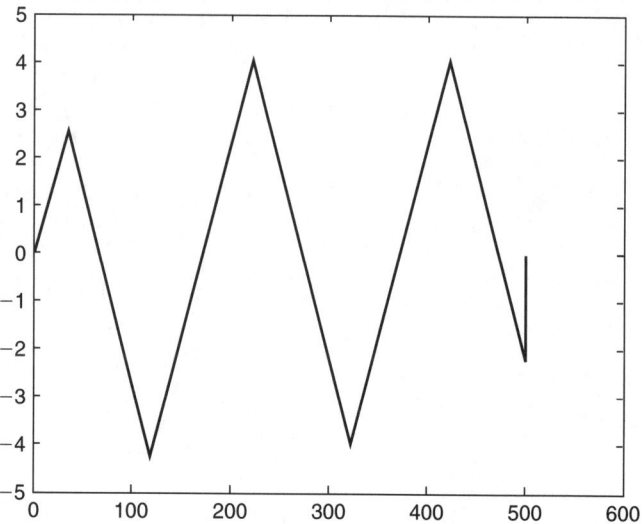

Figure A1.18 *Slope overload, step too small, amplitude and/or signal frequency too high and the delta modulator cannot keep up with the sinus signal*

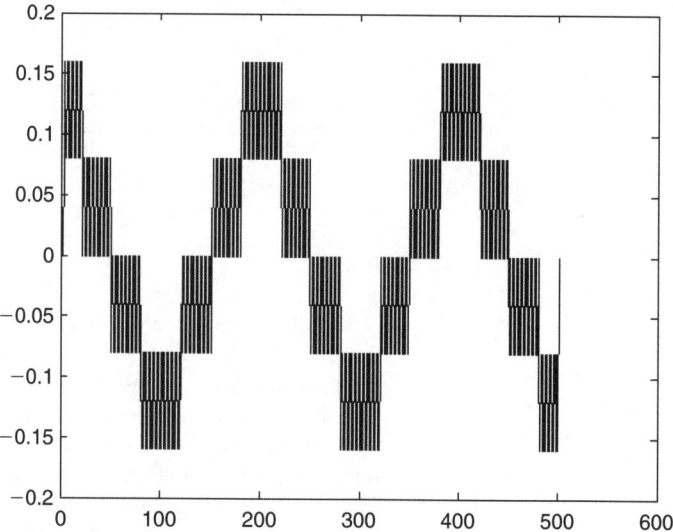

Figure A1.19 *Granularity error, step size too large or signal amplitude too small*

(f) The symbol error probability, i.e. the probability of transmission error when using error correction by majority vote decoding

$$P_{\text{err}} = \sum_{i=t+1}^{n} \binom{n}{i} p^i (1-p)^{(n-i)}$$

$$= \binom{5}{3} p^3 (1-p)^2 + \binom{5}{4} p^4 (1-p) + \binom{5}{5} p^5$$

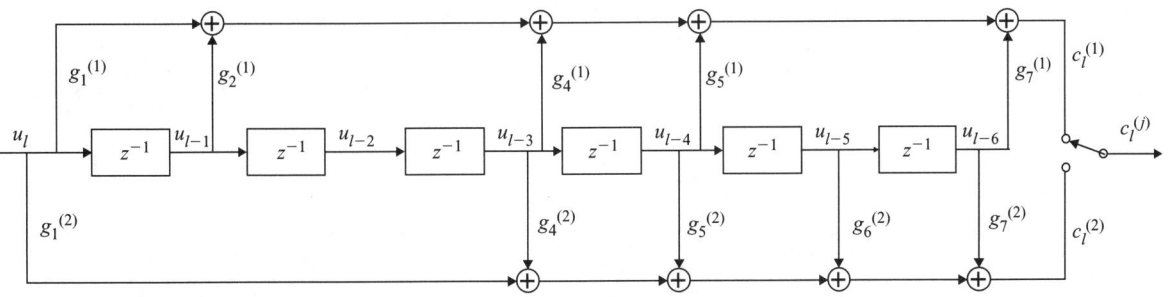

Figure A1.20 *Block diagram, convolution encoder for Jet Propulsion Lab (2,1,6)*

$$= 10 \cdot 0.05^3 \cdot 0.95^2 + 5 \cdot 0.05^4 \cdot 0.95 + 1 \cdot 0.05^5$$
$$= 1.16 \cdot 10^{-3}$$

P8-2

The notation for convolution codes is $(n, k, m) = (2, 1, 6)$ where $n = 2$ is the number of generators, $k = 1$ the number of data bits and $m = 6$ the number of delay elements in the encoder.

(a) Since there are $m = 6$ delay elements the number of states is $2^6 = 64$ states
(b) The length of the tail (on the output) is equal to the number of delay elements multiplied by the number of generators, i.e. $m \cdot n = 6 \cdot 2 = 12$ bits (corresponding to 6 bits on the input)
(c) The constraint length is given by $n_s = (m + 1)n = (6 + 1) \cdot 2 = 14$ bits
(d) The code rate is given by $R = L/((L + m)n) = 8/((8 + 6) \cdot 2) = 0.286$ where $L = 8$ bits, since we are transmitting a byte
(e) See Figure A1.20
(f) Using a diagram of the encoder as in (e), the output sequence can be determined given the input sequence 10110101. The generators are (see Section 8.2.4)

$$g^{(1)} = (1, 1, 0, 1, 1, 0, 1) \quad g^{(2)} = (1, 0, 0, 1, 1, 1, 1)$$

The output sequence will be (including tail):

```
111011100101010111001000011 1
```

P8-3

A suggested MATLAB™ program is found below:

```
% pr83

p=0:0.01:1;                          % error probability vector

                                     % calculate capacity
c=1+(p.*log(p)+(1-p).*log(1-p))/log(2);
plot(p,c)                            % plot
```

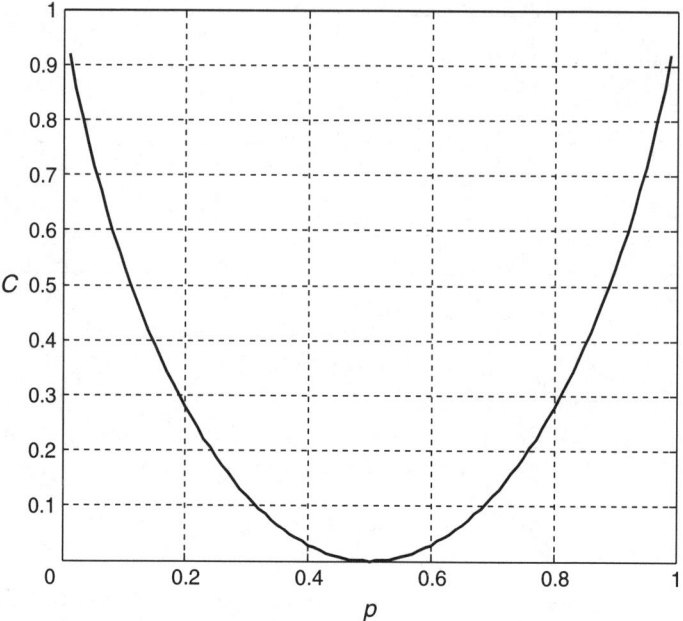

Figure A1.21 *Channel capacity of the BSC as a function of bit error probability*

The resulting plot is shown in Figure A1.21. It is interesting to note that if the error probability is 0, the channel, of course, works fine. But if the error probability is 1, i.e. all bits are wrong (inverted), it works as well. The worst case is for the error probability to be equal to 0.5 when no information passes through the channel. One could continue guessing.

P8-4

We have $C = W\text{lb}(1 + X/W)$ since $N_0 = 1$, the MATLAB™ program is found below:

```
% pr84

x=logspace(-18,-9);              % power
w=logspace(3,6);                 % bandwidth

N0=4e-21;                        % noise density

[X,W] = MESHGRID(x,w);           % prepare input values
z=W*log(1+X./(W*N0))/log(2);     % calculate capacity
surfl(x,w,z);                    % plot
colormap(gray);
```

The plot is shown in Figure A1.22.

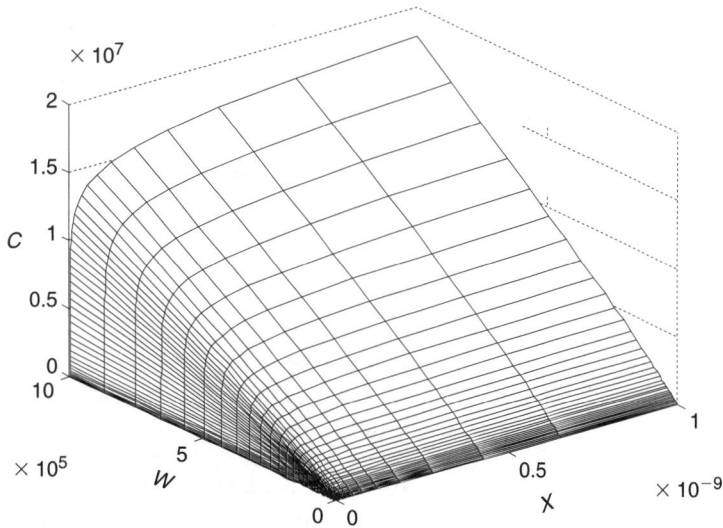

Figure A1.22 *Channel capacity of the Gaussian channel, capacity (C) versus bandwidth (W) and signal power (X)*

Chapter 9   P9-1

From the transfer function $H(z) = \sum_{k=0}^{N-1} b_k z^{-k}$ with $b_k = 1/N$ the difference equation can easily be obtained

$$H(z) = \frac{Y(z)}{X(z)} = \sum_{k=0}^{N-1} b_k z^{-k} = \frac{1}{N}\left(1 + z^{-1} + z^{-2} + \cdots + z^{N-1}\right)$$

$$\Rightarrow Y(z) = \frac{1}{N}\left(X(z) + X(z)z^{-1} + X(z)z^{-2} + \cdots + X(z)z^{N-1}\right)$$

$$\Rightarrow y(n) = \frac{1}{N}(x(n) + x(n-1) + x(n-2) + \cdots + x(n-N+1))$$

$$\text{(A1.9.1)}$$

Assume the worst case, i.e. every incoming sample $x(n)$ has maximum positive amplitude. Considering that the input bus width is 16 bits, the maximum positive amplitude would be $2^{16} - 1$, i.e. all ones. We have to add all the N samples in the accumulator before dividing the sum by N. Doing it the other way, i.e. dividing the samples before adding in the accumulator is not a good idea. Small sample values may be truncated (i.e. approximated to zero, due to the limited resolution), implying that a series of small sample values may result in an erroneous output signal.

Since the accumulator consisting of 24 bits has a maximum value of $2^{24} - 1$, we realize that

$$N(2^{16} - 1) \leq (2^{24} - 1) \Rightarrow N \leq \frac{2^{24} - 1}{2^{16} - 1} \approx \frac{2^{24}}{2^{16}} = 2^8 = 256$$

Now, the analysis above is very **pessimistic**. Firstly, it is probably not very likely that all incoming samples will have maximum positive amplitude. Secondly, if the DSP in use has wraparound arithmetic, and we know that the output average values should be within the given range, overflow in the accumulator does not matter.

Applying scaling to the input samples may be risky for the truncation reasons given above.

P9-2

If we assume that the filter length in P9-1 is $N = 256$ and that $b_k = 1/N$ it is obvious that using the MAC instruction is overkill, since we do not need to multiply by one. If there is a simpler type of instruction, like "add-and-accumulate" in the instruction repertoire of the DSP, it should preferably be used. Anyway, since the sampling frequency is 200 kHz, the sampling time is $t_s = 5\,\mu s$ which is also the maximum loop time since $t_p \leq t_s$. The loop typically contains instructions for: reading the input, adding and accumulating, increasing (or decreasing) and testing a loop counter and branching. Doing this in $5\,\mu s$ should not be a problem using a DSP chip. It would probably also be possible to use a fast standard microprocessor chip.

No data memory, except for processor registers, is needed if the filter is implemented for **batch mode** operation as outlined above. That is, 256 input samples are collected and the average is calculated. If, however, a **running average** is required, we have another situation. In such a case an output value should be delivered at **every** sampling instant, based on the past 256 samples. This means that we need to store the past 256 samples in a delay-line type list in the memory, hence requiring $256 \cdot 2 = 512$ bytes of memory. This is no big problem, but the allowed execution time is certainly a problem. The allowed processing time for the loop is now approximately $t_p = (5 \cdot 10^{-6})/256 = 19\,ns$ which is definitely not possible, even if using a top speed DSP of today. This would probably require a parallel hardware implementation, for instance using fast ASICs or FPGAs.

In this particular case, however, a trick can be used. Rewriting the original difference equation (A1.9.1) into a recursive form (see Chapter 1, Section 1.3.12) yields

$$y(n) - y(n-1) = \frac{1}{N}(x(n) + x(n-1) + x(n-2) + \cdots + x(n-N+1))$$

$$- \frac{1}{N}(x(n-1) + x(n-2) + x(n-3) + \cdots + x(n-N))$$

$$= \frac{1}{N}(x(n) - x(n-N))$$

which can be rewritten as (see also equation (1.49))

$$y(n) = y(n-1) + \frac{1}{N}(x(n) - x(n-N)) \qquad (A1.9.2)$$

This is a comb-type filter (see Chapter 2, Section 2.3.7) and is quite simple to implement. In this case, we need to make one subtraction, one multiplication

Table A1.2 *Transition table for the simplified Morse code receiver, indices are event numbers and old state numbers, values in the table are new state numbers*

Event	Old state						
	1	2	3	4	5	6	7
1	2	1	1	1	7	1	1
2	5	4	1	1	1	1	1
3	1	3	1	1	6	1	1

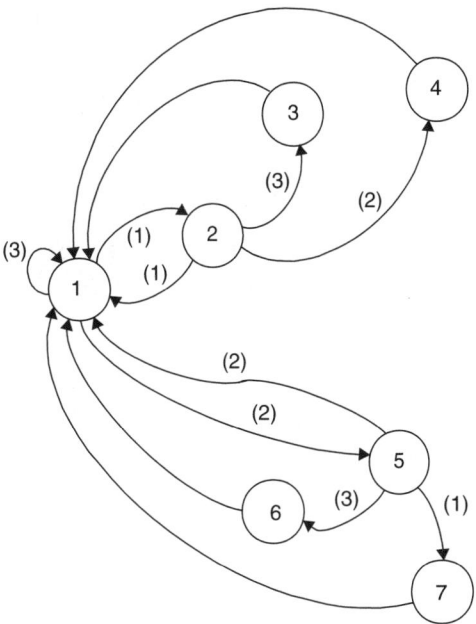

Figure A1.23 *State transition diagram for the simplified Morse code receiver*

and one addition and accumulation for each run through the loop, which should not be any problem to perform in $5\,\mu s$. The memory requirement would be 512 bytes, as above.

Conclusion: finding efficient and smart algorithms is important!

P9-3

From Table 9.1 we get $a = 0.225 \Rightarrow 0001$ and $b = -0.125 \Rightarrow 1111$, which gives the sum 0000 and multiplied by $c = 4$ gives $0000 \Rightarrow 0$. Doing the calculations the other way yields: $ca = 0.900 \Rightarrow 0111$ and $cb = -0.5 \Rightarrow 1100$, summing gives $0111 + 1100 = 0011 \Rightarrow 0.375$. The "true" answer is 0.400.

Obviously, the last method gives a better result. In the first case, the limited resolution caused a **cancellation**, since the difference was smaller than 1 LSB.

A potential problem with the latter method is of course the risk of overflow. Working with scaling and limited resolution is a delicate task.

P9-4

We have the following seven states:

State 1: wait for signal to arrive
State 2: wait for second part of letter or time out, i.e. symbol complete
State 3: print "e", symbol complete
State 4: print "a", symbol complete
State 5: wait for second part of letter or time out, i.e. symbol complete
State 6: print "t", symbol complete
State 7: print "n", symbol complete

Further, we can identify three events:

Event 1: dot received
Event 2: dash received
Event 3: time out, i.e. no signal received within 1 s, i.e. symbol complete

The state transition can be summarized by Table A1.2, and the transition state diagram is depicted in Figure A1.23.

Appendix 2 A MATLAB™/ Simulink™ primer

A2.1 Introduction

This primer is not intended to be a complete tutorial, but only a quick way of getting started with MATLAB™ if you are not an experienced user. There are many good books around on this particular topic (Buck *et al.*, 2002; Pratap, 2002). It is also quite common that an introductory chapter in many signal processing books deal with MATLAB™, for instance Denbigh (1998) and Tewari (2002). Anyway, here comes yet another introductory text to this well-known and versatile software package.

A2.1.1 The software

MATLAB™ is a software package for numerical computation and visualization and is a product of The MathWorks Inc., www.mathworks.com. The acronym MATLAB™ stands for MATrix LABoratory, since the basic data type in the system is the matrix. MATLAB™ not only has powerful built-in functions for basic matrix algebra and graphics, but also for many types of highly advanced scientific and technical computations. Further, the user can easily write his own functions in the MATLAB™ language. There are also a vast number of optional "toolboxes", i.e. collections of functions for special applications available. Such applications could be image processing, filter design, neural networks, fuzzy logic, spline, etc. In this text we assume that you have the basic software package and the signal processing and control system toolboxes.

Simulink™ is an extension of MATLAB™ and is a tool for modeling, analyzing, and simulating physical and mathematical dynamic systems. We will give a brief presentation and some examples using Simulink™ you can try as well, provided you have that extension installed on your computer.

A2.2 Basics

MATLAB™ is a command-based, interactive system, awaiting for you to type a proper command after the prompt >>. One of the most useful commands is `help`. Try typing this command and you will get a list of all the available built-in functions and commands in your system. For example:

```
>> help

HELP topics:

matlab\general        - General purpose commands.
```

```
matlab\ops              - Operators and special characters.
matlab\lang             - Programming language constructs.
matlab\elmat            - Elementary matrices and matrix
                          manipulation.
matlab\elfun            - Elementary math functions.
matlab\specfun          - Specialized math functions.
matlab\matfun           - Matrix functions - numerical linear
                          algebra.
matlab\datafun          - Data analysis and Fourier
                          transforms.
matlab\audio            - Audio support.
matlab\polyfun          - Interpolation and polynomials.
matlab\funfun           - Function functions and ODE solvers.
matlab\sparfun          - Sparse matrices.
matlab\graph2d          - Two-dimensional graphs.
matlab\graph3d          - Three-dimensional graphs.
matlab\specgraph        - Specialized graphs.
matlab\graphics         - Handle graphics.
matlab\uitools          - Graphical user interface tools.
matlab\strfun           - Character strings.
matlab\iofun            - File input/output.
matlab\timefun          - Time and dates.
matlab\datatypes        - Data types and structures.
matlab\verctrl          - Version control.
matlab\winfun           - Windows operating system interface
                          files (DDE/COM).
winfun\comcli           - No table of contents file.
matlab\demos            - Examples and demonstrations.
toolbox\local           - Preferences.
simulink\simulink       - Simulink.
simulink\blocks         - Simulink block library.
simulink\components     - Simulink components.
simulink\fixedandfloat  - No table of contents file.
simulink\simdemos       - Simulink 4 demonstrations and
                          samples.
simdemos\simfeatures    - Simulink: feature demonstrations
                          and samples.
simdemos\simgeneral     - Simulink: general model demon-
                          strations and samples.
simulink\dee            - Differential equation editor.
simulink\dastudio       - No table of contents file.
asap2\asap2             - No table of contents file.
asap2\user              - No table of contents file.
toolbox\compiler        - MATLAB compiler.
control\control         - Control system toolbox.
control\ctrlguis        - Control system toolbox - GUI support
                          functions.
control\ctrlobsolete    - Control system toolbox - obsolete
                          commands.
control\ctrlutil        - No table of contents file.
control\ctrldemos       - Control system toolbox - demos.
toolbox\sb2sl           - SB2SL (converts systembuild to
                          Simulink).
signal\signal           - Signal processing toolbox.
signal\sigtools         - Filter design and analysis tool
                          (GUI).
signal\sptoolgui        - Signal processing toolbox GUI.
```

```
signal\sigdemos        - Signal processing toolbox
                         demonstrations.

For more help on directory/topic, type "help topic".
For command syntax information, type "help syntax".

>>
```

If you type `help` followed by a function or command, you will obtain more information about that particular function. Try, for instance, `help abs` which will give you the help text for the `abs` function (giving the absolute value of the argument):

```
>> help abs

ABS Absolute value.
  ABS(X) is the absolute value of the elements of X. When
  X is complex, ABS(X) is the complex modulus (magnitude) of
  the elements of X.

  See also SIGN, ANGLE, UNWRAP.

Overloaded methods
  help iddata/abs.m
  help sym/abs.m

>>
```

Another useful command is `demo`. Try that for yourself.

A2.2.1 Some simple math

The easiest way of using MATLAB™ is as an interactive calculator. Try the following basic calculations and always end your command lines by pressing the "return" key:

```
>> 5*7

ans =

  35

>> pi/2

ans =

  1.5708

>> (45+7)*9/(12-3-90)

ans =

  -5.7778

>> 2^4

ans =

  16

>>
```

Table A2.1 *Some common MATLAB™ functions*

Common math	MATLAB™	Comment
e^x	`exp(x)`	
$\ln(x)$	`log(x)`	Natural logarithm
$\log(x)$	`log10(x)`	Logarithm base 10
$lb(x)$	`log2(x)`	Logarithm base 2
$\sqrt{x}$	`sqrt(x)`	
$\sin(x)$	`sin(x)`	x in radians
$\cos(x)$	`cos(x)`	x in radians
$\tan(x)$	`tan(x)`	x in radians
$\arcsin(x)$	`asin(x)`	Gives radians
$\arccos(x)$	`acos(x)`	Gives radians
$\arctan(x)$	`atan(x)`	Gives radians, see also `atan2(x)`
$x!$	`factorial(x)`	

As you can see, the result is assigned to a variable called `ans`. Further, the constant π is, not surprisingly, denoted `pi`. Further, the operand `^` corresponds to "raised to". All calculations in MATLAB™ are done in double precision, i.e. with 15 digit resolution. Using the command `format` you can, however, switch between different display formats. Try, for instance:

```
>> format long
>> pi/2

ans =

  1.57079632679490

>> format
>> pi/2

ans =

  1.5708

>>
```

Default is five digits and is obtained by typing `format` only. Try `help format` to find out what other display options are available. Table A2.1 shows some common mathematical functions in MATLAB™, however, there are many more. Remember that you can always use `help` to obtain more information about a particular function.

A2.2.2 Variables, scalars, vectors and matrices

A2.2.2.1 Scalars

The basic data type in MATLAB™ is the matrix. A row vector can be viewed as $[1 \times N]$ matrix and column vector as matrix with dimensions $[N \times 1]$. A scalar is of course a matrix $[1 \times 1]$. Variable names can be up to 19 characters. The first character must be a letter, but the remainder can be letters, digits or underscores. MATLAB™ is case sensitive, i.e. it distinguishes between small and capital letters. Assigning a number to a variable is straightforward:

```
>> A_variable=19

A_variable =

  19

>> another_variable=-4;
>>
```

Note that the value of a variable is printed immediately after the command is given, while if the command is ended with a **semicolon**, no message is printed. As soon as the variable name is typed, the value will be printed. The variables can, of course, be parts of equations, for instance:

```
>> 3*A_variable/another_variable

ans =

 -14.2500

>>
```

A2.2.2.2 Vectors and matrices

Above, we have assigned values to two scalar variables. Assigning values to a row vector $\mathbf{x} = [1 \ \ 2 \ \ 3]$ would look like:

```
>> x=[1, 2, 3];
>>
```

where the commas can be omitted. A column vector $\mathbf{y} = \begin{bmatrix} 4 \\ 5 \end{bmatrix}$ can be assigned by typing:

```
>>y=[4; 5];
>>
```

where we note that semicolons, in matrices, are used to mark the end of a row. Finally, let us define a matrix $\mathbf{B} = \begin{bmatrix} 10 & 11 & 12 \\ 13 & 14 & 15 \end{bmatrix}$

```
>> B=[10, 11, 12; 13, 14, 15];
>>
```

All the variables (scalars, vectors and matrices) defined so far are stored in the memory, called the **workspace** of MATLAB™. They are global and will stay in the workspace until MATLAB™ is shut down. One way to keep track of your variables is to use the command who or whos:

```
>> who

Your variables are:

A_variable    another_variable   x
B             ans                y
>> whos
Name                  Size      Bytes     Class

A_variable            1x1         8       double array
B                     2x3        48       double array
another_variable      1x1         8       double array
ans                   1x1         8       double array
x                     1x3        24       double array
y                     2x1        16       double array

Grand total is 14 elements using 112 bytes

>>
```

As you can see from the above, who only gives the names of your active variables, while whos gives a more detailed description. As mentioned before, once assigned, these variables will stay in the workspace. There is, however, a way to delete a variable, for instance, the vector **x**, by typing:

```
>> clear x
>>
```

Check this using who. Typing clear all will delete all variables in the workspace. To continue our experiments, we assume that you have restored your variables as they were, before we started fooling around with the clear command above. You can use the up-arrow and down-arrow keys on your keyboard to find and reissue old commands given. This certainly saves time.

A2.2.2.3 Vector and matrix computations

Doing calculations with vectors and matrices is a little bit more complicated than dealing with scalars, since we need to keep track of the dimensions. Further, the dimensions have to agree and in some algebraic operations, for instance, division is not defined for matrices. We must also remember that matrix calculations are not commutative in the general case, i.e. it does matter in which order, for instance, chain multiplication is computed. Let us do some experiments. For example, if you try to add vector **x** to vector **y** an error message will result, since the vectors do not have the same dimensions:

```
>> x+y
??? Error using ==> +
Matrix dimensions must agree.

>>
```

Define a new row vector $\mathbf{z} = 2\mathbf{x}$ which will have the same dimension as $\mathbf{x}$, hence it should be possible to add $\mathbf{z} + \mathbf{x}$:

```
>> z=2*x;
>> z+x

ans =

   3   6   9

>>
```

Let us now change the row vector $\mathbf{z}$ $[1 \times 3]$ into a column vector $\mathbf{w} = \mathbf{z}^{\mathrm{T}} [3 \times 1]$. Obtaining the transpose is done using the ' operator:

```
>> w=z'

w =

   2
   4
   6

>>
```

If we now multiply the two vectors, $\mathbf{xw}$ we should expect the dimensions of the product matrix to be $[1 \times 3][3 \times 1] = [1 \times 1]$:

```
>> x*w

ans =

   28

>>
```

This is basically a dot product or "inner product" between the two vectors $\mathbf{x}$ and $\mathbf{w}$, i.e. $\mathbf{x} \cdot \mathbf{w} = |\mathbf{x}||\mathbf{w}| \cos(\alpha)$, where α is the angle between the vectors $\mathbf{x}$ and $\mathbf{w}$. Now let us change the order of the vector multiplication to $\mathbf{wx}$. In this case, computing the "outer product", we expect to obtain a matrix with dimensions $[3 \times 1][1 \times 3] = [3 \times 3]$. Let us try:

```
>> w*x

ans =

   2    4    6
   4    8   12
   6   12   18

>>
```

So now the general rule is that if we type the operators *, ^, + or -, the actual action taken will depend on the type of the variables involved. For instance, if a and b are scalars the expression a*b will simply be a common multiplication, while typing w*x will result in a matrix since two vectors were involved (as above). The operator / requires special explanation. For scalars, this is nothing but a simple division, but division is not defined for matrices. Nevertheless, the / can actually be used with matrices in MATLAB™ but means multiplication from the right by the inverse of the second matrix, i.e. $\mathbf{A}/\mathbf{B} = \mathbf{A}\mathbf{B}^{-1}$; note that this is a definition in MATLAB™ and not proper mathematics notation. There is indeed an alternative as well, using a backslash $\mathbf{A}\backslash\mathbf{B} = \mathbf{A}^{-1}\mathbf{B}$. By the way, taking the inverse of a matrix $\mathbf{E}$ in MATLAB™ is written inv(E). Anybody who has inverted matrices by hand or with a calculator agrees that this MATLAB™ command is extremely valuable. It saves lots of tedious work.

Sometimes, one wishes to make operations between vectors or matrices on an element-to-element basis. For example, we would like to multiply the first element of vector $\mathbf{x}$ with the first element of vector $\mathbf{z}$, the second element of $\mathbf{x}$ with the second element of $\mathbf{z}$ and so on. In this case the * operator cannot be used since vector multiplication would be assumed, and that is not what we want to do. It is not even possible with $\mathbf{x}$ and $\mathbf{z}$, considering the dimensions. So, for this purpose, there are a number of special symbols in MATLAB™, starting with a dot, namely: .*, ./, .^ and so on. For example:

```
>> x*z
??? Error using ==> *
Inner matrix dimensions must agree.
```

but

```
>> x.*z

ans =

    2    8    18

>>
```

The first case did not work for the reasons given above, while the second one accomplished multiplication element by element, which was what we wanted. Note, even skilled MATLAB™ programmers forget the dots now and then. If you get mysterious error messages about dimensions where you do not expect any such problems, check your dots.

A2.2.2.4 Addressing vectors and matrices

So far we have been working with vectors and matrices as a whole, but sometimes there is a need to pick out a specific element of a vector or matrix, or maybe even a given set of elements. Hence, we need to be able to address a specific element. This is easily accomplished using parentheses as below. Assume that we want the second element of vector $\mathbf{x}$ or the third element in the second row of matrix $\mathbf{B}$:

```
> x(2)

ans =

    2
```

```
>> B(2,3)

ans =

   15

>>
```

Note, unlike other computer languages as, for instance, C++, the first element in a vector or matrix in MATLAB™, has the index number **1**. There is no element number 0. If we need to address a given set of elements, we may need to create a sequence of integers. For example, assume that we have a vector **d** consisting of 1027 elements, and we need to get a hold of elements 827–835. In this case, it would, of course, be quite simple to use the method outlined above, but a smarter way is to define a sequence of integers: $827, 828, \ldots, 835$ and use the sequence to point into the vector. Creating a sequence in MATLAB™ is easy:

```
>> 827:835

ans =

  827   828   829   830   831   832   833   834   835

>>
```

Getting the desired elements of vector **d** would look like

```
>> d(827:835)

ans =

Columns 1 through 8

  0.7439   0.8068   0.6376   0.2513   0.1443   0.6516   0.9461   0.8159

Column 9

  0.9302

>>
```

A2.2.2.5 Sequences

In Section A2.2.2.4, a simple example of a sequence was shown. We needed a sequence of integers between two given limits. This was adequate for addressing a vector, since only integers are accepted as index. In other situations, however, other types of sequences are needed. An example could be to get a proper axis for a plot. The first alternative sequence is an extension of what was shown above. Suppose we need a linear sequence t with a non-integer step size, for instance a time scale from 0 to 2 s, in increments of 0.25 s:

```
>> t=0:0.25:2

t =

 Columns 1 through 8

    0    0.2500    0.5000    0.7500    1.0000    1.2500    1.5000    1.7500

 Column 9

   2.0000

>>
```

An alternative could be the MATLAB™ function `linspace(X1, X2, N)`, giving a sequence starting with X1, ending with X2 and consisting of N steps:

```
>> linspace(0,2,9)

ans =

 Columns 1 through 8

    0    0.2500    0.5000    0.7500    1.0000    1.2500    1.5000    1.7500

 Column 9

   2.0000

>>
```

In certain applications a logarithmic sequence is required; in such cases, use the function `logspace(X1, X2, N)`. It generates a sequence between decades 10^{X1} and 10^{X2} in N steps:

```
>> logspace(1, 3, 8)

ans =

 1.0e+003 *

   0.0100    0.0193    0.0373    0.0720    0.1389    0.2683    0.5179    1.0000

>>
```

For some situations in which an all 0 or all 1 vector (sequence) is needed, try the commands:

```
>> zeros(1, 6)

ans =

    0    0    0    0    0    0
```

```
>> ones(1, 8)

ans =

  1   1   1   1   1   1   1   1

>>
```

In these examples, vectors were created. Generating matrices is equally as easy, by simply replacing the initial parameter 1 with the number of rows in the desired matrix.

A2.2.3 Basic input/output and graphics

A2.2.3.1 Input

To input values or strings from the keyboard, the input statement can be used. First it prompts the user and then awaits a numerical value, a defined variable or a string (see help input):

```
>> r=input('how old are you?')
how old are you? 19

r =

   19

>>
```

To read data from files, there are a number of functions depending on the type of file. Reading, for instance, a simple binary file, the commands fopen and fread can be used. These functions resemble much of the same commands used in C++:

```
>> filhand=fopen('test.dat');
>> data=fread(filhand);
>> data

data =

   1
   2
   3

>> fclose(filhand);
>>
```

The code above first opens a binary file named test.dat. A file handle is assigned to the variable filhand. The file handle is from here on used to identify this particular file, in case more than one file is open at the same time. The file is opened for **reading** data only (default); for more information see help fopen. The actual reading takes place on the second line, where the data read from the file using `fread(filhand)` is assigned to the vector variable `data`. Finally, calling this variable, we see the contents of the read data file, the three numbers: 1 2 3. After reading the file it should be closed, using `fclose(filhand)`. For more information about reading files, consult `help fileformats` and `help iofun`. There are functions for reading a number of common file types, for instance, a special function for reading audio files (.wav) `wavread`, making it easy to read and manipulate sound files.

A2.2.3.2 Numerical and character output

Printing the value of a variable on the computer screen is quite straightforward by typing the name of the variable as above. If we need to include a given text, the function `disp` can be used:

```
>> disp('the values in the variable data are'), disp(data)
the values in the variable data are
   1
   2
   3

>>
```

Another feature of MATLAB™ is demonstrated above; more than one command can be put on the same line, provided they are separated by a comma or semicolon. The first `disp` displays the given string, while the second `disp` prints the values of the variable `data`. Note, the name of the variable is not shown, as would have been the case typing `data` only.

Printing to files can be performed by using the functions `fopen` and `fwrite`:

```
>> file_id=fopen('test2.bin','w');
>> fwrite(file_id,x);
>> fclose(file_id);
>>
```

First we open the file `test2.bin` for writing (`'w'`). The file handle in this example is called `file_id`. On the second line, the actual writing of vector `x` takes place, and finally the file is closed. There are lots of alternative commands and arguments for these functions, so please consult the help texts.

A2.2.3.3 Graphic output

One of the great things with MATLAB™ is that it is easy to obtain nice plots in two or three dimensions and in color and so on. Here, we only demonstrate some basic types, but try the `demo` and `help` functions to obtain more information. Let us start by a simple example of plotting the function $y(t) = 2\sin(\omega t)$ for the frequency 5 kHz, where $\omega = 2\pi f = 2\pi \cdot 5 \cdot 10^3$. Note that a comment, i.e. text not being processed by MATLAB™, is preceded by a % sign:

```
>> f=5e3;                 % frequency 5 kHz
>> t=0:1e-6:1e-3;         % time sequence, 0-1 ms in 1 μs step
>> y=2*sin(2*pi*f*t);     % the function itself
>> plot(t,y)             % plot y versus t
>>
```

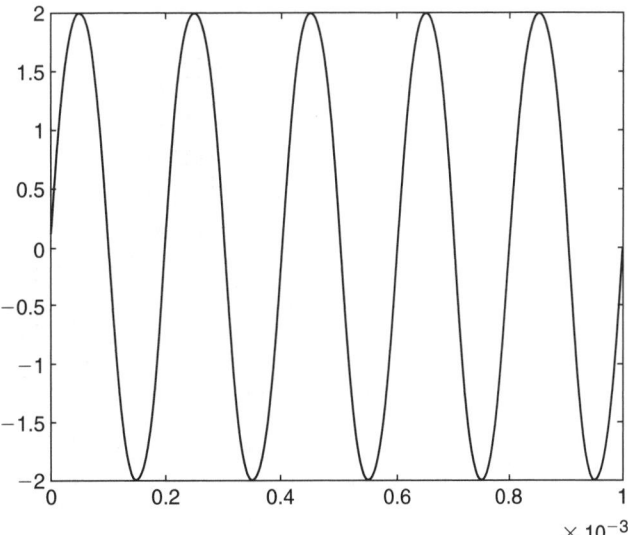

Figure A2.1 *An example of a simple two-dimensional, interpolated, continuous-time plot*

The resulting plot is shown in Figure A2.1. In this case we have not determined the scaling of the axes. It is made automatically. Nor have we assigned any labels to the axes of the plot. There are lots of functions for taking care of these issues, use `help plot`.

In the example above, we have approximated a continuous function in 1000 discrete points. The plot function interpolates in between the sample points, so the resulting plot looks continuous. Assume that we would like to instead plot a "digital" sequence, without interpolating. Then, the function `stem` can be used:

```
>> f=5e3;                % frequency 5 kHz
>> fs=1e5;               % sampling frequency 100 kHz
>> n=0:50;               % sample points 0-50
>> y=2*sin(2*pi*f*n/fs); % the function itself
>> stem(n,y)             % discrete plot
>>
```

In this case we are using a sampling frequency of 100 kHz (f_s) and we choose to plot the first 51 sample points. Figure A2.2 shows the result.

Making three-dimensional plots is a bit tricky, but gives excellent possibilities to visualize complicated mathematical functions. As in the previous cases above, we just give some simple examples, and there are numerous extensions and variations available. In this experiment we will make a three-dimensional plot of the function $f(x,y) = \cos(2\pi x) \cos(2\pi y)$ over the range $-0.5 < x < 0.5$ and $-0.5 < y < 0.5$:

```
>> [X,Y]=meshgrid(-.5:0.05:0.5,-0.5:0.05:0.5);
                            % create a matrix with all
                            % possible pairs of X and Y
>> f=cos(2*pi.*X).*cos(2*pi.*Y);  % compute the function f in
                            % all these points
```

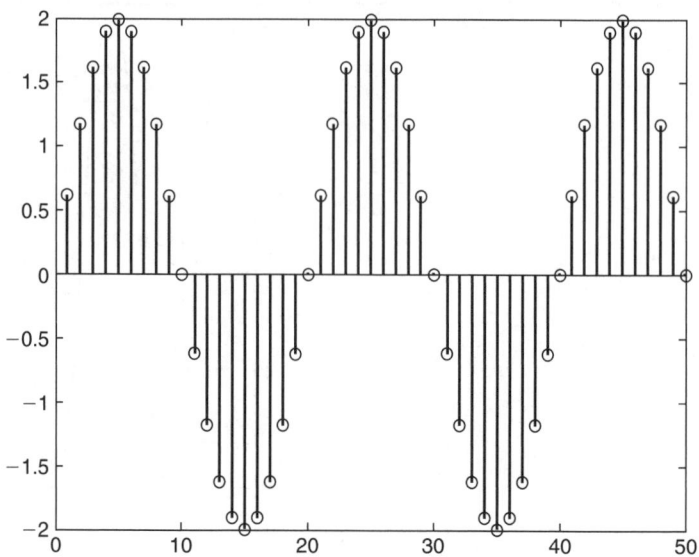

Figure A2.2 *A discrete-time plot obtained using the function stem*

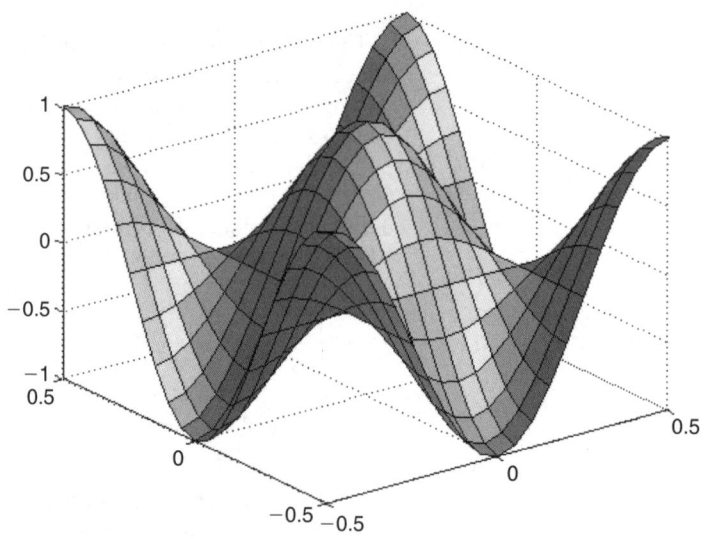

Figure A2.3 *A gray-scale three-dimensional plot*

```
>> surfl(X,Y,f)              % plot a three-dimensional
                             % surface
>> colormap(gray)            % keep to gray scale to
                             % avoid printing problems
>>
```

Note the use of dot operators to obtain elementwise computations. The final three-dimensional plot is shown in Figure A2.3. There are lots of possibilities using colors, shading, changing the viewing angle, etc.

A2.2.4 Other programming structures

In all programming languages there are instructions and functions for controlling the execution flow, as is the case in MATLAB™. Thanks to the simple handling of matrices and vectors, however, the need for loop structures is less than when using a general programming language such as, for instance, C++.

A2.2.4.1 Conditional execution

The basic conditional structure consists of the keywords `if`, `elseif`, `else` and `end`. Below, we have used these keywords to implement a soft limiter as in equation (4.10), repeated here for convenience:

$$f(x) = \begin{cases} a & x \geq a \\ x & b \leq x < a \\ b & x < b \end{cases} \tag{4.10}$$

The corresponding MATLAB™ code may look something like given below:

```
>> if x>a
      f=a;
   elseif x<b
      f=b;
   else
      f=x;
   end
>>
```

This type of structure does not need to have either `elseif` or `else` clauses, if not called for by the application. Note that the relational operator for equality is denoted `==` as in C++; the compatibility with C or C++ is, however, not total. Another structure with a strong resemblance to C++ is the `switch` structure. Below is an example testing the value of the variable `x`:

```
>>
switch x
  case 1
    disp('one');
  case 2
    disp('two');
  case 3
    disp('too much');
  otherwise
    disp('something else');
end
>>
```

Note that unlike C or C++, only the statements between the valid `case` and the next `case` (or `otherwise` or `end`) are executed. Program flow does not fall-through and hence, `break` statements are not necessary. The `otherwise` statement above is optional.

A2.2.4.2 Loop structures

Loops are in many cases not needed in MATLAB™, since vectors and sequences can often do the same job. However, in some cases loops are needed. The code example below will perform the same function as the vector-oriented approach in Section A2.2.3.3 above (i.e. this is an "unnecessary" loop application):

```
>> f=5e3;                          % frequency 5 kHz
>> fs=1e5;                         % sampling frequency 100 kHz
>> y=zeros(1,51);                  % assign space for vector y
>>for n=1:51                       % sample points 1-51
    y(n)=2*sin (2*pi*f*n/fs);      % the function itself
end
>>
```

The keywords here are for and end. There are two things that differ from the previous implementation of this task in Section A2.2.3.3. Firstly, we need to allocate memory space for the vector y in advance, unlike in the example in Section A2.2.3.3. This is accomplished by defining a zero vector y = zeros(1,51). Secondly, we must now use the sampling index 1–51, rather than 0–50, since 0 is not a valid addressing index for the y vector.

Another loop structure is the while to end model. The above code can be rewritten using this method:

```
>> f=5e3;                          % frequency 5 kHz
>> fs=1e5;                         % sampling frequency 100 kHz
>> y=zeros(1,51);                  % assign space for vector y
>> n=1;                            % initialize loop index
>>while n<=51                      % loop unit point 52
    y(n)=2*sin(2*pi*f*n/fs);       % the function itself
    n=n+1;                         % increment pointer
  end
>>
```

In this case we are responsible for initializing and incrementing the loop index n. Sometimes this approach is superior, if complicated conditions for exiting the loop exist. On the other hand, failing to update the loop counter properly may result in an infinite loop. (In some applications infinite loops are created deliberately.)

To skip out of an infinite loop, press **ctrl-c** on your keyboard.

A2.3 Workspace, scripts and functions

A2.3.1 The workspace

In MATLAB™, the workspace is the memory where all of your active variables are stored. The workspace can be saved to disk, and later be restored. There are two ways of saving your workspace:

```
>> save
>>
```

or

```
>> save('my_space.mat');
>>
```

In the first case, the workspace is saved to the default file "matlab.mat". In the second case, you can give the workspace file a specific name, in case you need to work with a number of different workspaces for different purposes.

To restore the workspace use:

```
>> load
>>
```

or

```
>> load('my_space.mat');
>>
```

As before, if no file name is given, the default file "matlab.mat" is used. If no file extension is given, .mat is assumed.

To navigate around your files and directories, many old Microsoft DOS commands apply such as `dir` for showing the files in the current directory and `cd` for changing directory and so on. The default start directory for MATLAB™ (where your workspace is stored) is named `work`.

A2.3.2 Scripts and m-files

Since MATLAB™ code consists of pure American standard code for information interchange (ASCII) characters, program code can easily be transferred between different operating system environments and different computers without any problems. Further, MATLAB™ program code can be written and edited in most text editors like Notepad™, Word™, Wordpad™ or the like. The code is saved as a text file. Such a saved code segment is called a script. If a script is saved as a text file with file suffix .m (so-called m-file), in the current directory, MATLAB™ will use this script as a new command having the same name as the file name (except the suffix). Let us have an example. First we write a script called `conus` as below and save it using the file name "conus.m":

```
% conus-a fantastic script for computing the cosine and sine
% input variable is: q (angle in radians)

disp('cosine is:'), disp(cos(q))
disp('sine is:'), disp(sin(q))
```

If we now type the following at the MATLAB™ prompt, the new command (m-file) will be directly invoked:

```
>> q=pi/2;
>> conus
cosine is:
 6.1232e-017

sine is:
   1

>>
```

As you can see, MATLAB™ now has been equipped with yet another command, `conus`. You may wonder why we have put all the comments over the script? Well, try the `help` command and you will understand:

```
>> help conus

conus - a fantastic script for computing the cosine and sine
input variable is: q (angle in radians)

>>
```

A drawback with scripts is that all variables are global. This means that if you are using a variable called, for instance, `test` (a very original name...?) in your current workspace and you run a script in which the very uncommon variable name `test` is also used, but for another purpose, problems may occur. The variable `test` in the script will overwrite the variable `test` of your workspace, which may not be what you want. On the other hand, the same mechanism is deliberately used to transfer data into and out of a script. This is a limitation with scripts. There is no way of transferring data into the script without using global variables.

A2.3.3 Functions

Functions are cousins to scripts, but now the problem with global variables has been solved, since functions have their own "private" workspace and hence private variables. Therefore, there is no risk that variables, which happen to have the same name as variables in the workspace, interfere with each other. To be able to transfer data into a function, we now need a formal variable transfer mechanism. The example below shows an example function VAT, which computes the VAT for a given amount. This code is stored as a text file with name VAT.m, as for m-files:

```
% VAT(am, trate)
% computes total amount including VAT
% am is amount without tax, trate is tax rate in %

function tot = VAT(am, trate)
tot=am+am*trate/100;
```

Running the function is now very simple. Assume we have sold digital filters for $120 and that the tax rate is 12%. Unfortunately, it has been 2 years since we used this function last, so we do not remember how to call it. Therefore, we first try the `help` command, before actually using the function:

```
>> help VAT

VAT(am, trate)
computes total amount including VAT
am is amount without tax, trate is tax rate in %

>> VAT(120, 12)
```

```
ans =

 134.4000

>>
```

A2.4 Some useful functions

In this section we will present some functions that are extra useful when working with digital signal processing and control. In this text we have assumed that you have the toolboxes signal processing and control systems installed. By installing more toolboxes, many new functions can be obtained. There are, for instance, special toolboxes for interactive filter design, neural networks, image processing, fuzzy logic and wavelets.

A2.4.1 Linear systems

There are four different model types within MATLAB™ to represent linear time invariant systems (LTI): these are transfer function (TF), zero–pole gain (ZPK), state–space (SS) and frequency response data (FRD). In MATLAB™ there are many functions to analyze and design linear systems. Some of these functions require the model to be of a specific type. For this reason, there are functions for converting one model type to another:

```
>>s=tf(s);          % converts model s to TF type
>>s=zpk(s);         % converts model s to ZPK type
>>s=ss(s);          % converts model s to SS type
>>s=frd(s,freq);    % converts model s to FRD type
                    % freq is a frequency vector
```

Note that FRD models cannot be converted to other types, and that conversion **to** FRD requires a frequency vector as input. Once the model has been converted to the desired type a number of useful functions are available. Some functions require the TF to be expressed as two vectors containing the coefficients of the numerator and denominator polynomials, e.g. **B** for numerator coefficients and **A** for denominator coefficients. If needed, these vectors can be converted to a TF model, e.g.:

```
>> B=[0, 1, 1];
>> A=[1, 3, 1];
>> H=tf(B,A)

Transfer function:
    s + 1
 -------------
s^2 + 3 s + 1

>>
```

For example, suppose we want a Bode plot of the TF $H(s) = (s+1)/(s^2 + 3s + 1)$. First let us generate a TF model as above or by using the method shown below, and after that we make a Bode plot:

```
>> s = tf('s'); H = (s+1)/(s^2+3*s+1)

Transfer function:
     s + 1
-------------
s^2 + 3 s + 1

>> Bode(H)
>>
```

The obtained Bode plot is shown in Figure A2.4.

An alternative function to plot the gain and phase shift versus frequency is freqs. In this case we need the coefficients of the numerator and denominator polynomials expressed as two vectors. Further, we need a vector stating frequencies for which the TF is to be plotted. Starting with the numerator coefficients, we have $B(s) = b_1 s + b_0 = s + 1, \Rightarrow b_1 = 1, b_0 = 1$ and these coefficients are stored in a row vector, starting with the coefficient corresponding to the highest order of s, $\mathbf{B} = [b_2 \ b_1 \ b_0] = [0 \ 1 \ 1]$. For the denominator we have

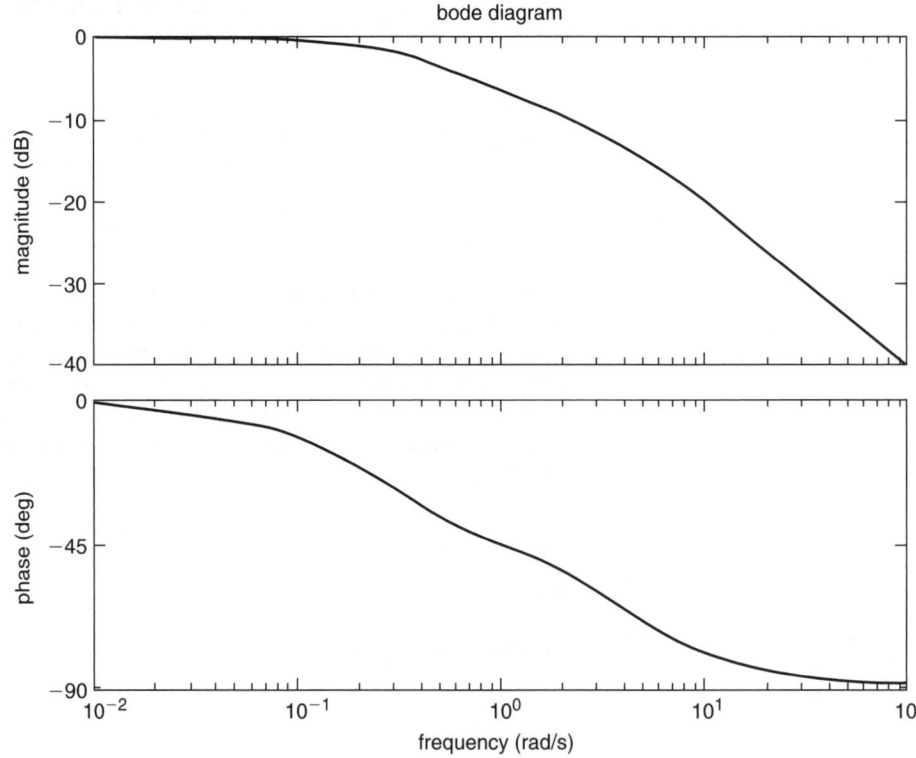

Figure A2.4 *Bode plot, in which the upper plot is the gain function and the lower plot is the phase shift function*

$A(s) = a_2 s^2 + a_1 s + a_0 = s^2 + 3s + 1, \Rightarrow a_2 = 1, a_1 = 3, a_0 = 1$ and the resulting vector will be $\mathbf{A} = [a_2 \ a_1 \ a_0] = [1 \ 3 \ 1]$. Note that it is recommended that these two vectors have the same dimensions, to avoid mixing up the order of the coefficients. That is why we added a seemingly "useless" coefficient $b_2 = 0$ in the vector representing the numerator polynomial. Finally, we need to determine a frequency vector, therefore we choose to use a logarithmic scale from 100 to 10 000 rad/s:

```
>> B=[0, 1, 1];        % numerator coefficients
>> A=[1, 3, 1];        % denominator coefficients
>> W=logspace(2,4);    % logarithmic frequency vector
>> freqs(B,A,W)        % plot gain and phase shift functions
>>
```

Suppose we need the pole–zero map of the TF above. That could easily be obtained by using the following:

```
>> pzmap(H)
>>
```

The pole–zero map is shown in Figure A2.5.

If we are working in the z-plane, rather than the Laplace domain, a corresponding function freqz is available. The arguments are basically the same, but the frequency vector can be expressed in Hz. Further, the sampling frequency has to be stated. The freqz function can handle a variety of input arguments, see help freqz. An example using freqz could be as follows. Assume we have a comb filter (see Chapter 2) with TF $H(z) = (1 - z^{-5})/(1 - z^{-1})$, the sampling frequency is 44 kHz and we are interested in the frequency response in the range 0–15 kHz. Plot the gain and phase shift functions:

```
>> B=[1  0 0 0 -1];    % numerator coefficients
>> A=[1 -1 0 0  0];    % denominator coefficients
>> F=0:0.5e3:15e3;     % linear frequency vector, Hz
>> freqz(B,A,F,44e3)   % plot gain and phase, sampling
                       % frequency 44 kHz
>>
```

Try this yourself and use help freqz to try other features of this function. **Note** that if the sampling frequency is not given, the frequency is given as fractions in the range 0–1 of the Nyquist frequency, i.e. half the sampling frequency. That is, 0 corresponds to 0 Hz while 1 corresponds to $f = f_s/2$ or $\Omega = \pi$ or $q = 1/2$.

A2.4.2 Filter design

A2.4.2.1 Analog filter design

There are a number of functions available for designing analog low-pass, high-pass, bandpass and bandstop filters, using different filter approximations (see Chapter 1). The outputs of these functions are, for instance, two vectors containing the polynomial coefficients of the numerator and denominator of the

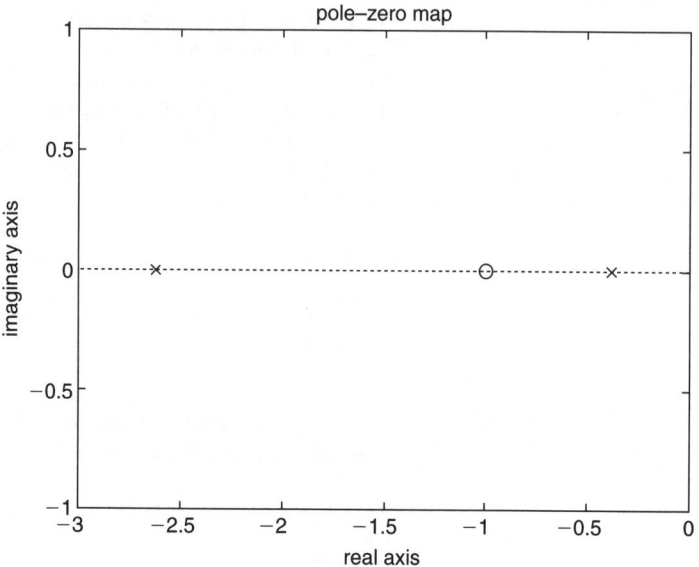

Figure A2.5 *An example of a pole–zero map*

TF. For example, we would like to design a Butterworth low-pass filter of order 5, with a cut-off frequency of 1 kHz:

```
>> [B,A]=butter(5,2*pi*1e3,'s');    % design low-pass
                                    % Butterworth filter
>> W=logspace(2,4)*2*pi;            % determine plot frequency
                                    % range
>> freqs(B,A,W)                     % plot gain and phase
>>
```

where the vector **B** contains the polynomial coefficients of the numerator and **A** contains the polynomial coefficients of the denominator. If a high-pass filter is needed, the first line is changed:

```
>> [B,A]=butter(5,2*pi*1e3,'high','s'); % design high-pass
                                        % Butterworth filter
>> W=logspace(2,4)*2*pi;                % determine plot
                                        % frequency range
>> freqs(B,A,W)                         % plot gain and phase
>>
```

For a bandpass filter an upper and a lower cut-off frequency is required. These frequencies are combined in a row vector. Suppose we want a bandpass filter of Butterworth type with lower cut-off frequency 700 Hz and upper cut-off frequency 1300 Hz. The filter should be of order 6. Then we would type as follows:

```
>> F=2*pi*[700, 1300];    % cut-off frequencies
>> [B,A]=butter(3,F,'s'); % design bandpass Butterworth
                          % filter
```

```
>> W=logspace(2,4)*2*pi;      % determine plot frequency range
>> freqs(B,A,W)               % plot gain and phase
>>
```

Note that the order of the filter will be 6, i.e. 2 times the parameter 3 stated in the `butter` function call. This goes for bandstop filters as well. Finally, a bandstop filter of Butterworth type is obtained if changing the first line above. We assume the same order and cut-off frequencies:

```
>> F=2*pi*[700, 1300];        % cut-off frequencies
>> [B,A]=butter(3,F,'stop','s');  % design bandstop
                              % Butterworth filter
>> W=logspace(2,4)*2*pi;      % determine plot frequency
                              % range
>> freqs(B,A,W)               % plot gain and phase
>>
```

In a similar way, Chebyshev filters can be designed. In this case we also have to state the allowed ripple. There are two functions for designing Chebyshev filters, one that gives ripple in the passband `cheby1` and one that gives ripple in the stopband, `cheby2`. The ripple specification is given in terms of peak-to-peak ripple expressed in dB. As an example, we need a Chebyshev bandstop filter of order 8, with cut-off frequencies 1.0 and 5.25 MHz, and max peak-to-peak ripple in the passband is 1 dB:

```
>> F=2*pi*[1, 5.25]*1e6;      % cut-off frequencies
>> [B,A]=cheby1(4,1,F,'stop','s');  % design bandstop
                              % Chebyshev filter 1 dB
                              % ripple
>> W=2*pi*logspace(5,7);      % determine plot frequency
                              % range
>> freqs(B,A,W)               % plot gain and phase
>>
```

There are also functions for designing elliptic filters (Cauer filters) `ellip` having ripple both in the passband and in the stopband and Bessel filters having linear-phase shift, `besself`.

A2.4.2.2 Transformations

As shown in Chapter 1, analog filters can be transformed into digital ones using transform techniques. In this section we will demonstrate MATLAB™ functions for doing impulse invariance and bilinear (Tustin) transformations.

The **impulse invariance transform** is performed using the function `impinvar`. Assume that we have designed an analog filter using, for instance, the function `butter` as above. The analog filter, specified by the coefficient vectors **B** and **A**, is then converted into a digital version having a TF of the type $G(z) = (b_0 + b_1 z^{-1} + b_2 z^{-2} + \cdots)/(a_0 + a_1 z^{-1} + a_2 z^{-2} + \cdots)$, which is expressed as the two coefficient vectors **Bz** and **Az**. To do the actual transformation, we also need to specify the sampling frequency, say, 10 kHz in this example. The last line is used to plot the gain and phase shift function of the resulting digital filter:

```
> >[B,A]=butter(5,2*pi*1e3,'s');      % design an analog low-pass
                                       % Butterworth filter
>> [Bz,Az]=impinvar(B,A,10e3);        % convert using impulse
                                       % invariance
>> freqz(Bz,Az)                       % plot gain and phase
>>
```

Now, let us repeat the same procedure but using the **bilinear transform (BLT)**. The corresponding MATLAB™ function is named bilinear. We first design an analog filter, transform it using the BLT and plot gain and phase. As in the previous example we are using the sampling frequency 10 kHz, but as can be seen below, there is a fourth argument of the function bilinear. The last parameter, 2 kHz, is the frequency at which **pre-warping** is done. If no frequency is given, no pre-warping takes place:

```
>> [B,A]=butter(3,2*pi*400,'s');      % design another
                                      % analog filter
>> [Bz,Az] = bilinear(B,A,10e3,2e3);  % convert using BLT
                                      % and pre-warping
>> freqz(Bz,Az)                       % plot gain and phase
>>
```

A2.4.2.3 Digital filter design

Digital filters can be designed **directly**, not taking the path via analog filters. The MATLAB™ functions, butter, cheby1, cheby2 and ellip, can be used to design digital filters directly. The only thing needed is to omit the parameter "s" when calling the functions and to remember that frequencies now are given in the interval 0–1, where 1 corresponds to $f = f_s/2$ (see Section A2.4.1 above). An example is shown below where we are designing a digital Butterworth filter, order 4 with cut-off frequency 0.4, i.e. $f_c = 0.4 f_s/2$:

```
>> [Bz,Az]=butter(4,0.4);     % direct digital filter design
>> freqz(Bz,Az)               % plot gain and phase
>>
```

There are many other methods for direct design of digital filters. A common method is the **Parks–McClellan optimal equiripple FIR filter design**, using the **Remez exchange algorithm** (see Chapter 1). There is a special function in MATLAB™ implementing this method. The function is denoted remez. Since this method is used for designing FIR filters only, the corresponding TF does not have a denominator polynomial. Hence, only the coefficient vector **Bz** for the numerator is generated. The designed filter has linear-phase response (symmetric, real weights) and, being a FIR filter, it is always stable. The input parameters to remez are the length of the filter and the desired gain function. The gain function is defined by two vectors **A** and **F**. Vector **F** defines the edges (in pair) between a set of frequency bands in the range 0–1, and vector **A** defines the desired gain in these frequency bands. For example, we need a low-pass filter with the gain specification ("wish list")

$$A = \begin{cases} 1 & 0 < f < 0.4 \dfrac{f_s}{2} \\ 0 & 0.5 \dfrac{f_s}{2} f < \dfrac{f_s}{2} \end{cases}$$

The undefined frequency region between $0.4(f_s/2) < f < 0.5(f_s/2)$ is a transition band, for the gain to drop from 1 to 0 (ideally). For execution time reasons, the length of the filter cannot be longer than 20 taps. The MATLAB™ code needed is shown below:

```
>> F=[0, 0.4, 0.5, 1];    % define frequency band edges
>> A=[1, 1, 0, 0];        % desired gain in the bands
>> Bz=remez(19,F,A);      % design filter, length 20
>> freqz(Bz,1)            % plot gain and phase
>>
```

Note that the order 19 given as the first parameter of `remez` corresponds to filter length 20. Please consult `help remez` for more information.

Using a special variation of this function, **Hilbert transformers** can be designed (see Chapter 5). The only modification needed is to add the parameter `'hilbert'` in the `remez` function call:

```
>> Bz=remez(19,F,A,'hilbert')    % design a Hilbert filter
>>
```

A2.4.3 Fast Fourier transform and convolution

Performing fast Fourier transform (FFT) using MATLAB™ is very convenient. Assume we have a vector **X** containing time domain samples. The FFT is done by simply calling the function `fft`:

```
>> n=1:500;               % sample points
>> X=cos(2*pi*0.1*n);     % create a signal vector
>> F=fft(X);              % FFT
>> plot(n,abs(F))         % plot the magnitude
>>
```

Since the output vector **F** of the FFT is a complex vector, we need to use the `abs` function to be able to plot the magnitude of the elements of **F**.

In the case above, we have implicitly used a **rectangular** windowing function. The input data simply begins and starts according to the length of the vector **X**. If we need to use other windowing functions, there are a number of windowing functions defined in MATLAB™, for instance, **Blackman**, **Hamming**, **Hanning**, **Kaiser**, **Hahn**, etc. Below, an example using the Hamming window is shown:

```
>> n=1:500;               % sample points
>> X=cos(2*pi*0.1*n);     % create a signal vector
>> X=X.*(hamming(500))';  % Hamming windowing
>> F=fft(X);              % FFT
>> plot(n,abs(F))         % plot the magnitude
>>
```

Note that the window sequence produced is a column vector, while X is a row vector, hence transpose of the window vector is needed.

Making an inverse FFT is also straightforward, using the function `ifft`. As an example, let us use the inverse FFT to restore the **X** vector from the vector **F**:

```
>> n=1:500;              % sample points
>> X=cos(2*pi*0.1*n);    % create a signal vector
>> F=fft(X);             % FFT
>> Xhat=ifft(F);         % inverse FFT
>> plot(n,real(Xhat))    % plot restored Xhat (real part)
>>
```

In the general case, the inverse FFT gives a complex output vector. The function `real` is used to plot only the real part. Since the original signal **X** was real, the imaginary parts found in **Xhat** are negligible and a result of rounding errors during the computations.

Yet another useful function is `conv`, used for convolution of signal vectors (see Chapter 1). For example, define two vectors **x** and **h** and evaluate the convolution sum. Plot the resulting vector **y**:

```
>> x=[1.2, -3.4, 2.3, 2.3, 5.6, -3.4];   % an example vector
>> h=[2, 1, -1];                         % another example
>> y=conv(x,h);                          % convolution of the
                                         % vectors
>> stem(y)                               % plot
>>
```

A2.5 Simulink™

Simulink™ (an extension to MATLAB™) is a tool for making simulations of dynamical systems, e.g. closed-loop control system. The user interface of Simulink™ is graphical and programming means putting block diagrams together. Working with Simulink™ is quite easy, so this presentation need not be long. "Learning by doing" is a good method of getting to know Simulink™.

Simulink™ can be started from the MATLAB™ prompt simply by typing `simulink`. After starting Simulink™ a library browser window will appear. A number of sub-groups will be presented: continuous, discrete, look-up table, etc. Open a new window by clicking the menu file/new/model. A new, blank window will appear. In this window, you can create your model by "drag-and-dropping" components from the library.

We will now simulate a simple closed-loop control system and analyze the step response as an example.

(1) From the sub-group *sources*, drag-and-drop a *step* to your new blank window. By means of this component, we will generate a step function as the reference signal to the closed-loop control system.
(2) From *math operations*, get *gain*. This will be used as an inverter.
(3) From *math operations*, get *sum*. This device will create the error signal by subtracting the measured output signal from the reference signal.
(4) From *sinks*, get *scope*. The scope is used to examine the output signal from the process.

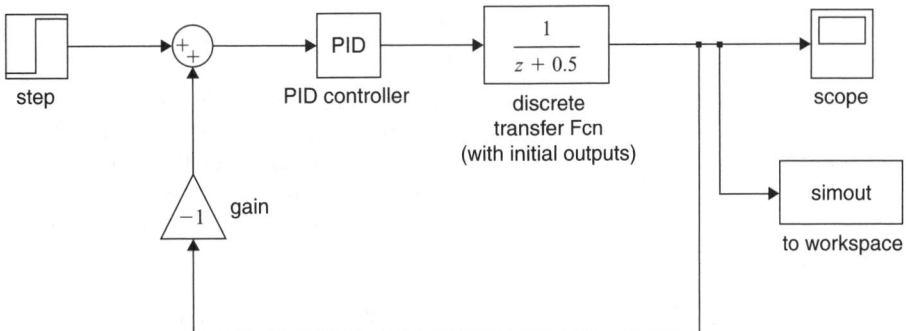

Figure A2.6 *A Simulink™ model of a simple, closed-loop control system*

(5) Open new sub-groups by clicking *Simulink extras/additional linear*. From this sub-group get *proportional-integral-derivative (PID) controller*. This is a basic PID controller.

(6) Open new sub-groups by clicking *Simulink extras/additional discrete*. From this sub-group get *discrete transfer Fcn (with initial outputs)*. This function will be used to simulate the process we would like to control.

(7) Move the components around so that your model looks something like Figure A2.6. The *gain* can be rotated by right clicking on it and choosing *format/rotate block*.

(8) Connect the blocks to each other as in Figure A2.6. Move the cursor to an input or an output. The shape of the cursor will change from an arrow to a cross. Press and hold the left mouse button while moving the cursor to the destination point (an input or output of another component). Release the mouse button and you should have a connection. A dashed line indicates that one of the ends is not properly connected. You can adjust your wiring by grabbing (press and hold the mouse button while moving) a wire and moving it. You can also move components and the wires will follow ("rubber banding").

(9) Double click on *gain*, a parameter window will open. Set the gain to -1, and click *OK*.

(10) Double click on *scope*, an oscilloscope-type window will open. Keep it open.

(11) Start the simulation by selecting *simulation/start* in your model window. You should get a graph on the oscilloscope, probably an oscillatory and unstable output signal.

(12) Double click on the *PID controller*, a parameter window for the controller will open. Try new setting of the *proportional, integral and derivative* parameters. Rerun the simulation to see if the output signal looks better. Repeat this procedure until you are happy with the output of the system, for an input step reference signal. (Be careful with the *derivative* parameter.) You can also experiment with changing the process model. Double click on the *discrete transfer Fcn* to open the corresponding parameter window.

As you have seen from the simple experiment above, Simulink™ is easy to use for simulations and experiments. Yet another interesting property is that Simulink can work in concert with MATLAB™. You can, for instance, send a signal to and from the workspace of MATLAB™, or to and from files. Try the following:

(13) In the sub-group *sinks*, get *to workspace*.
(14) Connect the *to workspace* component to the output of the process, i.e. to the same signal as shown on the *scope*.
(15) Double click on *to workspace* and in the parameter window, give the variable a desired name, the default is *simout*. Also, select *array* in the *save format* menu.
(16) Run the simulation as before. Go to the MATLAB™ window. Type whos, you will now find an array with the name you have chosen (default *simout*). The array contains the output signal from the simulation, i.e. the same signal that is shown in the *scope* window. Having the output signal as an array in the workspace, further analysis can be performed using MATLAB™ functions.

The demonstration above is just an example of the features of Simulink™. There is, for instance, a component *from workspace* in the *sources* sub-group which makes it possible to input signals from the MATLAB™ workspace to your model.

If you save your model, it will be stored as a file in the current directory with the file suffix .mdl.

References

Ahlin, L., Zander, J. (1998). *Principles of Wireless Communications*. Studentlitteratur. ISBN 91-44-00762-0.

Anderson, B.D.O., Moore, J.B. (1979). *Optimal Filtering*. Prentice-Hall. ISBN 0-13-638122-7.

Åström, K.J., Wittenmark, B. (1984). *Computer Controlled Systems, Theory and Design*. Prentice-Hall. ISBN 0-13-164319-3.

Becchetti, C., Ricotti, L.P. (1999). *Speech Recognition*. Wiley. ISBN 0-471-97730-6.

Bergh, J., Ekstedt, F., Lindberg, M. (1999). *Wavelets*. Studentlitteratur. ISBN 91-44-00938-0.

Bozic, S.M. (1994). *Digital and Kalman Filtering*. Edward Arnold. ISBN 0-340-61057-3.

Buck, J.R., Daniel, M.M., Singer, A.C. (2002). *Computer Explorations in Signals and Systems Using MATLAB*. Prentice-Hall. ISBN 0-13-042155-3.

Burrus, C.S., Parks, T.W. (1985). *DFT/FFT and Convolution Algorithms*. Wiley. ISBN 0-471-81932-8.

Carron Jr, L.P. (1991). *Morse Code: The Essential Language*. The American Relay League Inc. ISBN 0-87259-035-6.

Cavicchi, T.J. (2000). *Digital Signal Processing*. Wiley. ISBN 0-471-12472-9.

Chen, C.-T. (1999). *Linear System Theory and Design*. Oxford University Press Inc. ISBN 0-19-511777-8.

Chichocki, A., Unbehauen, R. (1993). *Neural Networks for Optimization and Signal Processing*. Wiley. ISBN 0-471-93010-5.

Cover, T.M., Thomas, J.A. (1991). *Elements of Information Theory*. Wiley. ISBN 0-471-06259-6.

Denbigh, P. (1998). *System Analysis & Signal Processing*. Addison-Wesley. ISBN 0-201-17860-5.

DSP56000/56001 Digital Signal Processor User's Manual (1989). Motorola literature distribution, Arizona. DSP56000UM/AD.

Esbensen, K., *et al.* (1994). *Multivariate Analysis – In Practice*. Computer-aided modelling AS, Norway. ISBN 82-993330-0-8.

Fitch, J.P., Doyle, E.J., Gallagher Jr, N.C. (1984). Median filtering by threshold decomposition. *IEEE Trans. Acoust. Speech, Signal Process.* **32**: 1183–1188.

Gallagher Jr, N.C., Wise, G.L. (1981). A theoretical analysis of the properties of median filters. *IEEE Trans. Acoust. Speech, Signal Process.* **29**: 1136–1141.

Gonzales, R.C., Woods, R.E. (2002). *Digital Image Processing*. Addison-Wesley. ISBN 0-201-18075-8.

Grossberg, S. (1987a). *The Adaptive Brain I*. Elsevier. ISBN 0-444-70413-2.

Grossberg, S. (1987b). *The Adaptive Brain II*. Elsevier. ISBN 0-444-70414-0.

Haykin, S. (2001). *Communication Systems*. Wiley. ISBN 0-471-17869-1.

Heller, S. (1997). *Introduction to C++*. Academic Press. ISBN 0-123-39099-0.

Hillis, W.D. (1987). *The Connection Machine*. MIT Press. ISBN 0-262-08157-1.

Hinton, G.E., Anderson, J.A. (1981). *Parallel Models of Associative Memory*. Lawrence Erlbaum. ISBN 0-89859-105-8.

Holmes, P.D. (1989). *The 8-Queens Problem and Task-Scheduling – A Neural Net Approach*. FOA, Linköping, March 1989. FOA Report D 30526-3.4.

Hopfield, J.J., Tanks, D.W. (1986). Computing with neural circuits: a model. *Science*. **253**: 625–633.

Hu, M.J.C. (1963). *A Trainable Weather-Forecasting System*. Stanford Electronics Laboratory, Stanford University, June 1963. Report No. ASD-TDR-63-636, Technical Report No. 6759-1.

Hunter, R., Robinson, H. (1980). International digital facsimile coding standards. *Proc. IEEE*, July 1980.

Lee, D.T.L., Yamamoto, A. (1994). *Wavelet Analysis: Theory and Application*. Hewlett-Packard J., December 1994.

Lippmann, R.P. (1987). An introduction to computing with neural nets. *IEEE ASSP Mag.* April 1987.

Lynn, P.A., Fuerst, W. (1998). *Digital Signal Processing with Computer Applications*. Wiley. ISBN 0-471-97631-8.

Marven, C., Ewers, G. (1993). *A Simple Approach to Digital Signal Processing*. Alden Press. ISBN 0-904-047-00-8.

McClelland, J.L., Rumelhart, D.E. (1986). *Parallel Distributed Processing, Part 2*. MIT Press. ISBN 0-262-13218-4.

Miller, G.M., Beasley, J.S. (2002). *Modern Electronic Communication*. Prentice-Hall. ISBN 0-13-016762-2.

Minsky, M.L., Papert, S.A. (1969). *Perceptrons* (expanded edition 1988). MIT Press. ISBN 0-262-63111-3.

Mitra, S.K., Kaiser, L.F. (1993). *Handbook for Digital Signal Processing*. Wiley. ISBN 0-471-61995-7.

Nalwa, V.S. (1993). *A Guided Tour of Computer Vision*. Addison-Wesley. ISBN 1-201-54853-4.

Oppenheimer, A.V., Schafer, R.W. (1975). *Digital Signal Processing*. Prentice-Hall. ISBN 0-13-214635-5.

Orfandis, S.J. (1985). *Optimum Signal Processing*. Macmillan. ISBN 0-02-949860-0.

Palm, R., Driankov, D., Hellendorrn, H. (1996). *Model Based Fuzzy Control*. Springer. ISBN 3-540-61471-0.

Papoulis, A., Pillai, S.U. (2001). *Probability, Random Variables and Stochastic Processes*. McGraw-Hill. ISBN 0-07-281725-9.

Passino, K.M., Yurkovich, S. (1998). *Fuzzy Control*. Addison-Wesley. ISBN 0-201-18074-X.

Pires, J. de Sousa (1989). *Electronics Handbook*. Studentlitteratur. ISBN 91-44-21021-3.

Pohlmann, K.C. (1989). *Principles of Digital Audio*. Howard W. Sams, Hayden Books. ISBN 0-672-22634-0.

Pratap, R. (2002). *Getting Started with MATLAB*. Oxford University Press. ISBN 0-19-515014-7.

Proakis, J.G. (1989). *Digital Communications*. McGraw-Hill. ISBN 0-07-100269-3.

Rabiner, L.R., Gold, B. (1975). *Theory and Application of Digital Signal Processing*. Prentice-Hall. ISBN 0-13-914101-4.

Rumelhart, D.E., McClelland, J.L. (1987). *Parallel Distributed Processing, Part 1*. MIT Press. ISBN 0-262-18120-7.

Schwartz, M., Shaw, L. (1975). *Signal Processing: Discrete Spectral Analysis, Detection, and Estimation*. McGraw-Hill. ISBN 0-07-055662-8.

Smith, S.W. (2003). *Digital Signal Processing*. Newnes. ISBN 0-750674-44-X.

Specht, D.F. (1964). *Vectorcardiographic Diagnosis Utilizing Adaptive Pattern-Recognition Techniques*. Stanford Electronics Laboratory, Stanford University, June 1964. Technical Report No. 6763-1.

Spiegel, M.R. (1971). *Calculus of Finite Differences and Difference Equations.* McGraw-Hill. ISBN 0-07-060218-2.

Stranneby, D. (1990). *Error Correction of Corrupted Binary Coded Data, Using Neural Networks.* KTH, Stockholm, 1990. Report No. TRITA-TTT9008.

Stranneby, D. (1996). *Power and Frequency Assignment in HF Radio Networks.* Radio Communication Systems Laboratory, KTH, Stockholm, April 1996. TRITA-S3-RTS-9605, ISSN 1400-9137.

Stranneby, D. (2001). *Digital Signal Processing: DSP & Applications.* Newnes. ISBN 0-7506-48112.

Tewari, A. (2002). *Modern Control Design with MATLAB and SIMULINK.* Wiley. ISBN 0-471-496790.

Widrow, B., Lehr, M.A. (1990). 30 years of adaptive neural networks: perceptron, madaline and backpropagation. *Proc. IEEE*, **78**(9).

Widrow, B., Stearns, S.D. (1985). *Adaptive Signal Processing.* Prentice-Hall. ISBN 0-13-004029-0.

Wilkie, J., Johnson, M., Katebi, R. (2002). *Control Engineering an Introductory Course.* Palgrave. ISBN 0-333-77129-X.

Wirth, N. (1976). *Algorithms + Data Structures = Programs.* Prentice-Hall. ISBN 0-13-022418-9.

Zwolinski, M. (2004). *Digital System Design with VHDL.* Prentice-Hall. ISBN 0-130-39985-X.

Glossary

A brief overview of some common abbreviations and buzz-words:

2PSK	2-ary phase shift keying (see BPSK)
4PSK	4-ary phase shift keying (see QPSK)
8PSK	8-ary phase shift keying
AC	Alternating current
A/D	Analog-to-digital
ADALINE	Adaptive linear neuron
ADC	Analog-to-digital converter
ADM	Adaptive delta modulation
ADPCM	Adaptive differential pulse code modulation
ADSL	Asymmetric digital subscriber loop
AGC	Automatic gain control
AI	Artificial intelligence
A-law	*Signal companding standard used in Europe (see μ-law)*
ALU	Arithmetic logic unit
AM	Amplitude modulation
AND	*Boolean function*
ANN	Artificial neural network
ANS	Artificial neural system
APC	Adaptive predictive coding
AR	Auto regressive
ARMA	Auto-regressive moving average
ARQ	Automatic repeat request
ASCII	American standard code for information interchange
ASIC	Application specific integrated circuit
ASK	Amplitude shift keying
ASR	Automatic speech recognition
AU	Address unit
AWGN	Additive white Gaussian noise
BCD	Binary-coded decimal
BCH	Bose, Chaudhuri, Hocquenghem (*class of error correcting codes*)
BCJR	Bahl, Cocke, Jelinke, Raviv
BFSK	Binary frequency shift keying
BIT	Binary digit
BLT	Bilinear transform
BP	Band-pass
BPSK	Binary phase shift keying
BSC	Binary symmetric channel
BU	Bus unit
C	*computer programming language (see C++)*

C++	*computer programming language (extension of C)*
CAM	Content addressable memory
CCITT	Comité Consultatif International Télegraphique et Télephonique
CD	Compact disc
CDMA	Code division multiple access
CD-ROM	Compact disc read only memory
CELP	Code excited linear prediction
CISC	Complex instruction set computer
CM	Connection machine
CMOS	Complementary metal oxide semiconductor
CODEC	Coder–decoder
CoA	Center of area
CoG	Center of gravity
CoM	Center of maximum
Compander	Compressor–expander
CRC	Cyclic redundancy check
CVSD	Continuously variable slope delta modulator
D/A	Digital-to-analog
DAB	Digital audio broadcasting
DAC	Digital-to-analog converter
DAT	Digital audio tape
DC	Direct current (*sometimes interpreted as a constant bias*)
DCC	Digital compact cassette
DCT	Discrete cosine transform
DFT	Discrete Fourier transform
DM	Delta modulator
DMA	Direct memory access
DPCM	Differential pulse code modulation
DPCM-AQB	Differential pulse code modulation adaptive quantization – backwards
DPCM-AQF	Differential pulse code modulation adaptive quantization – forward
DPSK	Differential phase shift keying
DSL	Digital subscriber loop
DSP	Digital signal processing
DSP	Digital signal processor
DTMF	Dual-tone multi-frequency
DVB	Digital video broadcasting
DVD	Digital video disc
ECC	Error-control code *or* error-correcting code
ECG	Electrocardiograph
EEG	Electroencephalogram
ENIAC	Electronic numerical integrator and computer
EPROM	Erasable programmable read only memory
EU	Execution unit
EXCEL™	*Spreadsheet type calculation software by Microsoft*
FFT	Fast Fourier transform
FIFO	First in first out (*a queue*)
FIR	Finite impulse response

FM	Frequency modulation
FOH	First-order hold
FORTRAN	Formula translation *(old computer programming language)*
FPGA	Field programmable gate array
FPU	Floating point unit
FRD	Frequency response data
FS	Full scale
FSK	Frequency shift keying
GMSK	Gaussian minimum shift keying
GPS	Global positioning system
GSM	Groupe speciale mobile *or* global system for mobile communication
HDTV	High definition television
IDFT	Inverse discrete Fourier transform
IEEE	The Institute of Electrical and Electronics Engineers
I^{2}C™	(IIC) Inter IC *(simple bidirectional 2-wire bus standard developed by Philips)*
IIR	Infinite impulse response
I/O	Input/output
I/Q	In phase/quadrature phase
ISDN	Integrated services digital network
ISI	Intersymbol interference
JAVA	*Computer programming language (subset of C++)*
JPEG	Joint photographics expert group
LDM	Linear delta modulation
LED	Light emitting diode
LIFO	Last in first out *(a stack)*
LMS	Least mean square
LPC	Linear predictive coding
LSB	Least significant bit
LTI	Linear time invariant *(commonly also implies causal)*
LUT	Look-up table
LZ	Lempel–Ziv
LZW	Lempel–Ziv–Welch
MA	Moving average
MAC	Multiply add accumulate
MAP	Maximum a posteriori probability
MathCad™	*Calculation software by MathSoft*
Mathematica™	*Calculation software by Wolfram Research*
MATLAB™	*Calculation software by MathWorks*
MDCT	Modified discrete cosine transform
MFLOPS	Million floating-point operations per second
MHC	Modified Huffman code
MIMD	Multiple instruction multiple data
MIMO	Multi-input multi-output
MIPS	Million instructions per second
MISO	Multi-input single-output
MIT	Massachusetts Institute of Technology
ML	Maximum likelihood

MLPC	Multipulse excited linear predictive coding
MoM	Mean of maximum
MP3	Layer-3 of MPEG-1 (*audio compression algorithm*)
MPEG	The moving pictures expert group
MRC	Modified Read code
MSB	Most significant bit
MSE	Minimum square error
MSK	Minimum shift keying
μ-law	*Signal companding standard used in USA (see A-law)*
NOP	No operation
NOT	*Boolean function*
NP	Non-polynomial
OCR	Optical character reading
OR	*Boolean function*
OSR	Oversampling ratio
PAM	Pulse amplitude modulation
Pascal	*Computer programming language*
PC	Personal computer
PC	Program counter
PCA	Principal component analysis
PCM	Pulse code modulation
PDM	Pulse density modulation
PDS	Parallel distributed system
PID	Proportional-integral-derivative
PM	Phase modulation
PNM	Pulse number modulation
PPM	Pulse position modulation
ppm	Parts per million
PROM	Programmable read only memory
PSK	Phase shift keying
PWM	Pulse width modulation
QAM	Quadrature amplitude modulation
QPSK	Quadrature phase shift keying
RELP	Residual excited linear prediction
RISC	Reduced instruction set computer
RLS	Recursive least square
RMS	Root mean square
ROM	Read only memory
RPM	Revolutions per minute
RS	Reed–Solomon (code)
RSC	Recursive systematic convolutional
RX	Receiver
SAR	Succesive approximation register
SBC	Sub-band coding
SC	Switched capacitor
SDR	Software defined radio
S&H	Sample-and-hold
S/H	Sample-and-hold
SIMD	Single instruction multiple data
SNR	Signal-to-noise ratio

SP	Stack pointer
SPI	Serial peripheral interface
SS	State–space
SSB	Single sideband
TF	Transfer function
TX	Transmitter
UWB	Ultra wideband
VHDL	Very high-speed integrated circuit hardware description language
VLSI	Very large scale integration
VSELP	Vector sum excited linear prediction
WCDMA	Wideband code division multiple access
XOR	*Exclusive OR, Boolean function*
ZOH	Zero-order hold
ZPK	Zero–pole gain

Index